AF352929

Prevention and Treatment of Periodontitis

Prevention and Treatment of Periodontitis

Editors

Yorimasa Ogata
Kenichi Imai

MDPI • Basel • Beijing • Wuhan • Barcelona • Belgrade • Manchester • Tokyo • Cluj • Tianjin

Editors
Yorimasa Ogata
Department of Periodontology
Nihon University School of
Dentistry at Matsudo
Matsudo
Japan

Kenichi Imai
Department of Microbiology
Nihon University School of
Dentistry
Tokyo
Japan

Editorial Office
MDPI
St. Alban-Anlage 66
4052 Basel, Switzerland

This is a reprint of articles from the Special Issue published online in the open access journal *Journal of Clinical Medicine* (ISSN 2077-0383) (available at: www.mdpi.com/journal/jcm/special_issues/Chronic_Periodontitis).

For citation purposes, cite each article independently as indicated on the article page online and as indicated below:

LastName, A.A.; LastName, B.B.; LastName, C.C. Article Title. *Journal Name* **Year**, *Volume Number*, Page Range.

ISBN 978-3-0365-1326-3 (Hbk)
ISBN 978-3-0365-1325-6 (PDF)

Contents

About the Editors . . . . **vii**

Preface to "Prevention and Treatment of Periodontitis" . . . . **ix**

Nobuhisa Furuse, Hideki Takai and Yorimasa Ogata
Effects of Initial Periodontal Therapy on Heat Shock Protein 70 Levels in Gingival Crevicular
Fluid from Periodontitis Patients
Reprinted from: *Journal of Clinical Medicine* 2020, *9*, 3072, doi:10.3390/jcm9103072 . . . . **1**

**Yuko Yamamoto, Toshiya Morozumi, Takahisa Hirata, Toru Takahashi, Shinya Fuchida,
Masami Toyoda, Shigeru Nakajima and Masato Minabe**
Effect of Periodontal Disease on Diabetic Retinopathy in Type 2 Diabetic Patients: A
Cross-Sectional Pilot Study
Reprinted from: *Journal of Clinical Medicine* 2020, *9*, 3234, doi:10.3390/jcm9103234 . . . . **9**

**Nanae Dewake, Yasuaki Ishioka, Keiichi Uchida, Akira Taguchi, Yukihito Higashi, Akihiro
Yoshida and Nobuo Yoshinari**
Association between Carotid Artery Calcification and Periodontal Disease Progression in
Japanese Men and Women: A Cross-Sectional Study
Reprinted from: *Journal of Clinical Medicine* 2020, *9*, 3365, doi:10.3390/jcm9103365 . . . . **21**

Norihisa Watanabe, Sho Yokoe, Yorimasa Ogata, Shuichi Sato and Kenichi Imai
Exposure to *Porphyromonas gingivalis* Induces Production of Proinflammatory Cytokine via
TLR2 from Human Respiratory Epithelial Cells
Reprinted from: *Journal of Clinical Medicine* 2020, *9*, 3433, doi:10.3390/jcm9113433 . . . . **33**

**Yuhei Takayanagi, Takeshi Kikuchi, Yoshiaki Hasegawa, Yoshikazu Naiki, Hisashi Goto,
Kousuke Okada, Iichiro Okabe, Yosuke Kamiya, Yuki Suzuki, Noritaka Sawada, Teppei
Okabe, Yuki Suzuki, Shun Kondo, Tasuku Ohno, Jun-Ichiro Hayashi and Akio Mitani**
Porphyromonas gingivalis Mfa1 Induces Chemokine and Cell Adhesion Molecules in Mouse
Gingival Fibroblasts via Toll-Like Receptors
Reprinted from: *Journal of Clinical Medicine* 2020, *9*, 4004, doi:10.3390/jcm9124004 . . . . **47**

**Takatoshi Nagano, Takao Yamaguchi, Sohtaro Kajiyama, Takuma Suzuki, Yuji Matsushima,
Akihiro Yashima, Satoshi Shirakawa and Kazuhiro Gomi**
Effect of Azithromycin on Proinflammatory Cytokine Production in Gingival Fibroblasts and
the Remodeling of Periodontal Tissue
Reprinted from: *Journal of Clinical Medicine* 2020, *10*, 99, doi:10.3390/jcm10010099 . . . . **61**

**Norio Aoyama, Toshiya Fujii, Sayuri Kida, Ichirota Nozawa, Kentaro Taniguchi, Motoki
Fujiwara, Taizo Iwane, Katsushi Tamaki and Masato Minabe**
Association of Periodontal Status, Number of Teeth, and Obesity: A Cross-Sectional Study in
Japan
Reprinted from: *Journal of Clinical Medicine* 2021, *10*, 208, doi:10.3390/jcm10020208 . . . . **71**

**Yuji Inagaki, Jun-ichi Kido, Yasufumi Nishikawa, Rie Kido, Eijiro Sakamoto, Mika Bando,
Koji Naruishi, Toshihiko Nagata and Hiromichi Yumoto**
Gan-Lu-Yin (Kanroin), Traditional Chinese Herbal Extracts, Reduces Osteoclast Differentiation
In Vitro and Prevents Alveolar Bone Resorption in Rat Experimental Periodontitis
Reprinted from: *Journal of Clinical Medicine* 2021, *10*, 386, doi:10.3390/jcm10030386 . . . . **81**

Yoshiaki Nomura, Toshiya Morozumi, Mitsuo Fukuda, Nobuhiro Hanada, Erika Kakuta, Hiroaki Kobayashi, Masato Minabe, Toshiaki Nakamura, Yohei Nakayama, Fusanori Nishimura, Kazuyuki Noguchi, Yukihiro Numabe, Yorimasa Ogata, Atsushi Saito, Soh Sato, Satoshi Sekino, Naoyuki Sugano, Tsutomu Sugaya, Fumihiko Suzuki, Keiso Takahashi, Hideki Takai, Shogo Takashiba, Makoto Umeda, Hiromasa Yoshie, Atsutoshi Yoshimura, Nobuo Yoshinari and Taneaki Nakagawa
Optimal Examination Sites for Periodontal Disease Evaluation: Applying the Item Response Theory Graded Response Model
Reprinted from: *Journal of Clinical Medicine* **2020**, *9*, 3754, doi:10.3390/jcm9113754 . . . . **99**

Yoshiaki Nomura, Toshiya Morozumi, Yukihiro Numabe, Yorimasa Ogata, Yohei Nakayama, Tsutomu Sugaya, Toshiaki Nakamura, Soh Sato, Shogo Takashiba, Satoshi Sekino, Nobuo Yoshinari, Nobuhiro Hanada, Naoyuki Sugano, Mitsuo Fukuda, Masato Minabe, Makoto Umeda, Koichi Tabeta, Keiso Takahashi, Kazuyuki Noguchi, Hiroaki Kobayashi, Hideki Takai, Fusanori Nishimura, Fumihiko Suzuki, Erika Kakuta, Atsutoshi Yoshimura, Atsushi Saito and Taneaki Nakagawa
Estimation of the Periodontal Inflamed Surface Area by Simple Oral Examination
Reprinted from: *Journal of Clinical Medicine* **2021**, *10*, 723, doi:10.3390/jcm10040723 . . . . **117**

Yoshiaki Nomura, Toshiya Morozumi, Atsushi Saito, Atsutoshi Yoshimura, Erika Kakuta, Fumihiko Suzuki, Fusanori Nishimura, Hideki Takai, Hiroaki Kobayashi, Kazuyuki Noguchi, Keiso Takahashi, Koichi Tabeta, Makoto Umeda, Masato Minabe, Mitsuo Fukuda, Naoyuki Sugano, Nobuhiro Hanada, Nobuo Yoshinari, Satoshi Sekino, Shogo Takashiba, Soh Sato, Toshiaki Nakamura, Tsutomu Sugaya, Yohei Nakayama, Yorimasa Ogata, Yukihiro Numabe and Taneaki Nakagawa
Prospective Longitudinal Changes in the Periodontal Inflamed Surface Area Following Active Periodontal Treatment for Chronic Periodontitis
Reprinted from: *Journal of Clinical Medicine* **2021**, *10*, 1165, doi:10.3390/jcm10061165 . . . . **129**

Codrina Ancuța, Rodica Chirieac, Eugen Ancuța, Oana Țănculescu, Sorina Mihaela Solomon, Ana Maria Fătu, Adrian Doloca and Cristina Iordache
Exploring the Role of Interleukin-6 Receptor Inhibitor Tocilizumab in Patients with Active Rheumatoid Arthritis and Periodontal Disease
Reprinted from: *Journal of Clinical Medicine* **2021**, *10*, 878, doi:10.3390/jcm10040878 . . . . **143**

Yuki Wada, Asami Suzuki, Hitomi Ishiguro, Etsuko Murakashi and Yukihiro Numabe
Chronological Gene Expression of Human Gingival Fibroblasts with Low Reactive Level Laser (LLL) Irradiation
Reprinted from: *Journal of Clinical Medicine* **2021**, *10*, 1952, doi:10.3390/jcm10091952 . . . . **155**

About the Editors

Yorimasa Ogata

Yorimasa Ogata, DDS, PhD, is Dean, Professor, and Chairman of the Department of Periodontology, Nihon University School of Dentistry at Matsudo, Japan. Dr. Ogata graduated from Nihon University School of Dentistry at Matsudo in 1984 and went on to complete his doctorate at the Graduate School of Tokyo Medical and Dental University, Department of Biochemistry, Japan in 1988. He then joined the staff of the Department of Periodontology, Nihon University School of Dentistry at Matsudo in 1988, first as an assistant professor and then as a lecturer before taking up his present position in 2001. He also spent a year (1992–1993) as a visiting scientist with the MRC Group in Periodontal Physiology, School of Dentistry, University of Toronto, Canada. Dr. Ogata is a member of a number of societies, including the International Association for Dental Research, the American Academy of Periodontology, and the Japanese Society of Periodontology (JSP). Now, he is a president of JSP.

Kenichi Imai

Kenichi Imai DDS, PhD, is Professor and Chairman of the Department of Microbiology, Nihon University School of Dentistry (NUSD). Dr. Imai graduated from Asahi University School of Dentistry in 1997 and went on to complete his doctorate at the Graduate School of Mekai University School of Dentistry, Japan in 2001. He then joined the group of Takashi Okamoto as a postdoctoral fellow to pursue his interest on the transcriptional regulation and silencing of HIV. He was appointed as an Assistant Professor of NCU-GSMS and NUSD in 2008 and 2010, respectively. Dr. Imai has been a recipient of a Young Investigator award from the Society for Microbial Ecology and Disease and the ECC Yamaguchi Memorial AIDS Award from the Japanese AIDS Society for his work on HIV. Currently, he studies on host-microbial interactions and infectious diseases with a focus on periodontitis-associated and other oral bacteria and viral latency.

Preface to "Prevention and Treatment of Periodontitis"

When I received an invitation from *J Clin Med* to be the guest editor, I was wondering whether to accept it, but as a result of accepting the role of guest editor together with Professor Kenichi Imai, Nihon University School of Dentistry, 13 papers were included in this book. I am very happy to be able to do this. I would like to thank all the authors for their cooperation.

Yorimasa Ogata
on behalf of:

Yorimasa Ogata, Kenichi Imai
Editors

Journal of
Clinical Medicine

Article

Effects of Initial Periodontal Therapy on Heat Shock Protein 70 Levels in Gingival Crevicular Fluid from Periodontitis Patients

Nobuhisa Furuse [1], Hideki Takai [1,2] and Yorimasa Ogata [1,2,*] 

[1] Department of Periodontology, Nihon University School of Dentistry at Matsudo, Matsudo,
     Chiba 271-8587, Japan; furuse-dental@jcom.zaq.ne.jp (N.F.); takai.hideki@nihon-u.ac.jp (H.T.)
[2] Research Institute of Oral Science, Nihon University School of Dentistry at Matsudo, Matsudo,
     Chiba 271-8587, Japan
*   Correspondence: ogata.yorimasa@nihon-u.ac.jp; Tel.: +81-47-360-9362

Received: 28 August 2020; Accepted: 22 September 2020; Published: 24 September 2020

Abstract: Periodontitis is an inflammatory disease of periodontium which is caused by periodontopathic bacteria. Moreover, various cytokines such as interleukin-1β (IL-1β), tumor necrosis factor-α (TNF-α), and IL-6 are expressed in the inflamed periodontium. Heat shock proteins (HSPs) protect cells from abnormal conditions including inflammation, microbial infection and diseases. The 70-kDa HSPs (HSP70s) are major HSPs that express in the inflamed tissues. In this study, an enzyme-linked immunosorbent assay was applied to measure the levels of HSP70 in gingival crevicular fluid (GCF) from two periodontal pockets in each of 10 patients with Stage III, Grade B periodontitis. Sites with probing pocket depth (PPD) of ≤3 mm were named the healthy control (HC) sites, and sites with PPD of ≥5 mm were named the diseased sites. HSP70 levels in GCF were expressed higher at diseased sites than at HC sites, and decreased after initial periodontal therapy at diseased sites. These results suggest the association of HSP70 with the stage of periodontitis.

Keywords: 70-kDa heat shock proteins; gingival crevicular fluid; heat shock protein; periodontitis

---

1. Introduction

Periodontitis, a common disease, is the inflammation of periodontium caused by oral microorganisms [1]. Numerous cytokines have been detected in the gingival crevicular fluid (GCF). Interleukin-1β (IL-1β) and tumor necrosis-α (TNF-α) levels have been found in increased GCF from inflamed gingival tissues [2,3]. IL-1β, matrix metalloproteinase-8, and IL-8 levels in GCF were significantly higher in diseased sites than in healthy subjects, and those levels in GCF were significantly decreased after periodontal treatment [4]. In a recent study, GCF volume and IL-1β levels in GCF reflected the disease severity, and these parameters were suggested to be better than probing pocket depth (PPD) and bleeding on probing (BOP) as markers of gingival inflammation [5]. Therefore, it could be helpful to investigate the inflammatory cytokine levels in GCF for diagnosis of the active phase of periodontal disease.

Heat shock proteins (HSPs) act as molecular chaperones, which are enhanced proteins for protecting cells immediately after cells are exposed to heat shock stress [6]. The 70-kDa HSPs (HSP70s) comprising various isoforms are involved in a number of human pathologies, ranging from cancer to neurodegenerative diseases [7,8]. HSP70 levels were higher in protein extract from inflamed gingiva collected during periodontal surgery, and HSP60 and HSP65 levels were higher in serum from the patients with periodontitis [9]. In addition, expression levels of IL-1β, IL-8, and HSP70 were increased in GCF from patients with periodontitis compared to healthy subjects [10]. Serum concentration of chitinase-3-like-1 (YKL-40), which is a novel marker of acute and chronic inflammation, was significantly higher in patients with periodontitis and diabetes compared to healthy groups. However, GCF concentration of YKL-40 was similar in patients with periodontitis, patients with diabetes, and healthy subjects. YKL-40 levels were significantly increased in diabetes patients with periodontitis [11]. Therefore, HSP70 can be considered potentially as a marker for the severity of periodontal disease. However, there is no report on changes in HSP70 concentration in GCF before and after initial periodontal therapy, or its correlation with clinical parameters. The aim of this study was to elucidate the HSP70 levels in GCF from patients with periodontitis, and to compare their concentrations in GCF at first visit, after initial periodontal therapy, and at a three month follow-up.

2. Materials and Methods

2.1. Study Population

Ten patients with Stage III, Grade B periodontitis (mean age, 53.9 ± 3 years) at Nihon University Hospital School of Dentistry at Matsudo, Japan were recruited and then received initial periodontal therapy, including oral hygiene instruction, scaling and root planing, and professional mechanical tooth cleaning. The stage of periodontitis was assessed clinically based on clinical attachment loss (CAL), radiographic bone loss (BL), periodontal history of tooth loss (PTL), PPD, BOP, furcation involvement (Class II or III), plaque index (PlI), and gingival index (GI) [12–14]. A CP11 probe (Hu-Friedy, Chicago, IL, USA) was used to measure the PPD and CAL. They were diagnosed when BL affected the middle third of the root or beyond and CAL was 5 mm or more. If PTL was four teeth or less, then the diagnosis was stage III. When the patient's previous periodontal records were available, the rate of periodontitis progression in the previous five years could be estimated. If progression was <2 mm, then the diagnosis was Grade B periodontitis [15]. GCF samples were taken from two periodontal PPD sites (shallow PPD of ≤3 mm were named the healthy control (HC) sites and deep PPD of ≥5 mm were named the diseased sites) for each patient.

This study was approved by the ethics committee at Nihon University School of Dentistry at Matsudo (EC17-017; 6Dec2017, EC18-17-017-1; 31, August, 2018). Written informed consent was obtained from all patients after all of the details of the investigative approach had been explained to them. Ten patients were confirmed physically healthy and had no experience of periodontal treatment or antibiotic treatment for at least three months prior to participation in this study. None of the patients were smokers.

2.2. Sample Size

Sample size was determined by the statistical software Easy R (EZR) on R commander Ver.1.27 (Tokyo, Japan) [16]. It was calculated that six samples (samples were taken from two sites per patient) were necessary for each group (HC and diseased sites) to reach 80% power at 5% level of significance. Therefore, the number of samples taken was appropriate.

2.3. Clinical Protocol

Clinical examinations were performed three times (first visit (1st), second examination (2nd) after initial periodontal therapy, and third examination (3rd) at a three month follow-up after initial periodontal therapy) by two periodontal specialists (H.T. and Y.O.). Average period of initial periodontal therapy was three months. At the point of each examination (1st, 2nd, and 3rd), GCF was collected using Periopaper subsequent to removal of supragingival plaque with a sterile curette. The GCF volume (μL) was measured using a Periotron 4000 (Oraflow, New York, NY, USA). GCF samples for the second visit data were taken before the periodontal therapy on the same day. Periopaper was then inserted into the pocket for 30 s each time. GCF samples were kept in Eppendorf tubes from two periodontal sites (≥5 mm (deep diseased site) and ≤3 mm (shallow healthy control site; HC)) from each patient and stored at −80 °C until measurement of HSP70 levels [17].

2.4. Enzyme-Linked Immunosorbent Assay (ELISA)

Concentrations of HSP70 in GCF samples were measured by ELISA using the Human HSPA4 (HSP70) ELISA kit (Invitrogen) according to the manufacturer's instructions. GCF samples were dissolved in 300 µL of Sample Diluent C in the ELISA kit. Diluted samples (100 µL) were incubated for 2.5 h in an anti-human HSP70 pre-coated 96-well strip plate. After a wash, biotinylated antibody was added to the wells for 1 h. After a wash, streptavidin-HRP solution was added to the wells for 45 min. After washing, TMB substrate was added to the wells for 30 min. Color development was stopped and the optical density at 450 nm of each well was measured within 30 min using a microplate reader. All measurements were performed in duplicate and the concentrations of HSP70 were expressed in ng/mL.

2.5. Statistical Analysis

Clinical parameters are shown as mean ± standard error (SE) in Table 1. The significant differences between each examination (1st, 2nd, and 3rd) for clinical parameters, GCF volume, and HSP70 levels among the groups were determined by one-way ANOVA. Difference in BOP at the 1st to 3rd examinations were analyzed by a Chi-squared test. The level of significance was adjusted at 5%. Four steps Ekuseru-Toukei (the publisher OMS Ltd.) was used for statistical analyses.

Table 1. Patient characteristics.

	HC Sites	Diseased Sites
Age		53.9 ± 3
Gender (male/female)		2/8
PPD	2.7 ± 0.2	6.5 ± 0.5
CAL	3.9 ± 0.4	7.7 ± 0.7
PlI	1.1 ± 0.2	1.1 ± 0.2
GI	0.3 ± 0.2	1.7 ± 0.2
BOP	0 (0%)	8 (80%)

HC, healthy control; PPD, probing pocket depth; CAL, clinical attachment loss; PlI, plaque index; GI, gingival index; BOP, bleeding on probing; mean ± SE ($n = 10$).

3. Results

The patient characteristics such as age, sex, PPD, CAL, PlI, GI, and BOP distributions for the 10 patients in this study are listed in Table 1. Average PPD and CAL at the HC sites (PPD ≤ 3 mm) were 2.7 ± 0.2 mm and 3.9 ± 0.4 mm, and at the diseased sites (PPD ≥ 5 mm) were 6.5 ± 0.5 mm and 7.7 ± 0.7 mm, respectively. GI and BOP scores at the diseased sites (1.7 ± 0.2 and 80%) were higher than those at the HC sites (0.3 ± 0.2 and 0%). PlI at the HC and diseased sites were the same score (1.1 ± 0.2).

The concentrations of HSP70 in GCF from the HC and diseased sites at each point of examination during initial periodontal therapy are shown in Table 2. The average HSP70 level at the 1st visit in GCF from diseased sites was significantly higher than HC sites. Moreover, the concentration of HSP70 at diseased sites was significantly decreased at 3rd examination (a three month follow-up after initial periodontal therapy) as compared to the 1st examination. The concentration of HSP70 at HC sites did not change through the periodontal therapy (1st, 2nd, and 3rd examinations) (Table 2).

Table 2. Changes in the concentrations of HSP70 in GCF collected from HC and diseased sites during the periodontal therapy.

Concentration (ng/mL)	1st	2nd	3rd
HC sites	18 ± 4.99 *	16.73 ± 6.37	6.64 ± 3.46
Diseased sites	64.36 ± 13.74	35.1 ± 6.67	5.69 ± 1.78 **

GCF, gingival crevicular fluid; HC, healthy control; 1st, first visit; 2nd, after initial periodontal therapy; 3rd, a three month follow-up after initial periodontal therapy; mean ± SE ($n = 10$), * $p < 0.05$, ** $p < 0.01$.

Changes in five kinds of clinical parameters (PPD, CAL, PlI, GI, and BOP scores) at the HC and diseased sites during initial periodontal therapy are listed in Tables 3 and 4. At HC sites, PlIs were significantly decreased at the 2nd and 3rd examinations as compared to the 1st examination (Table 3). On the other hand, PPD and PlI at diseased sites were significantly decreased at the 2nd and 3rd examinations compared to the 1st examination. Furthermore, GI and BOP scores at diseased sites were significantly decreased at the 3rd examination compared to the 1st examination. GCF volumes from HC and diseased sites were measured by Periotron 4000 during the course of periodontal therapy (Table 5). GCF volumes from HC and diseased sites did not change during the periodontal therapy, however, the volumes of GCF from the diseased sites were significantly higher than those in the HC sites at the 1st and 2nd visits (Table 5).

Table 3. Changes in clinical parameters at HC sites.

	1st	2nd	3rd
PPD	2.7 ± 0.2	2.5 ± 0.2	2.5 ± 0.2
CAL	3.9 ± 0.4	3.6 ± 0.5	3.6 ± 0.4
PlI	1.1 ± 0.2	0.6 ± 0.2 *	0.6 ± 0.2 *
GI	0.3 ± 0.2	0	0.4 ± 0.3
BOP	0 (0%)	0 (0%)	2 (20%)

HC, healthy control; 1st, first visit; 2nd, after initial periodontal therapy; 3rd, a three month follow-up after initial periodontal therapy; PPD, probing pocket depth; CAL, clinical attachment loss; PlI, plaque index; GI, gingival index; BOP, bleeding on probing; mean ± SE (n = 10), * p < 0.05.

Table 4. Changes in clinical parameters at diseased sites.

	1st	2nd	3rd
PPD	6.5 ± 0.5	4.3 ± 0.4 *	4.3 ± 0.5 *
CAL	7.7 ± 0.7	6.2 ± 0.9	5.9 ± 0.7
PlI	1.1 ± 0.2	0.5 ± 0.2 *	0.6 ± 0.2 *
GI	1.7 ± 0.2	1.2 ± 0.3	0.9 ± 0.3 *
BOP	8 (80%)	5 (50%)	4 (40%) *

1st, first visit; 2nd, after initial periodontal therapy; 3rd, a three month follow-up after initial periodontal therapy; PPD, probing pocket depth; CAL, clinical attachment loss; PlI, plaque index; GI, gingival index; BOP, bleeding on probing; mean ± SE (n = 10), * p < 0.05.

Table 5. Changes in GCF volume at HC and diseased sites during the periodontal therapy.

GCF (μL)	1st	2nd	3rd
HC sites	0.92 ± 0.2	0.47 ± 0.2	0.66 ± 0.2
Diseased sites	2.41 ± 0.5 *	1.6 ± 0.3 *	1.31 ± 0.3

GCF, gingival crevicular fluid; HC, healthy control; 1st, first visit; 2nd, after initial periodontal therapy; 3rd, a three month follow-up after initial periodontal therapy; mean ± SE (n = 10), * p < 0.05.

4. Discussion

In this study, we have shown that there was a significant difference in HSP70 concentration in GCF between HC and diseased sites at the first visit. The concentration of HSP70 in GCF from diseased sites was significantly decreased at the three month follow-up after initial periodontal therapy (3rd examination; Table 2). At HC sites, PlI was significantly decreased at the 2nd and 3rd examinations (Table 3). At diseased sites, PPD and PlI were significantly decreased at the 2nd and 3rd examinations, whereas GI and BOP were significantly decreased only at the 3rd examination as compared to the 1st visit (Table 4). These results suggest that initial periodontal therapy is effective in improving inflammation of periodontal tissues and there is an association between the level of HSP70 and periodontitis. In addition, improvements of GI, BOP, and HSP70 levels were found to take longer than improvements of PPD and PlI.

Intracellular HSP levels are elevated immediately after exposed to stresses such as high temperature. HSPs are involved in the maintenance of cellular homeostasis and protein repair in damaged cells [18].

However, there are several unclear points in the relationship between HSPs and periodontitis. Inflammatory periodontal pockets have a higher temperature than healthy pockets [19]. Inflammatory cytokines, such as IL-1, TNF-α, and INF-γ, are produced in inflamed periodontal tissues [20], and they might act as stressors to induce the expression of HSPs. Lipopolysaccharide (LPS) and IL-1 increased hyperthermia-induced HSP70 in monocyte/macrophage-like RAW264.7 cells [21]. However, one study described how HSP70 dramatically down-regulated in the inflamed periodontal tissues [22]. Another study showed that GCF volume at the first visit decreased significantly after initial periodontal therapy [23]. However, in this study, GCF volumes from HC and diseased sites did not change during periodontal therapy (Table 5), although the GCF volumes from diseased sites at the 1st and 2nd visits were significantly higher than the GCF volumes from HC sites (Table 5). Therefore, further study is necessary to elucidate the involvement of HSP70 in the onset and progression of periodontitis.

Stress and smoking are environmental factors for periodontitis [24,25]. Several studies have shown that smoking has an adverse effect on the incidence and progression of periodontitis [25]. In the synovial tissues of smokers with rheumatoid arthritis (RA), HSP70 levels were significantly higher than in the synovial tissues of non-smokers with RA [26]. Therefore, smoking could increase the expression of HSP70. There are several studies describing the association between HSP70 and cancer. Malignant cells, such as osteosarcoma derived cells, expressed higher levels of HSP70 during tumor progression compared to normal cells [27]. Moreover, HSP70 has been assessed as a marker for oral epithelial dysplasia such as oral leukoplakia [28]. Therefore, various studies have been conducted to develop the HSP70 inhibitors for cancer therapy [29]. We have previously shown that the anti-HSP70 antibody levels were significantly higher in GCF from HC sites than diseased sites, and the anti-HSP70 antibody levels were increased after initial periodontal therapy in both HC and diseased sites [30]. Therefore, these results suggest that anti-HSP70 antibody may reduce inflammation of periodontal tissues via decreasing the HSP70 levels.

In conclusion, GCF volumes from the diseased sites were significantly higher than those in the HC sites at the 1st and 2nd visits. HSP70 concentration in GCF from diseased sites was significantly higher than the concentration of HSP70 from HC sites at the 1st visit. Moreover, the HSP70 concentration at the 1st visit was significantly decreased at the three month follow-up after initial periodontal therapy together with clinical parameters, such as PPD, GI, PlI, and BOP. These results suggest that the HSP70 concentration could become an appropriate indicator for the healing process of periodontitis.

Author Contributions: Conceptualization, H.T. and Y.O.; methodology, N.F., H.T. and Y.O.; manuscript preparation, H.T. and Y.O. All authors have read and agreed to the published version of the manuscript.

Funding: This study was supported in part by Japan Society for the Promotion of Science KAKENHI Grants, Grants-in-Aid for Scientific Research (C), 18K09583 to H.T. and 17K11994 and 20K09945 to Y.O., and a Nihon University Multidisciplinary Research Grant (17-019) for 2017~2018 to Y.O.

Conflicts of Interest: The authors declare no conflict of interest.

Abbreviations

BOP	bleeding on probing
CAL	clinical attachment loss
GCF	gingival crevicular fluid
GI	gingival index
HC	healthy control
HSP	Heat shock protein
IL-1β	interleukin-1β
PlI	plaque index
PPD	probing pocket depth
SE	standard error
TNF-α	tumor necrosis factor-α

References

1.   Page, R.C.; Kornman, K.S. The pathogenesis of human periodontitis: An introduction. *Periodontology 2000* **1997**, *14*, 9–11. [CrossRef] [PubMed]
2.   Perozini, C.; Chibebe, P.C.; Leao, M.V.; Queiroz, C.S.; Pallos, D. Gingival crevicular fluid biochemical markers in periodontal disease: A cross-sectional study. *Quintessence Int.* **2010**, *41*, 877–883.
3.   Stashenko, P.; Jandinski, J.J.; Fujiyoshi, P.; Rynar, J.; Socransky, S.S. Tissue levels of bone resorptive cytokines in periodontal disease. *J. Periodontol.* **1991**, *62*, 504–509. [PubMed]
4.   Konopka, L.; Pietrzak, A.; Brzezińska-Błaszczyk, E. Effect of scaling and root planing on interleukin-1b, interleukin-8 and MMP-8 levels in gingival crevicular fluid from chronic periodontitis patients. *J. Periodontal Res.* **2012**, *47*, 681–688. [CrossRef]
5.   Oh, H.; Hirano, J.; Takai, H.; Ogata, Y. Effects of initial periodontal therapy on interleukin-1β level in gingival crevicular fluid and clinical periodontal parameters. *J. Oral Sci.* **2015**, *57*, 67–71. [CrossRef]
6.   Schlesinger, M.J. Heat shock proteins. *J. Biol. Chem.* **1990**, *265*, 12111–12114. [PubMed]
7.   Milićević, Z.T.; Petković, M.Z.; Drndarević, N.C.; Pavlović, M.D.; Todorović, V.N. Expression of Heat Shock Protein 70 (HSP70) in Patients with Colorectal Adenocarcinoma–Immunohistochemistry and Western Blot Analysis. *Neoplasma* **2007**, *54*, 37–45.
8.   Lackie, R.E.; Maciejewski, A.; Ostapchenko, V.G.; Marques-Lopes, J.; Choy, W.Y.; Duennwald, M.L.; Prado, V.F.; Prado, M.A.M. The Hsp70/Hsp90 Chaperone Machinery in Neurodegenerative Diseases. *Front. Neurosci.* **2017**, *11*, 254. [CrossRef]
9.   Ando, T.; Kato, T.; Ishihara, K.; Ogiuchi, H.; Okuda, K. Heat shock proteins in the human periodontal disease process. *Microbiol. Immunol.* **1995**, *39*, 321–327. [CrossRef]
10.   Tsybikov, N.N.; Baranov, S.V.; Kuznik, B.I. Serum, oral and gingival fluid levels of heat shock protein-70, cytokines and their autoantibodies by periodontal disease. *Stomatologiia (MosK)* **2014**, *93*, 16–18.
11.   Kumar, P.A.; Kripal, K.; Chandrasekaran, K.; Bhavanam, S.R. Estimation of YKL-40 Levels in Serum and Gingival Crevicular Fluid in Chronic Periodontitis and Type 2 Diabetes Patients among South Indian Population: A Clinical Study. *Contemp. Clin. Dent.* **2019**, *10*, 304–310. [PubMed]
12.   Loe, H.; Silness, J. Periodontal disease in pregnancy. I. Prevalence and severity. *Acta Odontol. Scand.* **1963**, *21*, 533–551. [CrossRef] [PubMed]
13.   Silness, J.; Loe, H. Periodontal disease in pregnancy. II. Correlation between oral hygiene and periodontal condition. *Acta Odontol. Scand.* **1964**, *22*, 122–135. [CrossRef]
14.   Tonetti, M.S.; Greenwell, H.; Kornman, K.S. Staging and grading of periodontitis: Framework and proposal of a new classification and case definition. *J. Periodontol.* **2018**, *89* (Suppl. 1), S15–S172. [CrossRef] [PubMed]
15.   Tonetti, M.S.; Sanz, M. Implementation of the new classification of periodontal diseases: Decision-making algorithms for clinical practice and education. *J. Clin. Periodontol.* **2019**, *46*, 398–405. [CrossRef] [PubMed]
16.   Kanda, Y. Investigation of the freely available easy-to-use software 'EZR' for medical statistics. *Bone Marrow Transpl.* **2013**, *48*, 452–458. [CrossRef]
17.   Ito, H.; Numabe, Y.; Sekino, S.; Murakashi, E.; Iguchi, H.; Hashimoto, S.; Sasaki, D.; Yaegashi, T.; Kunimatsu, K.; Takai, H.; et al. Evaluation of bleeding on probing and gingival crevicular fluid enzyme activity for detection of periodontally active sites during supportive periodontal therapy. *Odontology* **2014**, *102*, 50–56. [CrossRef]
18.   Rokutan, K.; Hirakawa, T.; Teshima, S.; Nakano, Y.; Miyoshi, M.; Kawai, T.; Konda, E.; Morinaga, H.; Nikawa, T.; Kishi, K. Implications of heat shock/stress proteins for medicine and disease. *J. Med. Investig.* **1998**, *44*, 137–147.
19.   Fedi, P.F., Jr.; Killoy, W.J. Temperature differences at periodontal sites in health and disease. *J. Periodontol.* **1992**, *63*, 24–27. [CrossRef]
20.   Alexander, M.B.; Damoulis, P.D. The role of cytokines in the pathogenesis of periodontal disease. *Curr. Opin. Periodontol.* **1994**, 39–53.
21.   Aditi, G.; Zachary, A.C.; Mohan, E.T.; Ratnakar, P.; Tapan, M.; Jeffrey, D.H.; Ishwar, S.S. Toll-like Receptor Agonists and Febrile Range Hyperthermia Synergize to Induce Heat Shock Protein 70 Expression and Extracellular Release. *J. Biol. Chem.* **2013**, *288*, 2756–2766.
22.   Seo, B.; Coates, D.E.; Seymour, G.J.; Rich, A.M. Unfolded protein response-related gene regulation in inflamed periodontal tissues with and without Russell bodies. *Arch. Oral Biol.* **2016**, *69*, 1–6. [CrossRef] [PubMed]

23. Rossi, V.; Romagna, R.; Angst, P.D.M.; Gomes, S.C. Gingival crevicular fluid response to protocols of non-surgical periodontal therapy: A longitudinal evaluation. *Ross. Indian J. Dent. Res.* **2019**, *30*, 736–741.

24. Alex, S.S.; Alessandra, N.P.; Tereza, A.D.V.S.; Jose, R.C.; Fernando, O.C.; Sheila, C.C.; Álvaro, F.B. Effects of two chronic stress models on ligature-induced periodontitis in Wistar rats. *Archs. Oral Biol.* **2012**, *57*, 66–72.

25. Leite, F.R.M.; Nascimento, G.G.; Scheutz, F.; López, R. Effect of Smoking on Periodontitis: A Systematic Review and Meta-regression. *Am. J. Prev. Med.* **2018**, *54*, 831–841. [CrossRef]

26. Ospelt, C.; Camici, G.G.; Engler, A.; Kolling, C.; Vogetseder, A.; Gay, R.E.; Michel, B.A.; Gay, S. Smoking induces transcription of the heat shock protein system in the joints. *Ann. Rheum. Dis.* **2014**, *73*, 1423–1426. [CrossRef] [PubMed]

27. Ciocca, D.R.; Calderwood, S.K. Heat shock proteins in cancer: Diagnostic, prognostic, predictive, and treatment implications. *Cell Stress Chaperones* **2005**, *10*, 86–103. [CrossRef]

28. Seoane, J.M.; Varela-Centelles, P.I.; Ramirez, J.R.; Cameselle-Teijeiro, J.; Romero, M.A.; Aguirre, J.M. Heat shock proteins (HSP70 and HSP27) as markers of epithelial dysplasia in oral leukoplakia. *Am. J. Dermatopathol.* **2006**, *28*, 417–422. [CrossRef]

29. Yun, C.W.; Kim, H.J.; Lim, J.H.; Lee, S.H. Heat Shock Proteins: Agents of Cancer Development and Therapeutic Targets in Anti-Cancer Therapy. *Cells* **2019**, *9*, 60. [CrossRef]

30. Takai, H.; Furuse, N.; Ogata, Y. Anti-heat shock protein 70 levels in gingival crevicular fluid of Japanese patients with chronic periodontitis. *J. Oral Sci.* **2020**, *62*, 281–284. [CrossRef]

Journal of
Clinical Medicine

Article

Effect of Periodontal Disease on Diabetic Retinopathy in Type 2 Diabetic Patients: A Cross-Sectional Pilot Study

Yuko Yamamoto [1], Toshiya Morozumi [2,*], Takahisa Hirata [2], Toru Takahashi [3], Shinya Fuchida [4], Masami Toyoda [5], Shigeru Nakajima [5] and Masato Minabe [2]

1   Department of Dental Hygiene, Kanagawa Dental University, Junior College, 82 Inaoka, Yokosuka 2388580, Kanagawa, Japan; yamamoto.yuko@kdu.ac.jp
2   Division of Periodontology, Department of Oral Interdisciplinary Medicine, Graduate School of Dentistry, Kanagawa Dental University, 82 Inaoka, Yokosuka 2388580, Kanagawa, Japan; t.hirata@kdu.ac.jp (T.H.); minabe@kdu.ac.jp (M.M.)
3   Department of Health and Nutrition, Faculty of Human Health, Kanazawa Gakuin University, 10 Sue-machi, Kanazawa 9201392, Ishikawa, Japan; t-takahasi@kanazawa-gu.ac.jp
4   Department of Disaster Medicine and Dental Sociology, Graduate School of Dentistry, Kanagawa Dental University, 82 Inaoka, Yokosuka 2388580, Kanagawa, Japan; fuchida@kdu.ac.jp
5   Nakajima Internal Medicine Clinic, 1-17 Yonegahamadori, Yokosuka 2380011, Kanagawa, Japan; toyoda@nakajima-naika.com (M.T.); nakajima@nakajima-naika.com (S.N.)
*   Correspondence: morozumi@kdu.ac.jp; Tel.: +81-46-822-8855

Received: 17 September 2020; Accepted: 7 October 2020; Published: 9 October 2020

Abstract: Both periodontal disease and diabetes are common chronic inflammatory diseases. One of the major problems with type 2 diabetes is that unregulated blood glucose levels damage the vascular endothelium and cause complications. A bidirectional relationship between periodontal disease and diabetic complications has been reported previously. However, whether periodontal disease affects the presence of diabetic complications has not been clarified. Therefore, we examined the effect of the periodontal disease status on diabetic complications in patients with type 2 diabetes. Periodontal doctors examined the periodontal disease status of 104 type 2 diabetic patients who visited a private diabetes medical clinic once a month between 2016 and 2018. The subject's diabetic status was obtained from their medical records. Bayesian network analysis showed that bleeding on probing directly influenced the presence of diabetic retinopathy in type 2 diabetes patients. In addition, bleeding on probing was higher in the diabetic retinopathy group ($n = 36$) than in the group without diabetic retinopathy ($n = 68$, $p = 0.006$, Welch's t-test). Bleeding on probing represents gingival inflammation, which might affect the presence of diabetic retinopathy in type 2 diabetes patients who regularly visit diabetic clinics.

Keywords: periodontal disease; diabetes; diabetic retinopathy; bleeding on probing; probing pocket depth; fasting blood sugar

1. Introduction

Both periodontal disease and type 2 diabetes are known to be common chronic inflammatory diseases [1,2]. In addition, many studies have reported a bidirectional relationship between periodontal disease and type 2 diabetes [3–5]. Epidemiological studies show that diabetics with poor glycemic control have an increased risk of periodontal disease [6]. On the other hand, it has been reported that severe periodontitis adversely affects glycemic control in type 2 diabetic patients [7]. It has become clear that the treatment of periodontal disease and type 2 diabetes has an effect on the pathophysiology of both diseases. Further, it has been reported that hemoglobin A1c (HbA1c) levels are reduced when type 2 diabetic patients are treated for periodontal disease [8–10]. Additionally, it was reported that when blood glucose was controlled in type 2 diabetic patients, the bleeding on probing (BOP) value, which represents the inflammation state of periodontal disease, improved [11]. Both periodontal disease and type 2 diabetes have a high prevalence worldwide [12]; however, the mechanism underlying the bidirectional relationship between them remains unclear.

Type 2 diabetes is a disorder of dysregulated blood glucose homeostasis due to a decrease in insulin action [13]. One of the major problems with type 2 diabetes is that unregulated blood glucose levels damage the vascular endothelium and cause complications in diabetics [14]. Diabetes complications include cardiovascular disease, diabetic nephropathy, neuropathy, leg amputation, and diabetic retinopathy [15]. Diabetic retinopathy is one of the most common diabetic complications and, as it progresses, can cause blindness and poor quality of life [16]. Although diabetic retinopathy is a major cause of blindness in the working population, only 35–55% of patients with diabetes undergo regular ophthalmic evaluations. This is because the disease progresses gradually and the patient barely notices the progression of diabetic retinopathy [16]. Therefore, it is important to prevent the development of diabetic retinopathy in diabetic patients.

It has recently been shown that there is a correlation between the severity of periodontal disease and diabetic retinopathy and that many diabetic retinopathy patients have periodontal disease [17,18]. However, it remains unclear as to whether periodontal disease affects the presence of diabetic retinopathy. Commisso et al. reported that poor oral health care was observed in the diabetic population [19]. It is known that effective oral health behaviors, such as brushing teeth two or more times a day, improve the condition of periodontal disease [20]. However, the relationship between the toothbrushing habits of type 2 diabetic patients and the pathological condition of diabetes has not been fully clarified. Therefore, the purpose of this study was to clarify the effect of periodontal disease status on diabetic complications, including diabetic retinopathy, in type 2 diabetic patients who visit our hospital once a month for diabetes management. Furthermore, this study also aimed to examine the relationship between tooth brushing habits and the pathophysiology of diabetes in patients with type 2 diabetes.

2. Methods

2.1. Study Population

We recruited 104 (45 men and 59 women) diabetic patients who visited a private diabetes medical clinic in Yokosuka City (Nakajima Internal Medicine Clinic) every month between October 2016 and August 2018. Patients with type 1 diabetes, toothless patients, patients who did not visit the diabetic department monthly, and those who did not consent to the study were excluded. The protocol of the present study was approved by the Ethics Committee of Kanagawa Dental University in 2016 (approval number: 359) and was conducted in accordance with the Helsinki Declaration of 1975, as revised in 2013. The study was registered with the Clinical Trial Registry of the University Hospital Medical Information Network Clinical Trials Registry (ID: UMIN000024627). On implementation, all participants were given written and verbal explanations regarding the purposes and methods of the study, any potential risks, potential benefits, protection of personal information, and freedom of consent and withdrawal. Subsequently, all participants signed the informed consent form.

2.2. Medical Examination

Data regarding age, height, weight, duration of diabetes, presence of diabetic retinopathy, and the presence of diabetic nephropathy were collected from the subject's medical records. For each subject, the diagnosis of diabetic retinopathy was made by an ophthalmologist based on the Davis classification as follows: no diabetic retinopathy (NDR); simple diabetic retinopathy (SDR); pre-proliferative diabetic retinopathy (pre-PDR); and proliferative diabetic retinopathy (PDR) [21]. Peripheral blood samples were collected by a nurse who specializes in diabetes during the routine monthly medical consultations. HbA1c levels were measured by an HbA1c analyzer (ADAMS^{TM} A1c HA-8180, ARKRAY Inc., Kyoto, Japan), and fasting blood sugar (FBS) levels were measured by a glucose analyzer (Glutest mint, SANWA KAGAKU KENKYUSHO Co., LTD., Aichi, Japan). The remaining peripheral blood was used to measure creatinine and triglyceride levels in a clinical laboratory (SRL, Inc., Tokyo, Japan).

2.3. Periodontal Examination

The periodontal disease status test was performed at the diabetic medical clinic on the day of the subject's visit by three trained periodontists. The probing pocket depth (PPD) and BOP were recorded with a manual probe (No.9550, YDM Corporation, Tokyo, Japan) at four points (mesial-buccal, mid-buccal, distal-buccal, and mid-lingual) for all teeth. Measurements of PPD and BOP were performed by three periodontists and then calibrated by a periodontist with the longest clinical experience (M.M.). Tooth mobility was measured for each tooth in the oral cavity. The Miller tooth mobility scale (0 degree, 1 degree, 2 degree, and 3 degree) was used to classify the degree of tooth mobility [22]. The presence of dental plaque on the surface adjacent to each tooth was determined by scratching the adjacent surface on the gingival margin with an explorer. Adherence of dental plaque to the adjacent surface even at one place led to the plaque adhesion on a surface that is adjacent to the tooth to be judged as "Yes". The number of times patients brushed their teeth per day and whether they underwent regular dental check-ups were confirmed by a questionnaire that was completed by the subjects.

2.4. Bayesian Network

A Bayesian network is a directed acyclic graph that is composed of a set of variables $\{X_1, X_2,...,X_N\}$ and a set of directed edges between the variables [23]. Bayesian networks are very successful in probabilistic knowledge representation and reasoning. In Bayesian networks, the joint probability distribution function of all nodes can be calculated as follows:

$$P\left(X_1, X_2, \ldots, X_N\right) = \prod_{i=1}^{N} P\left(Xi \mid Pai\right) \tag{1}$$

where *Pai* is the set of random variables whose corresponding nodes are parent nodes of *Xi*.

A Bayesian network contains two elements: structure and parameters. Each arc begins at a parent node and ends at a child node. *Pa* (*X*) represents the parent nodes of node *X*. X_1 is the root node because it has no input arcs. Root nodes have prior probabilities. Each child node has conditional probabilities based on the combination of states of its parent nodes.

Although this study was a cross-sectional study, it was analyzed by a Bayesian network with reference to previous studies by Tsuruta et al. [24].

2.5. Statistical Analysis

All statistical analyses were performed using JMP version 12 (SAS Institute Japan, Tokyo, Japan) and R version 3.2.0 (The R Project for Statistical Computing, Vienna, Austria, 2013). Results are expressed as the mean and the standard error of the mean (SEM). Comparisons between the two groups were analyzed using Welch's *t*-test. Spearman's rank correlation was employed to analyze the statistical significance of the correlation between two variables. Causal effects between variables were calculated using Bayesian network analysis. *p*-Values less than 0.05 were considered statistically significant.

3. Results

3.1. Subject Characteristics

The characteristics of the subjects included in this study are shown in Table 1. The mean age of the subjects was 70.0 ± 1.22 years (range: 23–86 years). In terms of dental characteristics, the mean number of teeth was 21.4 ± 0.735, and the number of teeth tended to decrease with age. The average BOP was 27.3%, and 13.2% of subjects had an average ratio of a PPD of 4 mm or greater, with most subjects having periodontal disease. Thirty-eight subjects underwent supportive periodontal therapy at the dental clinic, and 66 did not. In terms of medical characteristics, the average duration of diabetes was 13.6 ± 1.02 years. The mean HbA1c was 7.15%, the mean FBS was 147 mg/dL, and the mean serum creatinine concentration was 1.08 mg/dL. In addition, 36 subjects had diabetic retinopathy, 30 had SDR, 4 had pre-PDR, and 2 had PDR. Twelve subjects had diabetic nephropathy and two subjects had diabetic neuropathy. Most of the subjects in this study had a history of diabetes for more than 10 years, and some had diabetic complications.

Table 1. Characteristics of the subjects in this study.

Total number	$n = 104$
Sex (n): Female/male	59/45
Age (year): Mean ± SEM	70.0 ± 1.22
Age (year): Minimum/maximum	23/86
Number of teeth: Mean ± SEM	21.4 ± 0.735
BOP (%)	27.3 ± 1.74
Ratio of PPD, 4 mm or greater (%)	13.2 ± 1.61
Percentage of teeth with more than 1 degree of movement (%)	9.90 ± 1.74
Plaque adhesion on a surface that is adjacent to the tooth (n): Yes/no	69/35
Brushing the teeth 2 or more times a day: Yes/no	65/39
SPT in the dental clinic: Yes/no	38/66
Duration of diabetes (years): Mean ± SEM	13.6 ± 1.02
BMI (%): Mean ± SEM	29.4 ± 0.467
FBS (mg/dL): Mean ± SEM	147 ± 4.95
HbA1c (%): Mean ± SEM	7.15 ± 0.0948
Serum creatinine (mg/dL)	1.08 ± 0.178
Diabetic nephropathy: Yes/no	12/92
Diabetic retinopathy: Yes/no	36/68
Diabetic retinopathy: PDR/total number	30/36
Diabetic retinopathy: pre-PDR/total number	4/36
Diabetic retinopathy: PDR/total number	2/36
Diabetic neuropathy: Yes/no	2/102
Serum LDL cholesterol (mg/dL): Mean ± SEM	113 ± 3.09
Serum HDL cholesterol (mg/dL): Mean ± SEM	57.3 ± 1.48
Serum triglyceride (mg/dL): Mean ± SEM	161 ± 1.48

BOP, Bleeding on probing; PPD, probing pocket depth; SPT, supportive periodontal therapy; BMI, body mass index; FBS, fasting blood sugar; HbA1c, hemoglobin A1c; NDR, no diabetic retinopathy; SDR, simple diabetic retinopathy; PDR, proliferative diabetic retinopathy; LDL, low-density lipoprotein; HDL, high-density lipoprotein; SEM, standard error of the mean.

3.2. Hemoglobin A1c in Type 2 Diabetic Patients with Adjacent Dental Plaque Attached to the Tooth Surface

The subjects were divided into a group with dental plaque attached to the adjacent tooth surface ($n = 69$) and a group without adjacent dental plaque attachment ($n = 35$). HbA1c was higher in the group with dental plaque attachment compared to the group without ($p < 0.05$, Welch's *t*-test, Table 2). No other differences were found between the two groups (Welch's *t*-test, Table 2).

Table 2. Characteristics of the groups with and without adjacent dental plaque attachment.

Variable	Adjacent Dental Plaque Attachment ($n = 69$)	No Adjacent Dental Plaque Attachment ($n = 35$)	*p*-Value *
Duration of diabetes (years)	14.3 ± 8.44	12.3 ± 13.4	0.4
BMI (%)	29.5 ± 4.69	29.1 ± 4.97	0.7
FBS (mg/dL)	154 ± 49.1	134 ± 51.2	0.06
HbA1c (%)	7.27 ± 1.01	6.90 ± 0.835	<0.05
Creatinine (mg/dL)	1.25 ± 2.21	0.735 ± 0.190	0.06
LDL cholesterol (mg/dL)	112 ± 32.5	113 ± 29.7	0.9
HDL cholesterol (mg/dL)	55.6 ± 14.1	60.5 ± 11.6	0.1
Triglyceride (mg/dL)	164 ± 88.4	156 ± 112	0.8

Data are presented as the mean ± standard error of the mean. BMI, body mass index; FBS, fasting blood sugar; HbA1c, hemoglobin A1c; Creatinine, serum creatinine; LDL, low-density lipoprotein; HDL, high-density lipoprotein; Triglyceride: serum triglyceride. * Welch's *t*-test. *p*-values < 0.05 are considered statistically significant.

3.3. Fasting Blood Sugar of Type 2 Diabetic Patients Who Brushed Their Teeth more than Twice a Day

The subjects were divided into two groups: those who brushed their teeth less than once per day ($n = 39$) and those who brushed their teeth two or more times per day ($n = 65$). The FBS in the group who brushed their teeth more than twice per day was lower than that in the group who brushed once a day or less ($p < 0.0001$, Welch's *t*-test, Table 3). No other differences were found between the two groups (Welch's *t*-test, Table 3).

Table 3. Characteristics of the groups who brushed their teeth <1 and ≥2 times per day.

Variable	Group Who Brushed Their Teeth ≥ 2 Times per Day ($n = 65$)	Group Who Brushed Their Teeth < 1 Time per Day ($n = 39$)	*p*-Value *
Duration of diabetes (years)	12.7 ± 10.7	15.1 ± 9.70	0.2
BMI (%)	28.9 ± 4.66	30.3 ± 4.88	0.2
FBS (mg/dL)	128 ± 31.4	179 ± 60.0	<0.0001
HbA1c (%)	7.00 ± 0.906	7.39 ± 1.03	0.05
Creatinine (mg/dL)	0.945 ± 1.56	1.31 ± 2.17	0.4
LDL cholesterol (mg/dL)	115 ± 32.0	108 ± 30.4	0.3
HDL cholesterol (mg/dL)	59.1 ± 13.8	54.2 ± 16.8	0.1
Triglyceride (mg/dL)	153 ± 93.2	173 ± 102	0.3

Data are presented as the mean ± standard error of the mean. BMI, body mass index; FBS, fasting blood sugar; HbA1c, hemoglobin A1c; Creatinine, serum creatinine; LDL, low-density lipoprotein; HDL, high-density lipoprotein; Triglyceride, serum triglyceride. * Welch's *t*-test. *p*-values < 0.05 are considered statistically significant.

3.4. Bleeding on Probing in Type 2 Diabetic Patients with Diabetic Retinopathy (1)

The subjects were divided into the diabetic retinopathy group ($n = 36$) and the non-diabetic retinopathy group ($n = 68$). The BOP was higher in the group with diabetic retinopathy than in the group without diabetic retinopathy ($p = 0.006$, Welch's *t*-test, Table 4). No other differences were found between the two groups (Welch's *t*-test, Table 4).

Table 4. Characteristics of the groups with and without diabetic retinopathy.

Variable	Diabetic Retinopathy ($n = 36$)	No Diabetic Retinopathy ($n = 68$)	*p*-Value *
Number of teeth	20.7 ± 8.50	21.7 ± 6.94	0.5
BOP (%)	34.1 ± 18.1	23.7 ± 16.6	0.006
Ratio of PPD, 4 mm or greater (%)	17.2 ± 19.0	11.1 ± 14.6	0.09
Proportion of teeth with more than 1 degree of movement (%)	12.0 ± 24.6	8.76 ± 12.7	0.5
Number of times the subjects brushed their teeth per day	1.53 ± 0.774	1.78 ± 0.666	0.1

Data are presented as the mean ± standard error of the mean. BOP, bleeding on probing; PPD, probing pocket depth. * Welch's *t*-test. *p*-values < 0.05 are considered statistically significant.

3.5. Bleeding on Probing in Type 2 Diabetic Patients with Diabetic Retinopathy (2)

Using Spearman's rank correlation, we found that the FBS positively correlated with the ratio of PPD of 4 mm or greater (r_s = 0.29, p = 0.003, n = 104) and BOP (r_s = 0.21, p = 0.04, n = 104) (Table 2). In contrast, FBS was not correlated with the percentage of teeth with more than 1 degree of movement (r_s = −0.041, p = 0.7, n = 104, Table 5) or the number of teeth (r_s = 0.034, p = 0.7, n = 104, Table 5).

Table 5. Correlations among fasting blood sugar, the PPD ratio (4 mm or greater), and BOP.

Variable	Fasting Blood Sugar (mg/dL)		
	r_s *	*p*-Value	*n*
Ratio of PPD 4 mm or greater (%)	0.29	0.0029	104
BOP (%)	0.21	0.037	104
Proportion of teeth with more than 1 degree of movement (%)	−0.041	0.68	104
Number of teeth	0.034	0.73	104

PPD, Probing pocket depth; BOP, bleeding on probing. * Spearman's rank correlation coefficient. *p*-values < 0.05 are considered statistically significant.

3.6. Determination of Causal Effects Using Bayesian Network Analysis

Bayesian network analysis showed that the presence of diabetic retinopathy was directly affected by BOP. In addition, the number of times patients brushed their teeth per day was directly affected by HbA1c and FBS.

4. Discussion

4.1. Effect of Bleeding on Probing on Diabetic Retinopathy

In this study, the subjects with diabetic retinopathy had a higher BOP compared to those without diabetic retinopathy (Table 4). Furthermore, the Bayesian network analysis showed that the presence of diabetic retinopathy was directly affected by the BOP (Figure 1). The BOP indicates the gingival inflammatory conditions caused by periodontal disease [25]. Accordingly, these results indicate that gingival inflammation in diabetic patients might have affected the development of diabetic retinopathy.

The involvement of reactive oxygen species (ROS) is considered to be a factor in which inflammation of the gingiva affected diabetic retinopathy [26–33]. *Porphyromonas gingivalis* (*P. gingivalis*) and *Fusobacterium nucleatum* (*F. nucleatum*) are the major periodontal bacteria that cause gingival inflammation [26,27]. It has been reported that diabetic model rats infected with *P. gingivalis* have increased maxillofacial oxidative stress and decreased gingival microvascular reactivity [28]. Tothova et al. reported that gingival inflammation caused by periodontitis leads to the production of ROS by neutrophils [29]. Furthermore, ROS production by neutrophils against *F. nucleatum* has been shown to be higher than that of *P. gingivalis* [30]. Overproduced ROS in the oral cavity can cause oxidative stress and oxidative damage to cells, proteins, and lipids throughout the body, which can lead to systemic disease [31]. Oxidative stress has been shown to reduce neuroretinal function [32]. In addition, oxidative stress has been found to be associated with the accelerated onset of diabetic retinopathy [33]. Therefore, ROS produced during inflammation of the gingiva in type 2 diabetic patients might induce oxidative stress and contribute to the development of diabetic retinopathy. Suppressing the production of ROS by treating periodontal disease might reduce the development of diabetic retinopathy in diabetic patients.

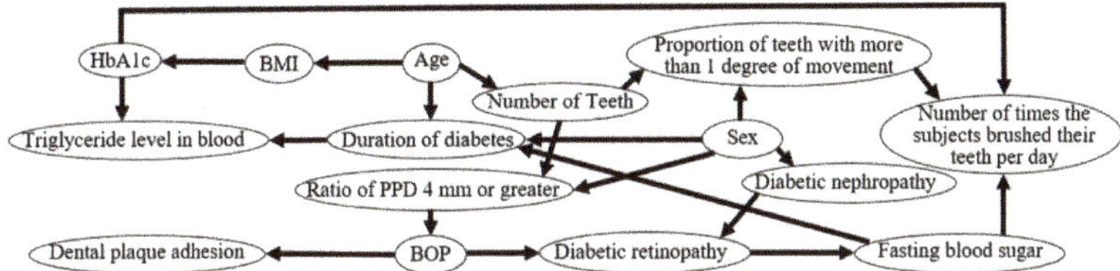

Figure 1. Bayesian network showing the causal effects among the following parameters: gender, age, number of teeth, bleeding on probing (BOP), ratio of probing pocket depth (PPD) of 4 mm or greater, proportion of teeth with more than 1 degree of movement, dental plaque adhesion, number of times the patients brushed their teeth per day, duration of diabetes, body mass index (BMI), fasting blood sugar, hemoglobin A1c (HbA1c), presence of diabetic retinopathy, presence of diabetic nephropathy, and triglyceride level in the blood. Causes and effects are indicated by arrowheads and lines, respectively.

Interleukin-17 (IL-17), which is secreted by IL-17-producing cells (T helper-17 cell: Th-17), causes inflammation and is involved in the development of chronic diseases such as autoimmune diseases [34]. IL-17 and Th-17 cells are present in human periodontal disease lesions and cause gingival inflammation [35]. Patients with a high BOP ratio have been found to have higher IL-17A concentrations in the saliva and gingival crevicular fluid compared with those without periodontal disease [36]. Furthermore, patients with periodontal disease have been found to have more Th-17 cells that produce IL-17 in the serum compared to healthy individuals [37]; in addition, the serum IL-17 concentration decreased due to a decrease in the BOP ratio in periodontal patients [38]. IL-17A has been reported to exacerbate diabetic retinopathy by impairing the function of Muller cells, which are the major glial cells in the retina [39]. Furthermore, IL-17A receptors have been found to be expressed in Muller glial cells, retinal endothelial cells, and photoreceptors [40]. Increased production of IL-17A in the periodontal lesions of patients with type 2 diabetes might have affected the development of diabetic retinopathy via the IL-17A receptor in the retina. Therefore, suppression of gingival inflammation with proper periodontal treatment might reduce IL-17 production in the gingiva and reduce the development of diabetic retinopathy in diabetic patients.

4.2. Relationship between Glycemic Control and Oral Hygiene Behavior

The Bayesian network analysis showed that the number of times patients brushed their teeth per day was directly affected by the HbA1c and FBS (Figure 1). In addition, HbA1c was higher in the group with dental plaque on the adjacent surfaces of the teeth compared to the group without adjacent dental plaque (Table 2). The FBS in subjects who brushed more than twice a day was lower than that of the subjects who brushed once a day or less (Table 3). It is well known that the amount of plaque on the tooth surface increases when the number of times toothbrushing is performed decreases [41]. Furthermore, HbA1c tended to be lower in the group who brushed more than twice a day than in the group who brushed once a day or less ($p = 0.05$, Table 3). From these findings, it is inferred that not only do FBS and HbA1c directly affect the number of times that diabetic patients brush their teeth per day, but also that oral hygiene habits and oral hygiene in diabetic patients are related to glycemic control.

Health literacy is the ability of an individual to obtain, understand, evaluate, and use health information to make decisions regarding the treatment of illnesses and overall health, and to maintain and improve the quality of life [42]. Sense of coherence (SOC) is a personality trait that allows individuals to adapt to and cope with stress to promote their health [43]. Recently, it has been reported that health literacy and SOC influence the pathophysiology of chronic diseases [42,44]. Patients with type 2 diabetes who have poor glycemic control have been shown to have poor health literacy [45]. Since SOC is indirectly involved in the glycemic control of patients with type 2 diabetes, it is necessary to increase SOC to improve glycemic control in such patients [46,47]. In addition, it has been reported that health literacy and SOC also affect oral hygiene behavior. Those who brush their teeth more than twice a day had a higher health literacy [48,49]. It has been reported that subjects who brush their teeth more than twice a day had higher SOC scores [49]. Inferring from these reports, the diabetic patients in this study with a higher health literacy and SOC would have been aware of not only blood glucose control, but also oral health improvement and behavior. Moreover, the ability to maintain and promote the health of diabetics would have been associated with glycemic control and oral hygiene habits. As a result, FBS and HbA1c would have had an effect on the number of times toothbrushing was performed per day, even in the Bayesian network analysis (Figure 1). Patients with type 2 diabetes who brush their teeth less often (per day) and have a large amount of plaque in the oral cavity might have poor glycemic control. Therefore, dentists and dental hygienists need to educate type 2 diabetics who are not interested in oral health in order to improve their health literacy and SOC, which might improve glycemic control in type 2 diabetics.

4.3. Relationship Between Periodontal Inflammation and Fasting Blood Sugar

In this study, there was a correlation between FBS and the ratio of both PPD of 4 mm or greater and BOP (Table 5). Similar to BOP, a PPD of 4 mm or greater indicates inflammation of the periodontal tissue [50]. Therefore, this relationship between periodontal inflammation and FBS in patients with type 2 diabetes was expected. There are many reports regarding the relationship between periodontal disease and FBS. Bleeding on probing (BOP) and PPD of 4 mm or greater were associated with fasting blood glucose [51]. For example, Joshipura et al. reported that the FBS of diabetic patients was reduced due to reduced periodontal pockets and improved BOP [52]. Further, Choi et al. reported that the FBS increased as the PPD increased [53]. Periodontal disease increases insulin resistance and raises blood sugar in diabetic patients [54]. Moreover, Colombo et al. reported that periodontal disease model rats had elevated plasma concentrations of tumor necrosis factor-α and increased insulin resistance [55]. The subjects in the present study might have also had high FBS levels due to periodontal inflammation causing insulin resistance. Treatment and education to reduce periodontal inflammation in dental clinics would help to control blood glucose levels in patients with type 2 diabetes.

4.4. Considerations

To the best of our knowledge, this study is the first to clarify the effect of periodontal disease on diabetic retinopathy in type 2 diabetic patients, as assessed by the diabetes department. However, this study has some limitations. First, we could not determine the smoking history of subjects because this was not described in the diabetes medical charts of the subjects. Second, we were unable to obtain data on the subjects' insulin injection status and medication status. As such, future studies will need to consider the association between these data and the effect of periodontal disease on diabetic retinopathy. Third, analysis of the Bayesian network revealed a causal relationship between multiple items. However, since the main purpose of this study was to clarify the effect of periodontal disease on diabetic complications, we considered only the causal relationship between the two diseases.

5. Conclusions

In conclusion, the BOP of subjects with diabetic retinopathy was higher than that of the subjects without diabetic retinopathy. Moreover, the gingival inflammation exhibited by BOP affected the presence of diabetic retinopathy in type 2 diabetics who regularly visit diabetic clinics. Controlling gingival inflammation by treating periodontal disease in dental clinics might suppress the development of diabetic retinopathy in patients with type 2 diabetes.

Author Contributions: T.M., Y.Y., and M.M. designed the study. T.H., Y.Y., T.M., M.M., M.T., and S.N. conducted the study. S.F. and T.T. performed the statistical analysis. Y.Y. and T.M. wrote the paper. M.M. and T.M. acquired the funding. All authors have read and agreed to the published version of the manuscript.

Funding: This study was supported by the Kanagawa Dental University Research Fund.

Acknowledgments: The authors thank the staff of the Nakajima Internal Medicine Clinic who cooperated with the data collection.

Conflicts of Interest: The authors declare no conflict of interest.

References

1. Doğan, Ş.B.; Ballı, U.; Dede, F.Ö.; Sertoğlu, E.; Tazegül, K. Chemerin as a novel crevicular fluid marker of patients with periodontitis and type 2 diabetes mellitus. *J. Periodontol.* **2016**, *87*, 923–933. [CrossRef] [PubMed]

2. Jourdain, M.L.; Velard, F.; Pierrard, L.; Sergheraert, J.; Gangloff, S.C.; Braux, J. Cationic antimicrobial peptides and periodontal physiopathology: A systematic review. *J. Periodontal Res.* **2019**, *54*, 589–600. [CrossRef] [PubMed]

3. Sanz, M.; Ceriello, A.; Buysschaert, M.; Chapple, I.; Demmer, R.T.; Graziani, F.; Herrera, D.; Jepsen, S.; Lione, L.; Madianos, P.; et al. Scientific evidence on the links between periodontal diseases and diabetes: Consensus report and guidelines of the joint workshop on periodontal diseases and diabetes by the International Diabetes Federation and the European Federation of Periodontology. *J. Clin. Periodontol.* **2018**, *45*, 138–149. [CrossRef] [PubMed]

4. Mohamed, H.G.; Idris, S.B.; Mustafa, M.; Ahmed, M.F.; Åstrøm, A.N.; Mustafa, K.; Ibrahim, S.O. Impact of chronic periodontitis on levels of glucoregulatory biomarkers in gingival crevicular fluid of adults with and without type 2 diabetes. *PLoS ONE* **2015**, *10*, e0127660. [CrossRef] [PubMed]

5. Casanova, L.; Hughes, F.J.; Preshaw, P.M. Diabetes and periodontal disease: A two-way relationship. *Br. Dent. J.* **2014**, *217*, 433–437. [CrossRef] [PubMed]

6. Mealey, B.L.; Ocampo, G.L. Diabetes mellitus and periodontal disease. *Periodontol. 2000* **2007**, *44*, 127–153. [CrossRef]

7. Taylor, G.W.; Burt, B.A.; Becker, M.P.; Genco, R.J.; Shlossman, M.; Knowler, W.C.; Pettitt, D.J. Severe periodontitis and risk for poor glycemic control in patients with non-insulin dependent diabetes mellitus. *J. Periodontol.* **1996**, *67*, 1085–1093. [CrossRef] [PubMed]

8. Madianos, P.N.; Koromantzos, P.A. An update of the evidence on the potential impact of periodontal therapy on diabetes outcomes. *J. Clin. Periodontol.* **2018**, *45*, 188–195. [CrossRef]

9. Darré, L.; Vergnes, J.N.; Gourdy, P.; Sixou, M. Efficacy of periodontal treatment on glycaemic control in diabetic patients: A meta-analysis of interventional studies. *Diabet. Metab.* **2008**, *34*, 497–506. [CrossRef]

10. Choi, S.E.; Sima, C.; Pandya, A. Impact of treating oral disease on preventing vascular diseases: A model-based cost-effectiveness analysis of periodontal treatment among patients with type 2 diabetes. *Diabet. Care* **2020**, *43*, 563–571. [CrossRef]

11. Katagiri, S.; Nitta, H.; Nagasawa, T.; Izumi, Y.; Kanazawa, M.; Matsuo, A.; Chiba, H.; Fukui, M.; Nakamura, N.; Oseko, F.; et al. Effect of glycemic control on periodontitis in type 2 diabetic patients with periodontal disease. *J. Diabet. Investig.* **2013**, *4*, 320–325. [CrossRef] [PubMed]

12. Sonnenschein, S.K.; Meyle, J. Local inflammatory reactions in patients with diabetes and periodontitis. *Periodontol. 2000* **2015**, *69*, 221–254. [CrossRef] [PubMed]

13. Merino, J.; Leong, A.; Liu, C.T.; Porneala, B.; Walford, G.A.; von Grotthuss, M.; Wang, T.J.; Flannick, J.; Dupuis, J.; Levy, D.; et al. Metabolomics insights into early type 2 diabetes pathogenesis and detection in individuals with normal fasting glucose. *Diabetologia* **2018**, *61*, 1315–1324. [CrossRef] [PubMed]

14. Forbes, J.M.; Cooper, M.E. Mechanisms of diabetic complications. *Physiol. Rev.* **2013**, *93*, 137–188. [CrossRef] [PubMed]

15. Deshpande, A.D.; Harris-Hayes, M.; Schootman, M. Epidemiology of diabetes and diabetes-related complications. *Phys. Ther.* **2008**, *88*, 1254–1264. [CrossRef]

16. Sinclair, S.H.; Schwartz, S.S. Diabetic retinopathy-an underdiagnosed and undertreated inflammatory, neuro-vascular complication of diabetes. *Front. Endocrinol. (Lausanne)* **2019**, *10*, 843. [CrossRef]

17. Veena, H.R.; Natesh, S.; Patil, S.R. Association between diabetic retinopathy and chronic periodontitis—A cross-sectional study. *Med. Sci.* **2018**, *6*, 104.

18. Song, S.J.; Lee, S.S.; Han, K.; Park, J.B. Periodontitis is associated with diabetic retinopathy in non-obese adults. *Endocrine* **2017**, *56*, 82–89. [CrossRef]

19. Commisso, L.; Monami, M.; Mannucci, E. Periodontal disease and oral hygiene habits in a type 2 diabetic population. *Int. J. Dent. Hyg.* **2011**, *9*, 68–73. [CrossRef]

20. Mizutani, S.; Ekuni, D.; Tomofuji, T.; Irie, K.; Azuma, T.; Iwasaki, Y.; Morita, M. Self-efficacy and progression of periodontal disease: A prospective cohort study. *J. Clin. Periodontol.* **2015**, *42*, 1083–1089. [CrossRef]
21. Tanaka, K.; Kawai, T.; Saisho, Y.; Meguro, S.; Harada, K.; Satoh, Y.; Kobayashi, K.; Mizushima, K.; Abe, T.; Itoh, H. Relationship between stage of diabetic retinopathy and pulse wave velocity in Japanese patients with type 2 diabetes. *J. Diabet. Res.* **2013**, *2013*, 193514. [CrossRef] [PubMed]
22. Miller, P.D. A classification of marginal tissue recession. *Int. J. Periodont. Restorat. Dent.* **1985**, *5*, 8–13.
23. Maglogiannis, I.; Zafiropoulos, E.; Platis, A.; Lambrinoudakis, C. Risk analysis of a patient monitoring system using Bayesian Network modeling. *J. Biomed. Inform.* **2006**, *39*, 637–647. [CrossRef] [PubMed]
24. Tsuruta, M.; Takahashi, T.; Tokunaga, M.; Iwasaki, M.; Kataoka, S.; Kakuta, S.; Soh, I.; Awano, S.; Hirata, H.; Kagawa, M.; et al. Relationships between pathologic subjective halitosis, olfactory reference syndrome, and social anxiety in young Japanese women. *BMC Psychol.* **2017**, *5*, 7. [CrossRef]
25. Killeen, A.C.; Harn, J.A.; Erickson, L.M.; Yu, F.; Reinhardt, R.A. Local minocycline effect on inflammation and clinical attachment during periodontal maintenance: Randomized clinical trial. *J. Periodont.* **2016**, *87*, 1149–1157. [CrossRef]
26. Huck, O.; Mulhall, H.; Rubin, G.; Kizelnik, Z.; Iyer, R.; Perpich, J.D.; Haque, N.; Cani, P.D.; de Vos, W.M.; Amar, S. Akkermansia muciniphila reduces *Porphyromonas gingivalis*-induced inflammation and periodontal bone destruction. *J. Clin. Periodont.* **2020**, *47*, 202–212. [CrossRef] [PubMed]
27. Yang, N.Y.; Zhang, Q.; Li, J.L.; Yang, S.H.; Shi, Q. Progression of periodontal inflammation in adolescents is associated with increased number of *Porphyromonas gingivalis*, *Prevotella intermedia*, *Tannerella forsythensis*, and *Fusobacterium nucleatum*. *Int. J. Paediatr. Dent.* **2014**, *24*, 226–233. [CrossRef] [PubMed]
28. Sugiyama, S.; Takahashi, S.S.; Tokutomi, F.A.; Yoshida, A.; Kobayashi, K.; Yoshino, F.; Wada-Takahashi, S.; Toyama, T.; Watanabe, K.; Hamada, N.; et al. Gingival vascular functions are altered in type 2 diabetes mellitus model and/or periodontitis model. *J. Clin. Biochem. Nutr.* **2012**, *51*, 108–113. [CrossRef]
29. Tóthová, L.; Celec, P. Oxidative stress and antioxidants in the diagnosis and therapy of periodontitis. *Front. Physiol.* **2017**, *8*, 1055. [CrossRef]
30. Wright, H.J.; Chapple, I.L.; Matthews, J.B.; Cooper, P.R. Fusobacterium nucleatum regulation of neutrophil transcription. *J. Periodont. Res.* **2011**, *46*, 1–12. [CrossRef] [PubMed]
31. Żukowski, P.; Maciejczyk, M.; Waszkiel, D. Sources of free radicals and oxidative stress in the oral cavity. *Arch. Oral Biol.* **2018**, *92*, 8–17. [CrossRef] [PubMed]
32. Suzumura, A.; Kaneko, H.; Funahashi, Y.; Takayama, K.; Nagaya, M.; Ito, S.; Okuno, T.; Hirakata, T.; Nonobe, N.; Kataoka, K.; et al. n-3 fatty acid and its metabolite 18-HEPE ameliorate retinal neuronal cell dysfunction by enhancing Müller BDNF in diabetic retinopathy. *Diabetes* **2020**, *69*, 724–735. [CrossRef] [PubMed]
33. Cecilia, O.M.; José Alberto, C.G.; José, N.P.; Ernesto Germán, C.M.; Ana Karen, L.C.; Luis Miguel, R.P.; Ricardo Raúl, R.R.; Adolfo Daniel, R.C. Oxidative stress as the main target in diabetic retinopathy pathophysiology. *J. Diabet. Res.* **2019**, *2019*, 8562408. [CrossRef] [PubMed]
34. Ren, S.; Hu, J.; Chen, Y.; Yuan, T.; Hu, H.; Li, S. Human umbilical cord derived mesenchymal stem cells promote interleukin-17 production from human peripheral blood mononuclear cells of healthy donors and systemic lupus erythematosus patients. *Clin. Exp. Immunol.* **2016**, *183*, 389–396. [CrossRef] [PubMed]
35. Cheng, W.C.; Hughes, F.J.; Taams, L.S. The presence, function and regulation of IL-17 and Th17 cells in periodontitis. *J. Clin. Periodont.* **2014**, *41*, 541–549. [CrossRef] [PubMed]
36. Apatzidou, D.A.; Iskas, A.; Konstantinidis, A.; Alghamdi, A.M.; Tumelty, M.; Lappin, D.F.; Nile, C.J. Clinical associations between acetylcholine levels and cholinesterase activity in saliva and gingival crevicular fluid and periodontal diseases. *J. Clin. Periodont.* **2018**, *45*, 1173–1183. [CrossRef]
37. Chen, X.T.; Chen, L.L.; Tan, J.Y.; Shi, D.H.; Ke, T.; Lei, L.H. Th17 and Th1 lymphocytes are correlated with chronic periodontitis. *Immunol. Investig.* **2016**, *45*, 243–254. [CrossRef]
38. Duarte, P.M.; da Rocha, M.; Sampaio, E.; Mestnik, M.J.; Feres, M.; Figueiredo, L.C.; Bastos, M.F.; Faveri, M. Serum levels of cytokines in subjects with generalized chronic and aggressive periodontitis before and after non-surgical periodontal therapy: A pilot study. *J. Periodont.* **2010**, *81*, 1056–1063. [CrossRef] [PubMed]
39. Qiu, A.W.; Liu, Q.H.; Wang, J.L. Blocking IL-17A alleviates diabetic retinopathy in rodents. *Cell Physiol. Biochem.* **2017**, *41*, 960–972. [CrossRef]

40. Sigurdardottir, S.; Zapadka, T.E.; Lindstrom, S.I.; Liu, H.; Taylor, B.E.; Lee, C.A.; Kern, T.S.; Taylor, P.R. Diabetes-mediated IL-17A enhances retinal inflammation, oxidative stress, and vascular permeability. *Cell Immunol.* **2019**, *341*, 103921. [CrossRef]

41. Kinane, D.F.; Zhang, P.; Benakanakere, M.; Singleton, J.; Biesbrock, A.; Nonnenmacher, C.; He, T. Experimental gingivitis, bacteremia and systemic biomarkers: A randomized clinical trial. *J. Periodont. Res.* **2015**, *50*, 864–869. [CrossRef] [PubMed]

42. Wang, W.; Zhang, Y.; Lin, B.; Mei, Y.; Ping, Z.; Zhang, Z. The urban-rural disparity in the status and risk factors of health literacy: A cross-sectional survey in Central China. *Int. J. Environ. Res. Public Health* **2020**, *17*, 3848. [CrossRef] [PubMed]

43. Tsiligianni, I.; Sifaki-Pistolla, D.; Gergianaki, I.; Kampouraki, M.; Papadokostakis, P.; Poulonirakis, I.; Gialamas, I.; Bempi, V.; Ierodiakonou, D. Associations of sense of coherence and self-efficacy with health status and disease severity in COPD. *NPJ Prim. Care Respir. Med.* **2020**, *30*, 27. [CrossRef] [PubMed]

44. Robat-Sarpooshi, D.; Mahdizadeh, M.; Alizadeh Siuki, H.; Haddadi, M.; Robatsarpooshi, H.; Peyman, N. The relationship between health literacy level and self-care behaviors in patients with diabetes. *Pat. Relat. Outcome Meas.* **2020**, *11*, 129–135. [CrossRef] [PubMed]

45. Saeed, H.; Saleem, Z.; Naeem, R.; Shahzadi, I.; Islam, M. Impact of health literacy on diabetes outcomes: A cross-sectional study from Lahore, Pakistan. *Public Health* **2018**, *156*, 8–14. [CrossRef] [PubMed]

46. Cohen, M.; Kanter, Y. Relation between sense of coherence and glycemic control in type 1 and type 2 diabetes. *Behav. Med.* **2004**, *29*, 175–183. [CrossRef]

47. Odajima, Y.; Sumi, N. Factors related to sense of coherence in adult patients with Type 2 diabetes. *Nagoya J. Med. Sci.* **2018**, *80*, 61–71.

48. Cepova, E.; Cicvakova, M.; Kolarcik, P.; Markovska, N.; Geckova, A.M. Associations of multidimensional health literacy with reported oral health promoting behaviour among Slovak adults: A cross-sectional study. *BMC Oral Health* **2018**, *18*, 44. [CrossRef]

49. Bernabé, E.; Kivimäki, M.; Tsakos, G.; Suominen-Taipale, A.L.; Nordblad, A.; Savolainen, J.; Uutela, A.; Sheiham, A.; Watt, R.G. The relationship among sense of coherence, socio-economic status, and oral health-related behaviours among Finnish dentate adults. *Eur. J. Oral. Sci.* **2009**, *117*, 413–418. [CrossRef]

50. Javed, F.; Al Amri, M.D.; Al-Kheraif, A.A.; Qadri, T.; Ahmed, A.; Ghanem, A.; Calvo-Guirado, J.L.; Romanos, G.E. Efficacy of non-surgical periodontal therapy with adjunct Nd:YAG laser therapy in the treatment of periodontal inflammation among patients with and without type 2 diabetes mellitus: A short-term pilot study. *J. Photochem. Photobiol. B* **2015**, *149*, 230–234. [CrossRef]

51. Jung, Y.S.; Shin, M.H.; Kweon, S.S.; Lee, Y.H.; Kim, O.J.; Kim, Y.J.; Chung, H.J.; Kim, O.S. Periodontal disease associated with blood glucose levels in urban Koreans aged 50 years and older: The Dong-gu study. *Gerodontology* **2015**, *32*, 267–273. [CrossRef] [PubMed]

52. Joshipura, K.J.; Muñoz-Torres, F.J.; Dye, B.A.; Leroux, B.G.; Ramírez-Vick, M.; Pérez, C.M. Longitudinal association between periodontitis and development of diabetes. *Diabet. Res. Clin. Pract.* **2018**, *141*, 284293. [CrossRef] [PubMed]

53. Choi, Y.H.; McKeown, R.E.; Mayer-Davis, E.J.; Liese, A.D.; Song, K.B.; Merchant, A.T. Association between periodontitis and impaired fasting glucose and diabetes. *Diabet. Care* **2011**, *34*, 381–386. [CrossRef] [PubMed]

54. Akutagawa, K.; Fujita, T.; Ouhara, K.; Takemura, T.; Tari, M.; Kajiya, M.; Matsuda, S.; Kuramitsu, S.; Mizuno, N.; Shiba, H.; et al. Glycyrrhizic acid suppresses inflammation and reduces the increased glucose levels induced by the combination of *Porphyromonas gulae* and ligature placement in diabetic model mice. *Int. Immunopharmacol.* **2019**, *68*, 30–38. [CrossRef] [PubMed]

55. Colombo, N.H.; Shirakashi, D.J.; Chiba, F.Y.; Coutinho, M.S.; Ervolino, E.; Garbin, C.A.; Machado, U.F.; Sumida, D.H. Periodontal disease decreases insulin sensitivity and insulin signaling. *J. Periodontol.* **2012**, *83*, 864–870. [CrossRef] [PubMed]

Journal of
Clinical Medicine

Article

Association between Carotid Artery Calcification and Periodontal Disease Progression in Japanese Men and Women: A Cross-Sectional Study

Nanae Dewake [1],*, Yasuaki Ishioka [1], Keiichi Uchida [2], Akira Taguchi [3], Yukihito Higashi [4], Akihiro Yoshida [5,6] and Nobuo Yoshinari [1,5]

1 Department of Operative Dentistry, Endodontology and Periodontology, School of Dentistry, Matsumoto Dental University, Shiojiri 399-0781, Japan; yasuaki.ishioka@mdu.ac.jp (Y.I.); nobuo.yoshinari@mdu.ac.jp (N.Y.)

2 Department of Oral Sciences, Matsumoto Dental University Hospital, Shiojiri 399-0781, Japan; keiichi.uchida@mdu.ac.jp

3 Department of Oral and Maxillofacial Radiology, School of Dentistry, Matsumoto Dental University, Shiojiri 399-0781, Japan; akira.taguchi@mdu.ac.jp

4 Department of Cardiovascular Regeneration and Medicine, Research Institute for Radiation Biology and Medicine, Hiroshima University, Hiroshima 734-0037, Japan; yhigashi@hiroshima-u.ac.jp

5 Department Oral Health Promotion, Graduate School of Oral Medicine, Matsumoto Dental University, Shiojiri 399-0781, Japan; akihiro.yoshida@mdu.ac.jp

6 Department of Oral Microbiology, School of Dentistry, Matsumoto Dental University, Shiojiri 399-0781, Japan

* Correspondence: nanae.dewake@mdu.ac.jp; Tel.: +81-26-352-3100

Received: 16 September 2020; Accepted: 19 October 2020; Published: 20 October 2020

Abstract: Objective: To evaluate the association between alveolar bone loss (ABL) detected on panoramic radiographs and carotid artery calcification (CAC) detected on computed tomography (CT). Methods: The study subjects included 295 patients (mean age ± SD: 64.6 ± 11.8 years) who visited the Matsumoto Dental University Hospital. The rate of ABL and the number of present teeth were measured on panoramic radiographs. Univariate analyses with t-tests and chi-squared tests were performed to evaluate the differences in age, gender, history of diseases, number of present teeth, and the ABL between subjects, with and without CAC. Moreover, multivariate logistic regression analysis, with forward selection and receiver operating characteristic curve (ROC) analysis, was performed. Results: The number of subjects without and with CAC was 174 and 121, respectively. Univariate analyses revealed that CAC was significantly associated with age, hypertension, osteoporosis, number of present teeth, and ABL. Multivariate logistic regression analysis adjusted for covariates revealed that the presence of CAC was significantly associated with ABL (OR = 1.233, 95% CI = 1.167–1.303). In the ROC analysis for predicting the presence of CAC, the the area under the ROC curve was the highest at 0.932 (95% CI = 0.904–0.960) for ABL, which was significant. Conclusions: Our results suggest that the measurement of ABL on panoramic radiographs may be an effective approach to identifying patients with an increased risk of CAC.

Keywords: carotid artery calcification; periodontal disease; alveolar bone loss; computed tomography; panoramic radiographs

---

1. Introduction

Japan is a super-aged society where 23.2% of the people die from heart and cerebrovascular diseases caused by arteriosclerosis [1]. According to the results of a patient survey conducted by the Ministry of Health, Labor, and Welfare in 2017, arteriosclerotic diseases have been recorded in people aged 30–90 years [2]. It is, therefore, important to determine asymptomatic arteriosclerosis as soon as possible for proactive prevention of vascular diseases in an effort to extend the life span of Japanese individuals.

Periodontitis, which develops principally from the age of 30 years, is a chronic oral inflammatory disease initiated by the accumulation of a bacterial biofilm on tooth surfaces and perpetuated by dysregulated local and systemic inflammatory immune responses [3]. Along with dental caries, different forms of periodontitis represent the two major global oral health burdens on our society [4]. In Japan, the percent of moderate periodontitis in individuals aged ≥65 years with a periodontal pocket depth (PD) of >4 mm was 57.5% in 2016. The presence of periodontitis is directly correlated with the number of present teeth [5]. In particular, periodontitis shares several risk factors with other noncommunicable diseases, including cardiovascular diseases.

The association between periodontal disease and atherosclerosis is well-known in Western countries. However, the epidemiological evidence linking these two phenomena is relatively poor in Japan. Several past studies have suggested an association between periodontal diseases and markers of subclinical atherosclerosis that are used to assess morphological abnormalities such as carotid intimae media thickness (c-IMT) and carotid plaque, as well as the functional abnormalities such as pulse-wave velocity (PWV) and flow-mediated vasodilation (FMD) of the brachial artery induced by reactive hyperemia [6–12]. Orlandi et al. conducted a systematic review and meta-analysis and found that the diagnosis of periodontal disease was associated with a mean increase in c-IMT of 0.08 mm (95% CI = 0.07–0.09) and a mean difference in FMD of 5.1% in comparison with those in the corresponding controls (95% CI = 2.08–8.11) [13]. They also described a beneficial effect of periodontal treatment on FMD, indicating an improvement in the endothelial function [13].

In vascular calcification, which is the final step in the development of atherosclerosis, the deposition of hydroxyapatite mineral in the arterial wall has been associated with an increased risk of heart disease, stroke, and atherosclerotic plaque rupture [14]. However, there is only limited information available on the association between periodontal diseases and carotid artery calcification (CAC). In fact, there are several imaging diagnostic tools available for assessing CAC. Of these, CAC can be easily and accurately detected by computed tomography (CT) [15]. On the other hand, periodontal disease progression can be assessed as an alveolar bone loss (ABL) on panoramic radiographs. Therefore, the purpose of the present study is to evaluate the association between the rate of ABL measured on panoramic radiographs and CAC detected by CT.

2. Methods

2.1. Patient Setting

A total of 295 patients (174 men and 121 women) who underwent both panoramic radiographs and CT for the examination of their lesions (such as benign tumors and cystic lesions) and implant placement and who visited the Matsumoto Dental University Hospital between 2014 and 2018 were enrolled in this study. Patients with lesions and alveolar bone destruction were excluded from this study. Moreover, patients who had undergone radiation therapy for their head and neck regions were also excluded. The mean age (SD) of the subjects was 64.6 (11.8) years. The information on age, gender, and medical history of the subjects was obtained from their respective medical records. We also extracted data on the potential risk factors for arteriosclerosis, namely, hypertension, hyperlipidemia, diabetes mellitus, and osteoporosis, from the records.

Since this study was a retrospective cross-sectional study, written informed consents were not obtained from all subjects at the time of taking both panoramic radiographs and CT. Instead, the results of this study were provided to all subjects in accordance with the ethical guidelines of the Ministry of Health, Labor, and Welfare and the Ministry of Education, Culture, Sports, Science, and Technology. The institutional review board for clinical research at Matsumoto Dental University reviewed and approved this study protocol (no. 152).

2.2. Assessment of Alveolar Bone Resorption Using Panoramic Radiographs

Panoramic radiographs of all subjects were obtained using the digital AZ3000 device (Asahi X-Ren Kogyo, Kyoto, Japan). The received ray system was a computed radiography system (Konica Minolta, Tokyo, Japan). A periodontist with 6 years of experience examined the number of present teeth and the rate of ABL. The supernumerary tooth and the third molar were excluded from the number of present teeth. The rate of ABL was measured on panoramic radiographs [16]. The distances between the cement–enamel junction (CEJ) and the alveolar crest (AC) and that between the CEJ and the root apex were measured at two sites (i.e., mesial and distal) of each of the present teeth. The apex was defined as the most apically located point of the root. In teeth restored with fillings or crowns, the most apical limit of the restoration was considered to be equivalent to the CEJ. Finally, ABL was calculated as the CEJ–AC/CEJ–apex [17]. The measurements were made for all of the present teeth, excluding implants, supernumerary teeth, and the third molars. The residual roots, without a cap for overdenture, were excluded. Teeth with caries or periapical lesions were not excluded. In the PROKRANK study, the radiographic periodontal condition based on the rate of ABL was classified into three groups: healthy (≥80% remaining bone), mild to moderate (66–79% remaining bone), and severe (<66% remaining bone) bone loss groups [18].

2.3. Detection of CAC on CT

CAC was detected on the axial CT images captured by using the multislice CT system Activation 16 (Canon Medical Systems, Tochigi, Japan) at the Matsumoto Dental University Hospital. The presence of CAC near the bifurcation of the common carotid artery (about 2-cm upper and lower) was evaluated independently by two oral and maxillofacial diplomate radiologists (one with >20 years of experience and another with 30 years of experience; Figure 1). In case of different outcomes (such as ectopic calcifications) between the two oral and maxillofacial radiologists, a consensus was reached via discussion.

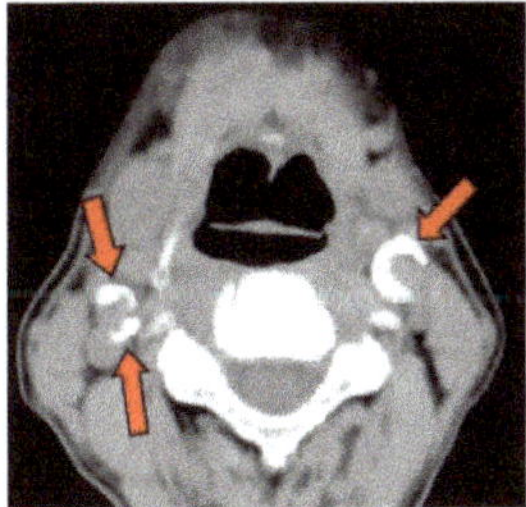

Figure 1. Observation of the carotid artery calcification. Arrow tip indicates the calcification.

2.4. Statistical Analysis

The patients' age was divided into 5 groups (30–49, 50–59, 60–69, 70–79, and 80–99 years). The number of present teeth was divided into 3 groups (1–9, 10–19, and ≥20). The rate of ABL (%) was divided into 3 groups (≤20%, 20–34%, and >34%) based on the radiographic periodontal classification [18]. Initially, univariate analyses with t-tests and chi-squared tests were used to evaluate the differences in age, gender (binary), history of diseases related to atherosclerosis, number of teeth present, and the rate of ABL (%) between the subjects, without and with CAC. Secondly, multivariate logistic regression analysis, with forward selection adjusted for age, gender (binary), and all other variables significant at $p < 0.20$ in the univariate analyses, were tested to calculate the adjusted odds ratio (OR) and 95% confidence interval (CI) of having CAC based on the ABL classification system. Furthermore, receiver operating characteristic (ROC) curve analysis was employed to clarify how asymptomatic CAC can be identified by age, presence of hypertension, number of present teeth, and the rate of ABL. According to the method suggested by Swets [19], the area under the ROC curve (AUROC) was determined as follows: less accurate ($0.5 <$ AUROC < 0.7), moderately accurate ($0.7 <$ AUROC < 0.9), highly accurate ($0.9 <$ AUROC < 1), and perfect tests (AUROC $= 1$). All comparisons were two-sided and performed at $p = 0.05$ level of significance. Statistical analysis was performed using SPSS ver. 24.0 for Windows (IBM Japan, Tokyo, Japan).

3. Results

The characteristics of all participants, according to their age groups, are shown in Table 1. The prevalence of calcification ($p < 0.001$), risk of osteoporosis ($p = 0.02$), number of present teeth ($p < 0.001$), and ABL ($p < 0.001$) were noted to significantly increase with advancing age. The number of subjects without and with CAC was 174 (99 men and 75 women; Group NC) and 121 (68 men and 53 women; group C), respectively. The mean ages (SD) of Group C and Group NC individuals were 72.0 (9.7) and 59.4 (10.3) years, respectively.

Table 1. The characteristics of all participants by age group.

	Age							
	30–39 y (*n* = 2)	40–49 y (*n* = 32)	50–59 y (*n* = 65)	60–69 y (*n* = 85)	70–79 y (*n* = 82)	80–89 y (*n* = 24)	90–95 y (*n* = 5)	*p*-Value
Male	1 (50.0)	16 (50.0)	40 (61.5)	54 (63.5)	42 (51.2)	11 (45.8)	3 (60.0)	0.40
Female	1 (50.0)	6 (50.0)	25 (38.5)	31 (36.5)	40 (48.8)	13 (54.2)	2 (40.0)	
Calcification	0 (0.0)	2 (6.3)	11 (16.9)	28 (32.9)	55 (67.1)	20 (83.3)	5 (100.0)	<0.001
Not Calcification	2 (100.0)	30 (93.8)	54 (83.1)	57 (67.1)	27 (32.9)	4 (16.7)	0 (0.0)	
Hypertension Yes	0 (0.0)	3 (9.4)	14 (21.5)	31 (36.5)	37 (45.1)	15 (62.5)	4 (80.0)	0.15
No	2 (100.0)	29 (90.6)	51 (78.5)	54 (63.5)	45 (54.9)	9 (37.5)	1 (20.0)	
Dyslipidemia Yes	0 (0.0)	1 (3.1)	10 (15.4)	12 (14.1)	10 (12.2)	2 (8.3)	1 (20.0)	0.68
No	2 (100.0)	31 (96.9)	55 (84.6)	73 (85.9)	72 (87.8)	22 (91.7)	4 (80.0)	
Diabetes mellitus Yes	0 (0.0)	1 (3.1)	6 (9.2)	14 (16.5)	13 (15.9)	2 (8.3)	2 (40.0)	0.58
No	2 (100.0)	31 (96.9)	59 (90.8)	71 (83.5)	69 (84.1)	22 (91.7)	3 (60.0)	
Osteoporosis Yes	0 (0.0)	0 (0.0)	2 (3.1)	1 (16.5)	13 (15.9)	2 (8.3)	1 (20.0)	0.02
No	2 (100.0)	32 (100.0)	63 (96.9)	84 (83.5)	69 (84.1)	22 (91.7)	4 (80.0)	
Cancer Yes	0 (0.0)	1 (3.1)	3 (9.2)	14 (1.2)	7 (8.5)	5 (20.8)	0 (0.0)	0.38
No	2 (100.0)	31 (96.9)	62 (90.8)	71 (98.8)	75 (91.5)	19 (79.2)	5 (100.0)	
Number of present teeth	27.5 ± 0.7	25.0 ± 4.6	23.9 ± 4.3	21.9 ± 5.7	17.7 ± 7.8	14.5 ± 6.5	11.4 ± 8.8	<0.001 [a]
1–9	0 (0.0)	1 (3.1)	1 (1.5)	3 (3.5)	16 (19.5)	7 (29.2)	2 (40.0)	
10–19	0 (0.0)	3 (9.4)	7 (10.8)	20 (23.5)	22 (26.8)	9 (37.5)	1 (20.0)	0.002
≥20	2 (100.0)	28 (87.5)	57 (87.7)	62 (72.9)	44 (53.7)	8 (33.3)	2 (40.0)	
ABL	19.8 ± 12.4	17.2 ± 7.7	19.4 ± 9.7	21.1 ± 9.3	29.5 ± 12.3	29.6 ± 9.2	35.7 ± 7.4	<0.001 [a]
≤20%	1 (50.0)	26 (81.3)	44 (67.7)	49 (57.6)	19 (23.2)	5 (20.8)	0 (0.0)	
>20%, ≤34%	1 (50.0)	5 (15.6)	15 (23.1)	28 (32.9)	38 (46.3)	11 (45.8)	2 (40.0)	<0.001
>34%	0 (0.0)	1 (3.1)	6 (9.2)	8 (9.4)	25 (30.5)	8 (33.3)	3 (60.0)	

[a] *t*-test: mean ± standard deviation; chi-square test: *n* (%); ABL: the rate of alveolar bone loss.

Significant differences were noted in age ($p < 0.001$), history of hypertension ($p < 0.001$), osteoporosis ($p = 0.004$), number of present teeth ($p < 0.001$), and ABL ($p < 0.001$) between Groups C and NC (Table 2). The rate of ABL in Group C was significantly greater than that in Group NC.

Table 2. The relationship between carotid artery calcification and other variables.

	Group C	Group NC	*p*-Value
	(*n* = 121) Male: 68, Female: 53	(*n* = 174) Male: 99, Female: 75	
Age (Year)	72.0 ± 9.7	59.4 ± 10.3	<0.001 [a]
30–49 y	2 (5.9)	32 (94.1)	
50–59 y	11 (16.9)	54 (83.1)	
60–69 y	28 (32.9)	57 (67.1)	<0.001
70–79 y	55 (67.1)	27 (32.9)	
80–95 y	25 (86.2)	4 (13.8)	
Male	68 (56.2)	99 (56.9)	0.91
Hypertension	69 (57.0)	35 (20.1)	<0.001
Dyslipidemia	14 (11.6)	22 (12.6)	0.78
Osteoporosis	12 (9.9)	4 (2.3)	0.004
Diabetes Mellitus	19 (15.7)	19 (10.9)	0.23
Cancer	13 (10.7)	19 (10.9)	0.96
Number of Present Teeth	17.1 ± 7.9	23.3 ± 4.8	<0.001 [a]
1–9	27 (90.0)	3 (10.0)	
10–19	31 (50.0)	31 (50.0)	<0.001
≥20	63 (31.0)	140 (69.0)	
ABL (%)	32.7 ± 9.7	17.2 ± 7.0	<0.001 [a]
≤20%	8 (5.6)	136 (94.4)	
>20%, ≤34%	67 (67.0)	33 (33.0)	<0.001
>34%	46 (90.2)	5 (9.8)	

[a] *t*-test: mean ± standard deviation; chi-square test: *n* (%); ABL: the rate of alveolar bone loss.

Multivariate logistic regression analysis, with forward selection adjusted for covariates, revealed that the presence of CAC was significantly associated with age (OR = 1.096, 95% CI = 1.051–1.143, $p < 0.001$), history of hypertension (OR = 3.748, 95% CI = 1.748–8.037, $p = 0.001$), and ABL (OR = 1.233, 95% CI = 1.167–1.303, $p < 0.001$; see Table 3). In the final model adjusted for the age group and the presence of hypertension, the OR of having CAC in subjects with 20% < ABL ≤ 34% and ABL > 34% was 23.676 (95% CI = 9.494–59.035, $p < 0.001$) and 111.848 (95% CI = 31.322–399.398, $p < 0.001$) in comparison with subjects with ABL ≤ 20%, respectively (Table 3).

In the ROC analysis predicting the presence of CAC, the AUROC was 0.932 (95% CI = 0.904–0.960, $p < 0.001$) for ABL, 0.815 (95% CI = 0.767–0.864, $p < 0.001$) for age, 0.685 (95% CI – 0.692–0.806, $p < 0.001$) for presence of hypertension, and 0.749 (95% CI = 0.621–0.748, $p < 0.001$) for the number of present teeth (Table 4, Figure 2).

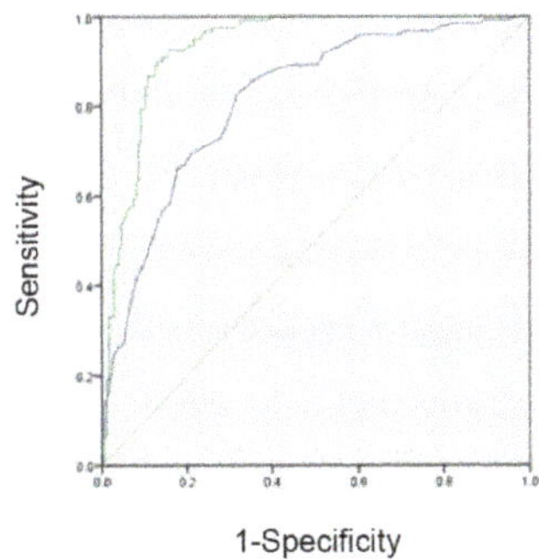

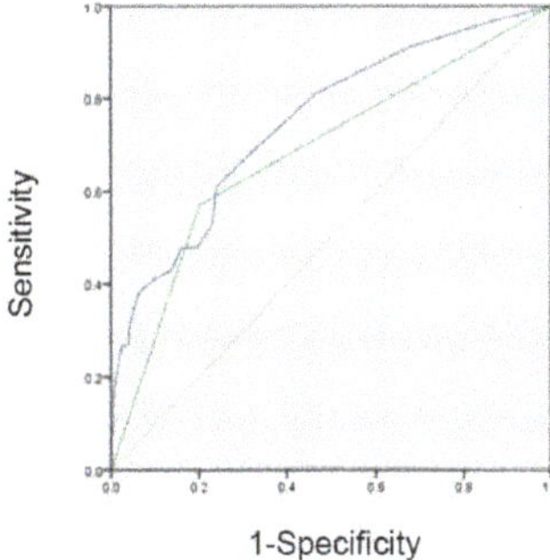

Figure 2. ROC curve of effective factors for the screening of carotid artery calcification. Green line in the left panel represents the rate of ABL. Blue line in the left panel represents the age. Green line in the right panel represents hypertension. Blue line in the right panel represents the number of present teeth.

Table 3. Factors associated with carotid artery calcification and the rate of alveolar bone loss evaluated by multivariate logistic regression analysis using forward selection.

		Partial Regression Coefficient	Standard Error	Odds Ratio (95% CI)	*p*-Value
Step 1	ABL	0.231	0.026	1.260 (1.197–1.325)	<0.001
	constant	−5.744	0.614	0.003	<0.001
Step 2	Age	0.099	0.020	1.105 (1.062–1.149)	<0.001
	ABL	0.214	0.028	1.239 (1.173–1.308)	<0.001
	Constant	−11.916	1.572	0.000	<0.001
Step 3	Age	0.092	0.021	1.096 (1.051–1.143)	<0.001
	Hypertension	1.321	0.389	3.748 (1.748–8.037)	0.001
	ABL	0.210	0.028	1.233 (1.167–1.303)	<0.001
	Constant	−11.883	1.658	0.000	<0.001
Step 1	ABL				
	≤20%			1.000	<0.001
	>20%, ≤34%	3.541	0.421	34.515 (15.112–78.833)	<0.001
	>34%	5.052	0.595	156.400 (48.722–502.053)	<0.001
	Constant	−5.744	0.614	0.003	<0.001
Step 2	Age				
	30–49 y			1.000	<0.001
	50–59 y	0.872	0.924	2.392 (0.391–14.617)	0.345
	60–69 y	1.937	0.883	6.940 (1.128–39.206)	0.028
	70–79 y	2.598	0.882	13.437 (2.387–75.654)	0.003
	80–95 y	4.188	1.102	65.902 (7.606–571.037)	<0.001
	ABL				
	≤20%			1.000	<0.001
	>20%, ≤34%	3.367	0.460	28.988 (11.759–71.456)	<0.001
	>34%	4.747	0.643	115.220 (32.686–406.154)	<0.001
	Constant	−4.705	0.905	0.009	<0.001
Step 3	Age				
	30–49 y			1.000	<0.001
	50–59 y	0.534	0.934	1.705 (0.273–10.638)	0.568
	60–69 y	1.498	0.893	4.471 (0.776–25.758)	0.094
	70–79 y	2.194	0.889	8.974 (1.571–51.274)	0.014
	80–95 y	3.747	1.130	42.410 (4.626–388.797)	0.001
	Hypertension	1.026	0.399	2.790 (1.275–6.104)	0.001
	ABL				
	≤20%			1.000	<0.001
	>20%, ≤34%	3.164	0.466	23.676 (9.494–59.035)	<0.001
	>34%	4.717	0.649	111.848 (31.322–399.398)	<0.001
	Constant	−11.883	1.658	0.000	<0.001

CI: confidence interval; ABL: the rate of alveolar bone loss.

Table 4. Effective factors for the screening of carotid artery calcification using ROC analysis.

Covariance	AUROC	Standard Error	*p*-Value (95% CI)
ABL	0.932	0.014	<0.001 (0.904–0.960)
Age	0.815	0.025	<0.001 (0.767–0.864)
Hypertension	0.685	0.032	<0.001 (0.621–0.748)
Number of Present Teeth	0.749	0.029	<0.001 (0.692–0.806)

CI: confidence interval; ABL: the rate of alveolar bone loss; AUROC: area under the receiver operating characteristic curve.

4. Discussion

This is the first study of its kind to investigate the association between the presence of CAC detected on CT and ABL that demonstrates the progression of periodontal diseases in a Japanese population.

Based on the report by Mattila et al. in 1989, no consensus has been attained on the presence of a causal relationship between periodontal diseases and ischemic heart diseases [20]. However, in their 2007 meta-analysis, Bahekar et al. noted that periodontal diseases were associated with an increased incidence of ischemic heart diseases [21]. In addition, in 2012, the American Heart Association submitted a systematic review of about 500 articles published between 1950 and 2011 that reported associations between periodontal diseases and atherosclerotic vascular disease (ASVD) [22]. Although periodontal interventions reduce the incidences of systemic inflammation and endothelial dysfunction in short-term studies, there is presently no evidence of them preventing ASVD or modifying its outcomes [22]. However, several common risk factors have been reported between periodontal diseases and ischemic heart diseases as confounding factors [23,24]. A 2014 meta-analysis reported that a periodontal disease increases the incidence of cerebrovascular disease [25]. In Japan, Taguchi et al. reported a significant association among the number of lacunar infarctions, a type of cerebral infarction, and ABL [26]. However, a cohort study found no such significant association [27].

In a recent study, periodontal bacterial flora was found to be associated with vascular diseases [28]. Advanced periodontitis in patients with ischaemic stroke is associated with a greater neurological deficit on admission [29], whereas epidemiological studies have reported no association between periodontal PD and myocardial infarction, stroke, or heart failure [30]. These confounding reports suggest that evidence linking periodontitis and cerebrovascular disease is inadequate, considering the lack of a consensus on the definition of a periodontal disease as well as on the objectives of clinical parameters employed in epidemiological studies and/or in interventional studies using standardized treatment protocols. On the other hand, periodontal disease affects short-term systemic inflammatory conditions and vascular endothelial cell functions. Therefore, long-term observational studies are required to clarify this point.

Periodontal diseases are associated with FMD and IMT (noninvasive assessments of vascular functions). The FMD refers to the rate of change in the blood vessel diameter; if the denominator vessel diameter is large, the FMD is relatively low even when the functions are normal. In addition, the image quality is poor in elderly or obese patients, as they tend to have larger upper arm diameters [31]. The diagnoses of early-stage atherosclerosis are usually performed via c-IMT with the detection of carotid artery echoes. Some past reviews have reported no significant association between c-IMT-indicated progression and increased risk for cardiovascular events. CT is slightly invasive, but it reliably detects calcification, which is the final stage of atherosclerosis, and CAC is associated with cardiovascular diseases [32–35]. Hence, CAC screening is considered more useful than the applications of FMD and c-IMT.

Since Friedlander et al. reported that CAC can be identified in panoramic radiographs [36], several researchers have evaluated the CAC-associated risks for cardiovascular lesions [32–35]. Accordingly, several positive associations have been reported, including that the CAC status is an effective indicator for cardiovascular lesions [37] as it can predict carotid stenosis [32], which is associated with peripheral arterial diseases (in Koreans aged ≥50 years) [33]. Peripheral arterial diseases are present in 84% of the patients with carotid artery stenosis [34] and are associated with carotid atherosclerosis [35]. Thus, a relationship exists between the assessment of CACs on digital panoramic radiographs and periodontitis [28,38,39]. However, Thanakun et al. reported that ABL was not associated with CAC [40]. In this study, we diagnosed CAC using CT because CT could be used to diagnose CAC with higher probability than panoramic radiography [41]. Moreover, we noted a significant association between the number of present teeth and CAC. However, ROC analyses revealed that ABL was a better screening factor than the number of present teeth.

We identified some limitations in this study. First, all subjects visited the Matsumoto Dental University Hospital and, therefore, probably lived in a specific region of Japan, which makes them nonrepresentative of the entire Japanese population. This factor may have introduced selection bias. Second, although we extracted the age and the status of hypertension, dyslipidemia, diabetes, osteoporosis, and cancer of the subjects from their medical records, we lacked data on their smoking status and the levels of C-reactive protein, total cholesterol, and high-density lipoprotein, which are all associated with CAC. In the Framingham heart study, 5573 first- and second-generation patients lacked cardiovascular diseases, high blood pressure, total cholesterol level, smoking, glucose-tolerance, and left ventricular hypertrophy, which are all risk factors for CVD progression [42]. Therefore, these factors may have also affected the CAC progression. We intend to explore this topic in the future. Third, we evaluated the periodontal disease status using a single method. Usually, periodontal PD, clinical attachment level (CAL), and bleeding on probing (BOP) are some of the factors used to diagnose periodontal diseases. However, as several of our patients underwent oral surgery, we lacked data on these points. We instead explored whether the CAC status was evident from the obtained panoramic images. We noted that ABL was more convenient to assess than PD, CAL, and BOP. Digital panoramic radiograph grades were also in good agreement with other radiographic, periodontal-disease-grading methods [43,44] and compatible with other measurement methods of periodontal diseases and other diagnostic methods [45–47]. We plan to develop a tool that measures the ABL rate in panoramic radiographs. Fourth, as this was a cross-sectional study, we could not clarify the presence of any causal relationship between CAC and ABL. This point requires further longitudinal study.

The strength of our study is that we enrolled 295 patients of different ages, unlike in previous related studies. In the ROC analyses, the AUROCs for ABL, age, hypertension, and the number of teeth were 0.932, 0.815, 0.685, and 0.749, respectively, suggesting that ABL and age may be used to screen for individuals with CAC. To the best of our knowledge, only a few epidemiological surveys have been conducted to evaluate the association between CAC and ABL in Japan.

5. Conclusions

The measurement of alveolar bone resorption in panoramic radiographs may effectively identify patients at increased risk for CAC. Further cross-sectional and longitudinal studies on a large scale, involving the exploration of important covariates such as smoking, are warranted in the future.

Author Contributions: Study design: N.Y., K.U., and A.T.; data collection: K.U. and A.T.; formal analysis and writing: A.T., N.Y., Y.I., and N.D.; data interpretation: A.T., N.Y., and N.D.; critical review: K.U., Y.H., and A.Y. All authors have read and agreed to the published version of the manuscript.

Funding: This study was supported by the Japan Society for the Promotion of Science, Grants-in-Aid for Scientific Research (18K09758).

Acknowledgments: The authors thank the study participants and the staff of Matsumoto Dental University Hospital.

Conflicts of Interest: The authors declare no conflict of interest.

References

1. Vital Statistics of Japan—The Latest Trends. 2018. Available online: https://www.mhlw.go.jp/toukei/saikin/hw/jinkou/kakutei18/dl/11_h7.pdf (accessed on 14 September 2020).

2. Data of Patient Survey 2017. Available online: https://www.e-stat.go.jp/stat-search/database?page=1&layout=datalist&toukei=00450022&tstat=000001031167&cycle=7&tclass1=000001124800&tclass2=000001124803&toukei_kind=6&statdisp_id=0003315901&result_page=1 (accessed on 14 September 2020).

3. Pihlstrom, B.L.; Michalowicz, B.S.; Johnson, N.W. Periodontal diseases. *Lancet* **2005**, *366*, 1809–1820. [CrossRef]

4. Dye, B.A. Global periodontal disease epidemiology. *Periodontol 2000* **2012**, *58*, 10–25. [CrossRef]

5. Data of Survey of Dental Diseases 2016. Available online: https://www.mhlw.go.jp/toukei/list/dl/62-28-02.pdf (accessed on 14 September 2020).

6. Beck, J.D.; Elter, J.R.; Heiss, G.; Couper, D.; Mauriello, S.M.; Offenbacher, S. Relationship of periodontal disease to carotid artery intima-media wall thickness: The atherosclerosis risk in communities (ARIC) study. *Arter. Thromb. Vasc. Biol.* **2001**, *21*, 1816–1822. [CrossRef]

7. Beck, J.D.; Eke, P.; Lin, D.; Madianos, P.; Couper, D.; Moss, K.; Elter, J.; Heiss, G.; Offenbacher, S. Associations between IgG antibody to oral organisms and carotid intima-medial thickness in community-dwelling adults. *Atherosclerosis* **2005**, *183*, 342–348. [CrossRef] [PubMed]

8. Desvarieux, M.; Schwahn, C.; Volzke, H.; Demmer, R.T.; Ludemann, J.; Kessler, C.; Jacobs, D.R., Jr.; John, U.; Kocher, T. Gender differences in the relationship between periodontal disease, tooth loss, and atherosclerosis. *Stroke* **2004**, *35*, 2029–2035. [CrossRef] [PubMed]

9. Desvarieux, M.; Demmer, R.T.; Rundek, T.; Boden-Albala, B.; Jacobs, D.R., Jr.; Sacco, R.L.; Papapanou, P.N. Periodontal microbiota and carotid intima-media thickness: The Oral Infections and Vascular Disease Epidemiology Study (INVEST). *Circulation* **2005**, *111*, 576–582. [CrossRef] [PubMed]

10. Seinost, G.; Wimmer, G.; Skerget, M.; Thaller, E.; Brodmann, M.; Gasser, R.; Bratschko, R.O.; Pilger, E. Periodontal treatment improves endothelial dysfunction in patients with severe periodontitis. *Am. Heart J.* **2005**, *149*, 1050–1054. [CrossRef] [PubMed]

11. Elter, J.R.; Hinderliter, A.L.; Offenbacher, S.; Beck, J.D.; Caughey, M.; Brodala, N.; Madianos, P.N. The effects of periodontal therapy on vascular endothelial function: A pilot trial. *Am. Heart J.* **2006**, *151*, 47. [CrossRef]

12. Tonetti, M.S.; D'Aiuto, F.; Nibali, L.; Donald, A.; Storry, C.; Parkar, M.; Suvan, J.; Hingorani, A.D.; Vallance, P.; Deanfield, J. Treatment of periodontitis and endothelial function. *N. Engl. J. Med.* **2007**, *356*, 911–920. [CrossRef]

13. Orlandi, M.; Suvan, J.; Petrie, A.; Donos, N.; Masi, S.; Hingorani, A.; Deanfield, J.; D'Aiuto, F. Association between periodontal disease and its treatment, flow-mediated dilatation and carotid intima-media thickness: A systematic review and meta-analysis. *Atherosclerosis* **2014**, *236*, 39–46. [CrossRef]

14. Nicoll, R.; Henein, M.Y. The predictive value of arterial and valvular calcification for mortality and cardiovascular events. *Int. J. Cardiol. Heart Vessel.* **2014**, *3*, 1–5. [CrossRef] [PubMed]

15. Ibrahimi, P.; Jashari, F.; Nicoll, R.; Bajraktari, G.; Wester, P.; Henein, M.Y. Coronary and carotid atherosclerosis: How useful is the imaging? *Atherosclerosis* **2013**, *231*, 323–333. [CrossRef] [PubMed]

16. Kim, T.S.; Obst, C.; Zehaczek, S.; Geenen, C. Detection of bone loss with different X-ray techniques in periodontal patients. *J. Periodontol.* **2008**, *79*, 1141–1149. [CrossRef]

17. Bassiouny, M.A.; Grant, A.A. The accuracy of the Schei ruler: A laboratory investigation. *J. Periodontol.* **1975**, *46*, 748–752. [CrossRef]

18. Ryden, L.; Buhlin, K.; Ekstrand, E.; de Faire, U.; Gustafsson, A.; Holmer, J.; Kjellstrom, B.; Lindahl, B.; Norhammar, A.; Nygren, A.; et al. Periodontitis Increases the Risk of a First Myocardial Infarction: A Report From the PAROKRANK Study. *Circulation* **2016**, *133*, 576–583.

19. Swets, J.A. Measuring the accuracy of diagnostic systems. *Science* **1988**, *240*, 1285–1293. [CrossRef]

20. Mattila, K.J.; Nieminen, M.S.; Valtonen, V.V.; Rasi, V.P.; Kesaniemi, Y.A.; Syrjala, S.L.; Jungell, P.S.; Isoluoma, M.; Hietaniemi, K.; Jokinen, M.J. Association between dental health and acute myocardial infarction. *BMJ* **1989**, *298*, 779–781. [CrossRef] [PubMed]

21. Bahekar, A.A.; Singh, S.; Saha, S.; Molnar, J.; Arora, R. The prevalence and incidence of coronary heart disease is significantly increased in periodontitis: A meta-analysis. *Am. Heart J.* **2007**, *154*, 830–837. [CrossRef]

22. Lockhart, P.B.; Bolger, A.F.; Papapanou, P.N.; Osinbowale, O.; Trevisan, M.; Levison, M.E.; Taubert, K.A.; Newburger, J.W.; Gornik, H.L.; Gewitz, M.H.; et al. Periodontal disease and atherosclerotic vascular disease: Does the evidence support an independent association?: A scientific statement from the American Heart Association. *Circulation* **2012**, *125*, 2520–2544. [CrossRef]

23. Hujoel, P.P.; Drangsholt, M.; Spiekerman, C.; DeRouen, T.A. Periodontal disease and coronary heart disease risk. *JAMA* **2000**, *284*, 1406–1410. [CrossRef]

24. Peacock, M.E.; Carson, R.E. Frequency of self-reported medical conditions in periodontal patients. *J. Periodontol.* **1995**, *66*, 1004–1007. [CrossRef]

25. Lafon, A.; Pereira, B.; Dufour, T.; Rigouby, V.; Giroud, M.; Bejot, Y.; Tubert-Jeannin, S. Periodontal disease and stroke: A meta-analysis of cohort studies. *Eur. J. Neurol.* **2014**, *21*, 1155-e67. [CrossRef] [PubMed]

26. Taguchi, A.; Miki, M.; Muto, A.; Kubokawa, K.; Migita, K.; Higashi, Y.; Yoshinari, N. Association between oral health and the risk of lacunar infarction in Japanese adults. *Gerontology* **2013**, *59*, 499–506. [CrossRef] [PubMed]

27. Howell, T.H.; Ridker, P.M.; Ajani, U.A.; Hennekens, C.H.; Christen, W.G. Periodontal disease and risk of subsequent cardiovascular disease in U.S. male physicians. *J. Am. Coll Cardiol.* **2001**, *37*, 445–450. [CrossRef]

28. Boillot, A.; Demmer, R.T.; Mallat, Z.; Sacco, R.L.; Jacobs, D.R.; Benessiano, J.; Tedgui, A.; Rundek, T.; Papapanou, P.N.; Desvarieux, M. Periodontal microbiota and phospholipases: The Oral Infections and Vascular Disease Epidemiology Study (INVEST). *Atherosclerosis* **2015**, *242*, 418–423. [CrossRef]

29. Slowik, J.; Wnuk, M.A.; Grzech, K.; Golenia, A.; Turaj, W.; Ferens, A.; Jurczak, A.; Chomyszyn-Gajewska, M.; Loster, B.; Slowik, A. Periodontitis affects neurological deficit in acute stroke. *J. Neurol. Sci.* **2010**, *297*, 82–84. [CrossRef]

30. Holmlund, A.; Lampa, E.; Lind, L. Oral health and cardiovascular disease risk in a cohort of periodontitis patients. *Atherosclerosis* **2017**, *262*, 101–106. [CrossRef]

31. Guidelines for Non-Invasive Vascular Function Test (JCS 2013). 2013. Available online: https://www.j-circ.or.jp/old/guideline/pdf/JCS2013_yamashina_h.pdf (accessed on 14 September 2020).

32. Almog, D.M.; Horev, T.; Illig, K.A.; Green, R.M.; Carter, L.C. Correlating carotid artery stenosis detected by panoramic radiography with clinically relevant carotid artery stenosis determined by duplex ultrasound. *Oral. Surg. Oral. Med. Oral. Pathol. Oral. Radiol. Endod.* **2002**, *94*, 768–773. [CrossRef]

33. Lee, J.S.; Kim, O.S.; Chung, H.J.; Kim, Y.J.; Kweon, S.S.; Lee, Y.H.; Shin, M.H.; Yoon, S.J. The prevalence and correlation of carotid artery calcification on panoramic radiographs and peripheral arterial disease in a population from the Republic of Korea: The Dong-gu study. *Dentomaxillofac Radiol.* **2013**, *42*, 29725099. [CrossRef]

34. Garoff, M.; Johansson, E.; Ahlqvist, J.; Jaghagen, E.L.; Arnerlov, C.; Wester, P. Detection of calcifications in panoramic radiographs in patients with carotid stenoses >/=50%. *Oral. Surg. Oral. Med. Oral. Pathol. Oral. Radiol.* **2014**, *117*, 385–391. [CrossRef]

35. Lee, J.S.; Kim, O.S.; Chung, H.J.; Kim, Y.J.; Kweon, S.S.; Lee, Y.H.; Shin, M.H.; Yoon, S.J. The correlation of carotid artery calcification on panoramic radiographs and determination of carotid artery atherosclerosis with ultrasonography. *Oral. Surg. Oral. Med. Oral. Pathol. Oral. Radiol.* **2014**, *118*, 739–745. [CrossRef] [PubMed]

36. Friedlander, A.H.; Lande, A. Panoramic radiographic identification of carotid arterial plaques. *Oral. Surg. Oral. Med. Oral. Pathol.* **1981**, *52*, 102–104. [CrossRef]

37. Cohen, S.N.; Friedlander, A.H.; Jolly, D.A.; Date, L. Carotid calcification on panoramic radiographs: An important marker for vascular risk. *Oral. Surg. Oral. Med. Oral. Pathol. Oral. Radiol. Endod.* **2002**, *94*, 510–514. [CrossRef] [PubMed]

38. Bilgin Çetin, M.; Sezgin, Y.; Nisancı Yilmaz, M.N.; Köseoğlu Seçgin, C. Assessment of carotid artery calcifications on digital panoramic radiographs and their relationship with periodontal condition and cardiovascular risk factors. *Int. Dent. J.* **2020**. [CrossRef] [PubMed]

39. Paju, S.; Pietiäinen, M.; Liljestrand, J.M.; Lahdentausta, L.; Salminen, A.; Kopra, E.; Mäntylä, P.; Buhlin, K.; Hörkkö, S.; Sinisalo, J.; et al. Carotid artery calcification in panoramic radiographs associates with oral infections and mortality. *Int. Endod. J.* **2020**. [CrossRef]

40. Thanakun, S.; Pornprasertsuk-Damrongsri, S.; Izumi, Y. C-reactive protein levels and the association of carotid artery calcification with tooth loss. *Oral. Dis.* **2017**, *23*, 69–77. [CrossRef]

41. Otawara, H.; Jinbu, Y.; Shinozaki, Y.; Hayasaka, J.; Tsuchiya, Y.; Itoh, H.; Noguchi, T.; Mori, Y.; Iida, Y.; Katsumata, A. Carotid calcification in panoramic Radiography. *J. Jpn.Oral. Med.* **2015**, *21*, 33–37. [CrossRef]

42. Anderson, K.M.; Odell, P.M.; Wilson, P.W.; Kannel, W.B. Cardiovascular disease risk profiles. *Am. Heart J.* **1991**, *121*, 293–298. [CrossRef]

43. Kaimenyi, J.T.; Ashley, F.P. Assessment of bone loss in periodontitis from panoramic radiographs. *J. Clin. Periodontol.* **1988**, *15*, 170–174. [CrossRef]

44. Rohlin, M.; Kullendorff, B.; Ahlqwist, M.; Henrikson, C.O.; Hollender, L.; Stenström, B. Comparison between panoramic and periapical radiography in the diagnosis of periapical bone lesions. *Dento Maxillo Facial Radiol.* **1989**, *18*, 151–155. [CrossRef]

45. Kiliç, A.R.; Efeoglu, E.; Yilmaz, S.; Orgun, T. The relationship between probing bone loss and standardized radiographic analysis. *Periodontal. Clin. Investig.* **1998**, *20*, 25–32. [PubMed]

46. Eickholz, P.; Hausmann, E. Accuracy of radiographic assessment of interproximal bone loss in intrabony defects using linear measurements. *Eur. J. Oral. Sci.* **2000**, *108*, 70–73. [CrossRef] [PubMed]

47. Graetz, C.; Plaumann, A.; Wiebe, J.-F.; Springer, C.; Sälzer, S.; Dörfer, C.E. Periodontal probing versus radiographs for the diagnosis of furcation involvement. *J. Periodontol.* **2014**, *85*, 1371–1379. [CrossRef] [PubMed]

Publisher's Note: MDPI stays neutral with regard to jurisdictional claims in published maps and institutional affiliations.

Journal of
Clinical Medicine

Article

Exposure to *Porphyromonas gingivalis* Induces Production of Proinflammatory Cytokine via TLR2 from Human Respiratory Epithelial Cells

Norihisa Watanabe [1,2], Sho Yokoe [1,2], Yorimasa Ogata [3], Shuichi Sato [1] and Kenichi Imai [2,*]

1   Department of Periodontology, Nihon University School of Dentistry, Chiyoda-ku, Tokyo 101-8310, Japan; watanabe.norihisa@nihon-u.ac.jp (N.W.); desh18040@g.nihon-u.ac.jp (S.Y.); satou.shuuichi@nihon-u.ac.jp (S.S.)
2   Department of Microbiology, Nihon University School of Dentistry, 1-8-13 Kanda-Surugadai, Chiyoda-ku, Tokyo 101-8310, Japan
3   Department of Periodontology, Nihon University School of Dentistry at Matsudo, Chiba 271-8587, Japan; ogata.yorimasa@nihon-u.ac.jp
*   Correspondence: imai.kenichi@nihon-u.ac.jp; Tel.: +81-3-3219-8115; Fax: +81-3-3219-8317

Received: 18 September 2020; Accepted: 22 October 2020; Published: 26 October 2020

Abstract: Aspiration pneumonia is a major health problem owing to its high mortality rate in elderly people. The secretion of proinflammatory cytokines such as interleukin (IL)-8 and IL-6 by respiratory epithelial cells, which is induced by infection of respiratory bacteria such as *Streptococcus pneumoniae*, contributes to the onset of pneumonia. These cytokines thus play a key role in orchestrating inflammatory responses in the lower respiratory tract. In contrast, chronic periodontitis, a chronic inflammatory disease caused by the infection of periodontopathic bacteria, typically *Porphyromonas gingivalis*, is one of the most prevalent microbial diseases affecting humans globally. Although emerging evidence has revealed an association between aspiration pneumonia and chronic periodontitis, a causal relationship between periodontopathic bacteria and the onset of aspiration pneumonia has not been established. Most periodontopathic bacteria are anaerobic and are therefore unlikely to survive in the lower respiratory organs of humans. Therefore, in this study, we examined whether simple contact by heat-inactivated *P. gingivalis* induced proinflammatory cytokine production by several human respiratory epithelial cell lines. We found that *P. gingivalis* induced strong IL-8 and IL-6 secretion by BEAS-2B bronchial epithelial cells. *P. gingivalis* also induced strong IL-8 secretion by Detroit 562 pharyngeal epithelial cells but not by A549 alveolar epithelial cells. Additionally, Toll-like receptor (TLR) 2 but not TLR4 was involved in the *P. gingivalis*-induced proinflammatory cytokine production. Furthermore, *P. gingivalis* induced considerably higher IL-8 and IL-6 production than heat-inactivated *S. pneumoniae*. Our results suggest that *P. gingivalis* is a powerful inflammatory stimulant for human bronchial and pharyngeal epithelial cells and can stimulate TLR2-mediated cytokine production, thereby potentially contributing to the onset of aspiration pneumonia.

Keywords: aspiration pneumonia; chronic periodontitis; *Porphyromonas gingivalis*; proinflammatory cytokines; TLR2

---

1. Introduction

Aspiration pneumonia is a major health problem because of its high mortality rates in elderly people who are frequently immunocompromised [1]. This occurs because of inflammation in the lower respiratory tract that is stimulated by mis-swallowed saliva entering through the pharynx [1]. Pneumonia is characterized by increased number of infiltrated inflammatory cells such as neutrophils and elevated levels of proinflammatory cytokines, including chemokines, in the lower respiratory tract [2,3]. The so-called respiratory bacteria, such as *Streptococcus pneumoniae*, are exogenous; hence, they initially infect lower respiratory epithelia and subsequently induce proinflammatory cytokine production [3]. In turn, secreted cytokines such as interleukin (IL)-8 recruit neutrophils to infected tissues, and IL-6 exerts proinflammatory effects on respiratory epithelial cells [3–7]. These cytokines thus play a key role in orchestrating inflammatory responses in the lower respiratory tract in the development of pneumonia.

Chronic periodontitis is defined as an endogenous microflora-associated chronic inflammatory disease that develops as a result of poor oral hygiene [8]. The disease which causes the destruction of the periodontium, including alveolar bone, is one of the most prevalent infectious diseases worldwide [8]. Chronic periodontitis is caused by increased inflammatory responses, including proinflammatory cytokine production executed by host local immunity that is triggered by and amplified in relation to an increase in the number of periodontopathic bacteria [8]. Among them, the most pathogenic bacterium is *Porphyromonas gingivalis*, a Gram-negative black-pigmented anaerobe that predominantly colonizes periodontal pockets and is detected on the tongue dorsum and in the saliva of patients with advanced chronic periodontitis [9]. *P. gingivalis* exerts its virulence by interacting with several host pattern recognition receptors such as Toll-like receptor (TLR) 2 and TLR4 [10].

Accumulating evidence indicates that chronic periodontitis is a risk factor for several systemic diseases such as pre-term birth, heart diseases, diabetes, and atherosclerosis [8]. In this regard, we have reported that a major metabolite of *P. gingivalis* possibly induces the reactivation of latently infected viruses, namely human immunodeficiency virus and Epstein–Barr virus [11,12]. Over the last two decades, chronic periodontitis has also been identified as a risk factor for aspiration pneumonia in the elderly [13,14]. *P. gingivalis* is isolated from bronchoalveolar lavage fluid (BALF) or sputum from patients with pneumonia [15–18]. In fact, an increase in teeth with periodontal pockets in the elderly is associated with increased mortality from aspiration pneumonia [19]. Furthermore, periodontal interventions such as oral hygiene instruction reduce the occurrence of aspiration pneumonia, even among high-risk individuals [20]. Further, in the elderly, there is an increased risk of aspiration because these individuals have reduced laryngopharyngeal sensitivity [21]. Thus, these observations suggest that periodontopathic bacteria present in saliva are aspirated through the pharynx into the lower respiratory tract, thereby contributing to the onset of aspiration pneumonia. Despite its importance, a causal relationship between *P. gingivalis* and aspiration pneumonia remains unexamined.

Based on the aforementioned observations, we have hypothesized that increased number of *P. gingivalis* in the aspiration may induce proinflammatory cytokine production by human respiratory epithelia. However, *P. gingivalis is* anaerobic, and it is therefore unlikely to exhibit stable virulence in the respiratory tract. In addition, the duration of *P. gingivalis* survival in vitro and in vivo remains unclear. Therefore, in this study, we used heat-inactivated bacterial cells to examine whether *P. gingivalis* induced the production of proinflammatory cytokines such as IL-8 and IL-6 by human bronchial, alveolar, and pharyngeal epithelial cells via TLR2 or TLR4. To the best of our knowledge, we have delineated for the first time a putative mechanism by which *P. gingivalis* induces inflammation in human respiratory epithelia, thereby potentially contributing to the onset of aspiration pneumonia.

2. Materials and Methods

2.1. Cell Culture and Reagents

Human bronchial (BEAS-2B), pharyngeal (Detroit 562), and alveolar (A549) epithelial cells were purchased from ATCC (Manassas, VA, USA) and maintained at 37 °C in Dulbecco's modified Eagle's medium (Sigma, St. Louis, MO, USA) containing 10% heat-inactivated fetal bovine serum (Thermo Fisher Scientific, Rockford, IL, USA), penicillin (100 U/mL), and streptomycin (100 µg/mL), as described previously. HEK293 human embryonic kidney cells stably transfected with an expression plasmid for either human TLR2 (293-TLR2; InvivoGen, San Diego, CA, USA) or human TLR4 (293-TLR4; InvivoGen) were purchased and maintained at 37 °C in Dulbecco's modified Eagle's medium (Sigma) containing 10% heat-inactivated fetal bovine serum (Thermo Fisher Scientific) in the presence of the antibiotic blasticidin (10 µg/mL) to maintain selection for these transfectants. Neutralizing antibodies against human TLR2 and TLR4 were obtained from R&D Systems (Minneapolis, MN, USA). Lipoteichoic acid (LTA) derived from *Staphylococcus aureus* as a TLR2 ligand and lipopolysaccharide (LPS) derived from *Escherichia coli* as a TLR4 ligand were also purchased from Sigma.

2.2. Bacterial Culture and Sample Adjustment

P. gingivalis ATCC 33,277 was cultured in brain heart infusion broth (BHIB: Becton, Dickinson and Company, Sparks, MD, USA) supplemented with 5 µg/mL hemin and 0.5 µg/mL menadione. *S. pneumoniae* ATCC 6303 was cultured in BHIB. The cultures were incubated at 37 °C for 24–72 h and grown in an anaerobic chamber (Te-Her Anaerobox, Hirasawa Co. Ltd., Tokyo, Japan) under an anaerobic condition of 10% H_2, 10% CO_2, and 80% N_2. The bacterial cell density was adjusted to 1.0×10^{10} CFU/mL, and the bacterial suspension was heat-inactivated at 60 °C for 1 h and stored at −80 °C until use.

2.3. mRNA Preparation and Real-Time Polymerase Chain Reaction (PCR)

The experimental procedures for RNA purification and real-time PCR were performed as previously described [22]. Briefly, the cells were washed once with ice-cold phosphate-buffered saline and homogenized using QIAshredder (QIAGEN, Alameda, CA, USA), and total RNA was purified using an RNeasy Mini Kit (QIAGEN). For cDNA synthesis, total RNA (1 µg) was reverse-transcribed using an RNA PCR kit (PrimeScript; Takara Bio, Shiga, Japan). The resulting cDNA mixture was subjected to real-time PCR analysis using Premix Ex *Taq* solution (Takara Bio) containing 5 µM sense and antisense primers. The primer sequences used for the amplification of each gene were as follows: IL-8, forward (5′-CTT GTC ATT GCC AGC TGT GT-3′) and reverse (5′-TGA CTG TGG AGT TTT GGC TG-3′); IL-6, forward (5′-TTC GGT CCA GTT GCC TTC TC-3′) and reverse (5′-GAG GTG AGT GGC TGT CTG TG-3′); and glyceraldehyde-3-phosphate dehydrogenase (GAPDH), forward (5′-ACC AGC CCC AGC AAG AGC ACA AG-3′) and reverse (5′-TTC AAG GGG TCT ACA TGG CAA CTG-3′). PCR assays were performed using a TP-800 Thermal Cycler Dice Real-Time System (Takara Bio) and analyzed using the software provided by the device manufacturer. The thermal cycling conditions were 40 cycles of 95 °C for 5 s, 60 °C for 30 s, and 72 °C for 1 min. All real-time PCR experiments were performed in triplicate, and the specificity of each product was verified via melting curve analysis. Calculated gene expression levels were normalized to GAPDH mRNA levels.

2.4. Cytokine Measurements

IL-8 and IL-6 concentrations in the cell culture supernatants were measured using an enzyme-linked immunosorbent assay (ELISA) kit (R&D Systems) according to the manufacturer's recommendations. All experiments were performed in triplicate, and data are presented as the mean ± SD.

2.5. Transfection and Luciferase Assay

For NF-κB reporter assays, HEK293, 293-TLR2, or 293-TLR4 cells were plated in 12-well plates (4×10^5 cells/mL) and grown overnight. These cells were transfected with reporter plasmids using a Lipofectamine 2000 transfection reagent (Thermo Fisher Scientific) according to the manufacturer's instructions. Further, 200 ng of 5xκB-luc, a plasmid in which luciferase gene expression is under the control of NF-κB, and 10 ng of an internal control plasmid pRL-TK, which expresses *Renilla reniformis* luciferase under the control of the TK promoter, were used for each transfection. Twenty-four hours after transfection, the cells were incubated in the presence of heat-inactivated *P. gingivalis*, LTA, and LPS for 24 h. Cells were harvested using Passive Lysis Buffer (Promega, Madison, WI, USA), and the extracts were assessed for luciferase activity using a Dual-Luciferase Assay System (Promega) as described previously [12]. Luciferase activity was normalized to *R. reniformis* luciferase activity, which acted as an as an internal control for transfection efficiency.

3. Results

3.1. P. gingivalis Induced IL-8 and IL-6 mRNA Expression and Protein Production by Human Bronchial Epithelial Cells

P. gingivalis is considered to orchestrate dysbiosis of periodontal flora, which triggers inflammatory responses including cytokine production in the periodontium and in turn causes alveolar bone-destruction as a major symptom of chronic periodontitis [8]. We therefore first examined the mRNA induction of IL-8 and IL-6 by BEAS-2B bronchial epithelial cells and pertinent protein release using real-time PCR and ELISA, respectively.

As shown in Figure 1A, *P. gingivalis* markedly upregulated IL-8 and IL-6 mRNA expression in BEAS-2B cells. The levels of both mRNAs rose after a 1 h exposure to *P. gingivalis* and peaked at 3 h of incubation (151 ± 21- and 24 ± 3- fold, respectively). As shown in Figure 1B, *P. gingivalis* elicited the release of both cytokines between 1 and 12 h in a time-dependent manner. To clarify the cause-and-effect relationship between the density of bacteria and the extent of cytokine mRNA expression and protein production levels, we used *P. gingivalis* at densities equivalent to 1×10^7–1×10^8 CFU/mL. As shown in Figure 1C,D, *P. gingivalis* induced cytokine production as well as mRNA expression in a density-dependent manner.

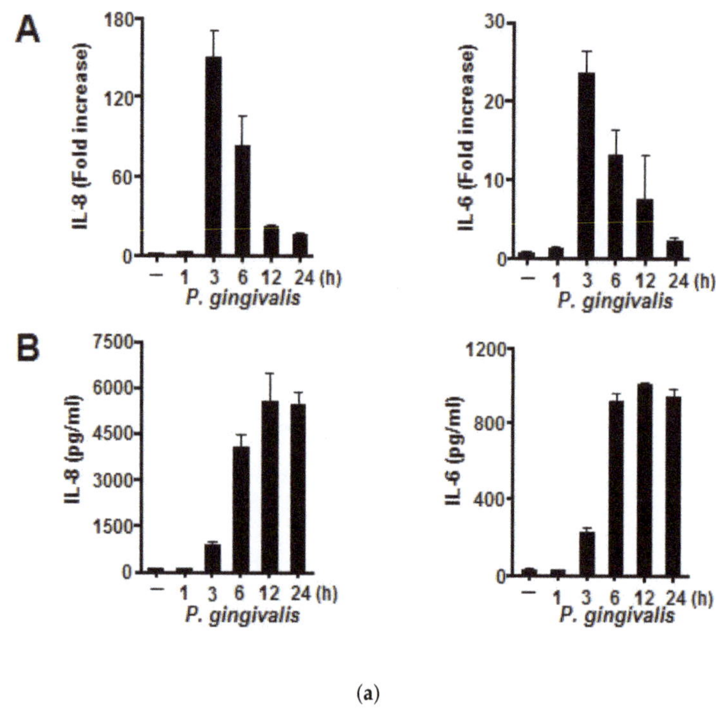

(a)

Figure 1. *Cont.*

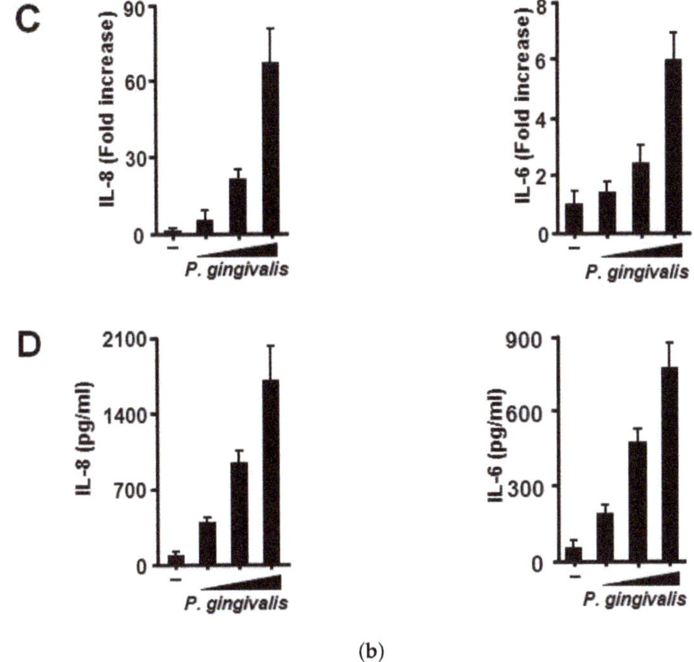

(**b**)

Figure 1. *Porphyromonas gingivalis*-induced mRNA expression and protein production of proinflammatory cytokines by human bronchial epithelial cells. BEAS-2B cells were exposed to heat-inactivated *P. gingivalis* (1×10^8 CFU/mL) for the indicated times (**A,B**) and at different bacterial cell densities (1×10^7, 0.5×10^8, or 1×10^8 CFU/mL) for 3 h (**C**) or 12 h (**D**). The cells were harvested, and IL-8 and IL-6 mRNA levels were measured using real-time polymerase chain reaction with specific primers. The mRNA levels were normalized to the GAPDH mRNA level and expressed as fold increases. IL-8 and IL-6 protein levels were determined by enzyme-linked immunosorbent assay and expressed as pg/mL. These experiments were performed in triplicate, and data are presented as the mean ± SD.

3.2. P. gingivalis Induced IL-8 and IL-6 Production by Human Pharyngeal Epithelial Cells but not by Human Alveolar Epithelial Cells

Because *P. gingivalis*-induced IL-8 and IL-6 production by BEAS-2B bronchial epithelial cells was observed, we examined whether pharyngeal epithelial cell line was also reactive to the stimulation with *P. gingivalis* because every material to be swallowed is retained in the pharynx for a while. As expected, stronger IL-8 production and much weaker IL-6 production than those by BEAS-2B cells were observed in Detroit 562 pharyngeal epithelial cells, which was in a density-dependent manner (Figure 2A). In contrast, proinflammatory cytokine production by *P. gingivalis* was not observed in A549 alveolar epithelial cells (Figure 2B).

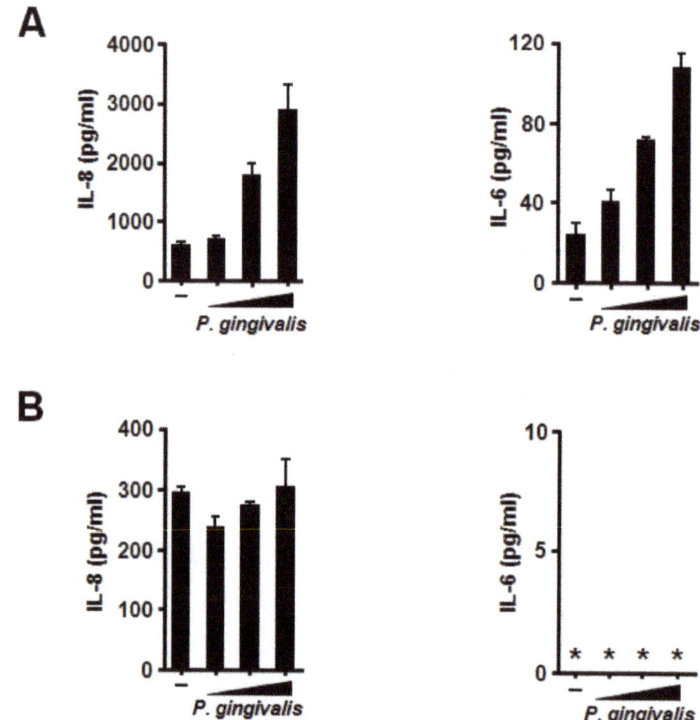

Figure 2. Effects of *Porphyromonas gingivalis* on interleukin (IL)-8 and IL-6 production by two human respiratory epithelial cells. Human pharyngeal (Detroit 562) (**A**) or alveolar (A549) (**B**) epithelial cells were incubated with heat-inactivated *P. gingivalis* (1×10^8 CFU/mL) for 12 h. IL-8 and IL-6 protein levels in the cell culture supernatants were determin ed by enzyme-linked immunosorbent assay. These experiments were performed in triplicate, and data are presented as the mean ± SD. *; below the detection limit.

3.3. P. gingivalis More Strongly Induced IL-8 and IL-6 Production than S. pneumoniae by Human Bronchial Epithelial Cells

S. pneumoniae contributes to the onset of aspiration pneumonia by inducing proinflammatory cytokines release. We therefore compared the inducibility of heat-inactivated bacterial cells between *P. gingivalis* and *S. pneumoniae* by human bronchial and pharyngeal epithelial cells. As shown in Figure 3A, *S. pneumoniae* induced IL-8 and IL-6 release by BEAS-2B cells but at much lower levels than *P. gingivalis*. Much stronger IL-8 production and much weaker IL-6 production were observed with Detroit 562 cells (Figure 3B), whereas no significant cytokine release was observed with A549 cells (Figure 3C).

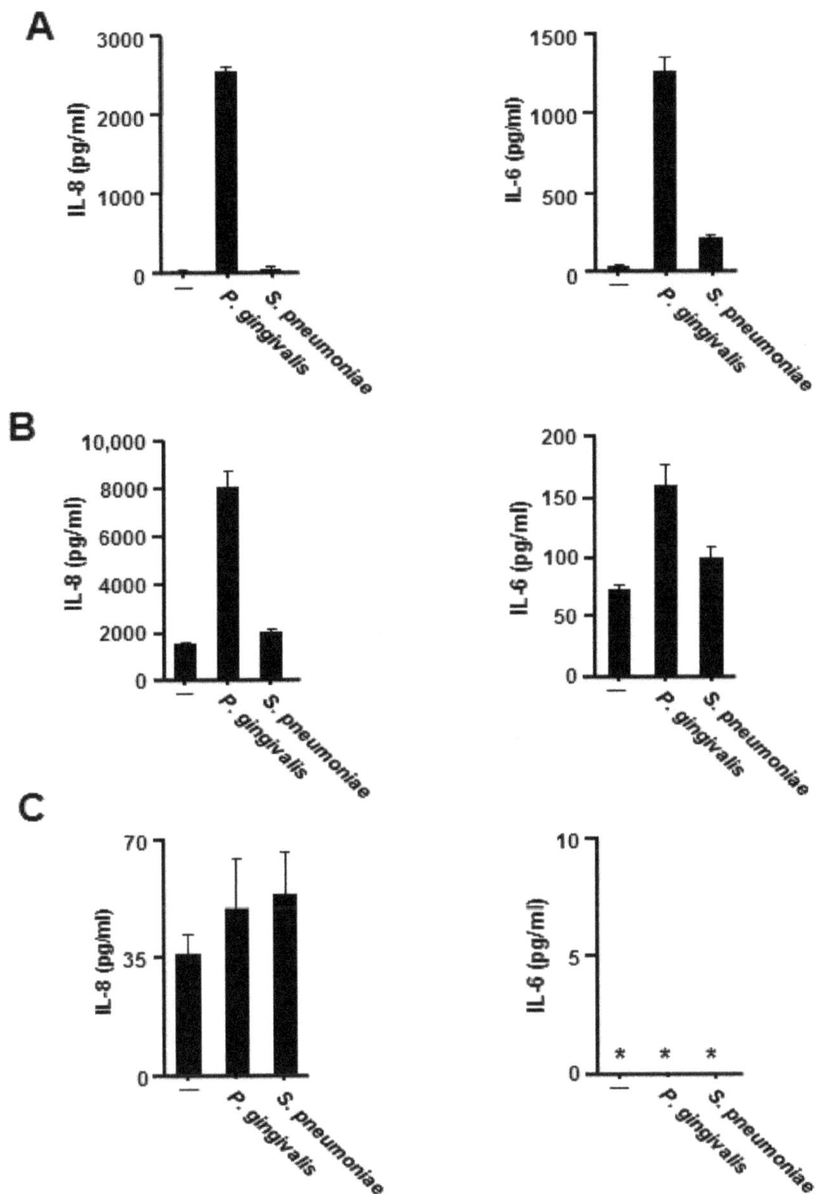

Figure 3. Comparison between the induction of proinflammatory cytokines by *Porphyromonas gingivalis* and by *Streptococcus pneumoniae*. BEAS-2B (**A**), Detroit 562 (**B**) or A549 (**C**) cells were incubated in the presence of heat-inactivated cells (1×10^8 CFU/mL) of *P. gingivalis* or *S. pneumoniae* for 12 h. IL-8 and IL-6 levels in the culture supernatants were then determined using enzyme-linked immunosorbent assay. These experiments were performed in triplicate, and data are presented as the mean ± SD. *; below the detection limit.

3.4. Involvement of TLR2 in P. gingivalis-Induced Proinflammatory Cytokine Production

Several studies have demonstrated TLR2 and TLR4 are important host receptors for recognizing *P. gingivalis* [10]. Eventually, these TLRs are reported to express on BASE-2B cells as well as on human bronchial epithelium and on Detroit 562 cells [23–25]. However, no study has yet examined whether human respiratory epithelial cells recognize *P. gingivalis* with their TLR2 or TLR4. In parallel, for producing IL-8 and IL-6, NF-κB is known to be an important transcription factor [26]. We therefore first examined whether these TLRs were involved in the NF-κB activation in *P. gingivalis*-stimulated epithelial cells using a luciferase assay. For this purpose, TLR-null HEK293 cells and HEK293 cells stably expressing either TLR2 or TLR4 cells were transfected with NF-κB reporter plasmids. As expected, NF-κB in 293-TLR2 cells were activated only by the stimulation with its specific ligand LTA from *S. aureus*, and likewise, 293-TLR4 cells were activated only by LPS from *E. coli*, whereas HEK293 cells were not activated by the both ligands (Figure 4A). When these cells were stimulated with *P. gingivalis*, 293-TLR2 cells alone were activated in a density-dependent manner (Figure 4A). These findings suggest that the observed cytokine production by BEAS-2B and Detroit 562 cells is induced via TLR2 but not via TLR4. To confirm this, BEAS-2B and Detroit 562 cells were treated with anti-TLR2 or anti-TLR4 antibodies prior to the stimulation with *P. gingivalis*. As shown in Figure 4B,C, anti-TLR4 antibodies had no effect on IL-8 and IL-6 production by either cell line following exposure to *P. gingivalis*, whereas the anti-TLR2 antibody significantly abrogated the production of both cytokines.

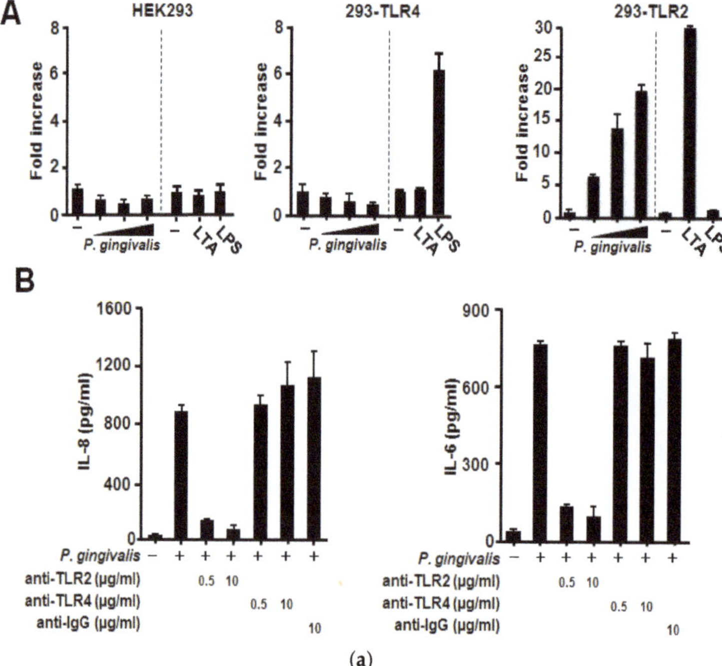

(a)

Figure 4. *Cont.*

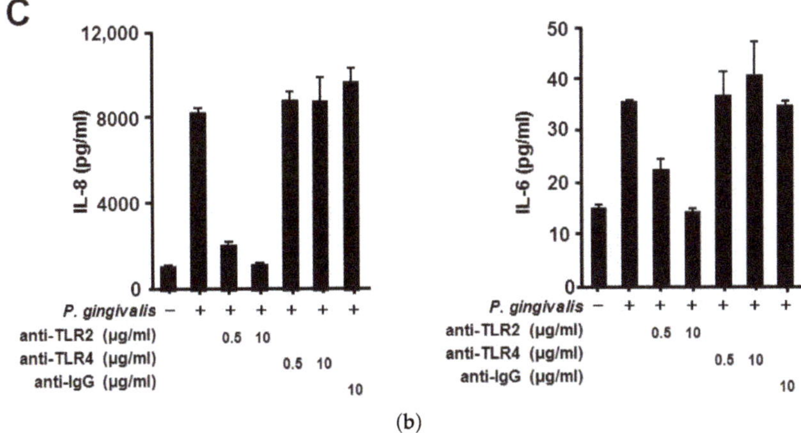

(b)

Figure 4. Involvement of Toll-like receptor (TLR)2 but not TLR4 in *Porphyromonas gingivalis*-induced proinflammatory cytokine production. (**A**) HEK293 cells, 293-TLR4 cells, or 293-TLR2 cells were transfected with a 5×κB-luc reporter plasmid together with an internal pRL-TK control plasmid. Twenty-four hours after transfection, these cells were stimulated with heat-inactivated *P. gingivalis* (1×10^7, 0.5×10^8, or 1×10^8 CFU/mL), *S. aureus* LTA (5 µg/mL), or *E. coli* LPS (200 ng/mL) for 24 h. Luciferase activity in the whole cell lysate was then determined. The data are presented as the fold increase in luciferase activity relative to that in control transfected cells (no stimulation). Data are presented as the mean ± SD of three independent transfections.BEAS-2B (**B**) and Detroit 562 (**C**) cells were pretreated with the indicated concentrations of anti-TLR2, anti-TLR4, or control antibodies for 1 h and then stimulated with or without *P. gingivalis* (1×10^8 CFU/mL). After 12 h incubation, culture supernatants were collected, and IL-8 and IL-6 levels were measured using enzyme-linked immunosorbent assay. These experiments were performed in triplicate, and data are presented as the mean ± SD.

4. Discussion

Because bacteria in saliva are aspirated into the lower respiratory tract, many studies have revealed an association between chronic periodontitis and aspiration pneumonia. In fact, *P. gingivalis* has been isolated from BALF or sputum in patients with aspiration pneumonia [15–18]. However, the mechanism by which the bacterium triggers the development of aspiration pneumonia has not been delineated. By contrast, *S. pneumoniae* is a major exogenous respiratory pathogen that infects the bronchial epithelium. This subsequently induces the release of proinflammatory cytokines, thereby contributing to the development of pneumonia [2,3]. In the present study, we found that *P. gingivalis* is a significant inflammatory stimulant for bronchial and pharyngeal epithelial cells.

IL-8, a potent neutrophil chemoattractant and activator, is associated with the pathogenesis of pneumonia. It accumulates and subsequently degranulates neutrophils, resulting in the destruction of normal tissues [2,3]. In fact, IL-8 levels are markedly elevated in BALF or sputum from patients with pneumonia in relation to an increase of neutrophil counts [4–6]. Interestingly, the extent of IL-8 release by respiratory epithelia is positively correlated with the bacterial load in the lower respiratory tract, which consequently contributes to lung injury [27,28]. In this regard, secreted IL-6 is involved in the stimulation of acute-phase protein synthesis, leukocyte recruitment, B-cell differentiation, and T-cell activation in many chronic inflammatory diseases [2,3]. IL-6 levels are also increased in the plasma and BALF of patients with pneumonia [7]. We observed that the release of IL-8 and IL-6 by bronchial and pharyngeal epithelial cells, both of which were induced by the exposure by *P. gingivalis*, was stronger than that by *S. pneumoniae*. Because IL-8 and IL-6 exert their proinflammatory effects in paracrine and autocrine manners, they must elicit inflammatory responses in neighboring host cells.

Our results thus suggest that inflammation in the lower respiratory epithelial cells is caused by an exposure to *P. gingivalis* itself and additionally by the absorption of paracrine proinflammatory cytokines released from the pharyngeal epithelium, which is induced by a precedent exposure to *P. gingivalis*. Therefore, these results may provide the novel insight that *P. gingivalis* contributes to the onset of aspiration pneumonia by inducing proinflammatory cytokine-production by bronchial and pharyngeal epithelial cells without infection.

From the aforementioned viewpoint, it should be also noted that elderly people have an increased risk of salivary aspiration because of their reduced laryngopharyngeal sensitivity and swallowing reflex impairment, both of which are ascribable to mild cognitive impairment [21]. In addition, an average person generates and ingests up to 1.5 L of saliva per day, which contains 1×10^8/mL bacteria released from oral biofilms, indicating more than 1×10^{11} bacteria are swallowed daily [29,30]. Moreover, a high incidence of silent aspiration during sleep is eventually recorded in the elderly [31,32]. These observations along with our findings support our hypothesis that increased number of *P. gingivalis* in saliva because of poor oral hygiene raises the risk of the onset of aspiration pneumonia.

Respiratory epithelia represent the first line of defense against exogenous respiratory pathogens. Respiratory epithelia are surely exposed to aspirated *P. gingivalis*, after which the bacterium presumably attaches to epithelial surfaces via several interactions between epithelial receptors and bacterial ligands such as fimbriae and LPS [10]. However, recognition of bacterial attachment by the epithelium alone might induce cytokine production. The present findings of proinflammatory secretion by bronchial and pharyngeal epithelial cells exposed to heat-inactivated *P. gingivalis* indicate that infection is not necessarily required for the observed induction of cytokine production. In this connection, several studies have indicated that heat-inactivated *P. gingivalis* appears capable of modulating the expression of inflammatory cytokines in several types of human cells other than respiratory epithelial cells, such as monocytes, whole blood cells, and gingival epithelial cells [33–36]. These observations are reminiscent of the manner how *S. pneumoniae* induces pneumonia.

Significant involvement of TLR2 but not TLR4 of several types of human cells, such as periodontal ligament cells, gingival fibroblasts, and monocytes, gingival epithelial cells, in proinflammatory cytokine production induced by *P. gingivalis* is also reported [35–37]. Although both BEAS-2B and Detroit 562 cells express TLR2 and TLR4, to the best of our knowledge, we found for the first time that only anti-TLR2 antibody significantly and concentration-dependently reduced *P. gingivalis*-induced IL-8 and IL-6 production by these cells. In parallel with these findings, *P. gingivalis* can elicit high levels of proinflammatory cytokine production in wild-type or TLR4-deficient mice, but not TLR2-deficient mice [38,39]. This suggests that without heat inactivation, TLR2 is necessary for *P. gingivalis* to induce proinflammatory cytokine production by host cells. These observations along with our findings suggest that bronchial and pharyngeal epithelia secrete some proinflammatory cytokines via TLR2 stimulation and become inflamed quickly when they are exposed to *P. gingivalis* irrespective of bacterial vitality.

Although *P. gingivalis* LPS and Fim A, a major fimbriae of *P. gingivalis*, have been widely accepted as TLR2 ligands [40–42], the stimulation of neither BEAS-2B nor Detroit 562 cells by these bacterial surface structures alone elicited IL-8 and IL-6 protein secretion. Therefore, the necessity of other surface structures or the whole *P. gingivalis* cell for the observed cytokine production is currently under study. In addition, in the human phagocytes, TLR2 is an essential molecule for sensing *P. gingivalis* and appears to be paired often by TLR1 and sometimes by TLR6 for arranging subsequent intracellular signaling due to imminent immunological necessity [40,43]. Because this has not been investigated in human respiratory epithelia, it would be interesting to see how blocking of TLR1 or TLR6 influences the *P. gingivalis*-induced IL-8 and IL-6 production. Further studies are therefore needed. However, it would be important that our findings have revealed a putatively causal relationship between *P. gingivalis* as a proinflammatory stimulant to several human respiratory epithelia and the onset of aspiration pneumonia.

Author Contributions: N.W., S.Y., and K.I. performed the experiments, analyzed the data, contributed reagents/materials/analytical tools, prepared the figures, and reviewed drafts of the article. Y.O. and S.S. contributed to the discussion, analyzed the data, and reviewed the drafts of the article. K.I. conceived, designed the experiments, authored or reviewed drafts of the article. All authors have read and agreed to the published version of the manuscript.

Funding: This work was supported by JSPS KAKENHI, and Uemura Fund, Dental Research Center, Nihon University School of Dentistry.

Conflicts of Interest: The authors confirm that there is no conflict of interests.

References

1. Okazaki, T.; Ebihara, S.; Mori, T.; Izumi, S.; Ebihara, T. Association between sarcopenia and pneumonia in older people. *Geriatr. Gerontol. Int.* **2019**, *20*, 7–13. [CrossRef]

2. Strieter, R.M.; Belperio, J.; Keane, M.P. Host innate defenses in the lung: The role of cytokines. *Curr. Opin. Infect. Dis.* **2003**, *16*, 193–198. [CrossRef]

3. Gomez, M.I.; Prince, A. Airway epithelial cell signaling in response to bacterial pathogens. *Pediatr. Pulmonol.* **2007**, *43*, 11–19. [CrossRef] [PubMed]

4. Boutten, A.; Dehoux, M.S.; Seta, N.; Ostinelli, J.; Venembre, P.; Crestani, B.; Dombret, M.C.; Durand, G.; Aubier, M. Compartmentalized IL-8 and elastase release within the human lung in unilateral pneumonia. *Am. J. Respir. Crit. Care Med.* **1996**, *153*, 336–342. [CrossRef] [PubMed]

5. Schütte, H.; Lohmeyer, J.; Rosseau, S.; Ziegler, S.; Siebert, C.; Kielisch, H.; Pralle, H.; Grimminger, F.; Morr, H.; Seeger, W. Bronchoalveolar and systemic cytokine profiles in patients with ARDS, severe pneumonia and cardiogenic pulmonary oedema. *Eur. Respir. J.* **1996**, *9*, 1858–1867. [CrossRef] [PubMed]

6. Chollet-Martin, S.; Montravers, P.; Gibert, C.; Elbim, C.; Desmonts, J.M.; Fagon, J.Y.; A Gougerot-Pocidalo, M. High levels of interleukin-8 in the blood and alveolar spaces of patients with pneumonia and adult respiratory distress syndrome. *Infect. Immun.* **1993**, *61*, 4553–4559. [CrossRef] [PubMed]

7. Montón, C.; Ewig, S.; Torres, A.; El-Ebiary, M.; Filella, X.; Rañó, A.; Xaubet, A. Role of glucocorticoids on inflammatory response in nonimmunosuppressed patients with pneumonia: A pilot study. *Eur. Respir. J.* **1999**, *14*, 218–220. [CrossRef]

8. Pihlstrom, B.L.; Michalowicz, B.S.; Johnson, N.W. Periodontal diseases. *Lancet* **2005**, *366*, 1809–1820. [CrossRef]

9. Bostanci, N.; Belibasakis, G.N. Porphyromonas gingivalis: An invasive and evasive opportunistic oral pathogen. *FEMS Microbiol. Lett.* **2012**, *333*, 1–9. [CrossRef] [PubMed]

10. Iii, F.C.G.; Genco, C.A. Porphyromonas gingivalis Mediated Periodontal Disease and Atherosclerosis: Disparate Diseases with Commonalities in Pathogenesis Through TLRs. *Curr. Pharm. Des.* **2007**, *13*, 3665–3675. [CrossRef]

11. Imai, K.; Ochiai, K.; Okamoto, T. Reactivation of Latent HIV-1 Infection by the Periodontopathic BacteriumPorphyromonas gingivalisInvolves Histone Modification. *J. Immunol.* **2009**, *182*, 3688–3695. [CrossRef] [PubMed]

12. Imai, K.; Inoue, H.; Tamura, M.; Cueno, M.E.; Inoue, H.; Takeichi, O.; Kusama, K.; Saito, I.; Ochiai, K. The periodontal pathogen Porphyromonas gingivalis induces the Epstein–Barr virus lytic switch transactivator ZEBRA by histone modification. *Biochimie* **2012**, *94*, 839–846. [CrossRef] [PubMed]

13. Van Der Maarel-Wierink, C.D.; Vanobbergen, J.N.O.; Bronkhorst, E.M.; Schols, J.M.G.A.; De Baat, C. Oral health care and aspiration pneumonia in frail older people: A systematic literature review. *Gerodontology* **2012**, *30*, 3–9. [CrossRef] [PubMed]

14. Terpenning, M.S.; Taylor, G.W.; Lopatin, D.E.; Kerr, C.K.; Dominguez, B.L.; Loesche, W.J. Aspiration Pneumonia: Dental and Oral Risk Factors in an Older Veteran Population. *J. Am. Geriatr. Soc.* **2001**, *49*, 557–563. [CrossRef]

15. Nagaoka, K.; Yanagihara, K.; Harada, Y.; Yamada, K.; Migiyama, Y.; Morinaga, Y.; Izumikawa, K.; Kohno, S. Quantitative detection of periodontopathic bacteria in lower respiratory tract specimens by real-time PCR. *J. Infect. Chemother.* **2017**, *23*, 69–73. [CrossRef] [PubMed]

16. Yatera, K.; Noguchi, S.; Mukae, H. The microbiome in the lower respiratory tract. *Respir. Investig.* **2018**, *56*, 432–439. [CrossRef]

17. Bartlett, J.G.; Finegold, S.M. Anaerobic infections of the lung and pleural space. *Am. Rev. Respir. Dis.* **1974**, *110*, 56–77. [CrossRef]

18. Bartlett, J.G. Anaerobic bacterial infection of the lung. *Anaerobe* **2012**, *18*, 235–239. [CrossRef]

19. Awano, S.; Ansai, T.; Takata, Y.; Soh, I.; Akifusa, S.; Hamasaki, T.; Yoshida, A.; Sonoki, K.; Fujisawa, K.; Takehara, T. Oral Health and Mortality Risk from Pneumonia in the Elderly. *J. Dent. Res.* **2008**, *87*, 334–339. [CrossRef]

20. Scannapieco, F.A.; Bush, R.B.; Paju, S. Associations Between Periodontal Disease and Risk for Nosocomial Bacterial Pneumonia and Chronic Obstructive Pulmonary Disease. A Systematic Review. *Ann. Periodontol.* **2003**, *8*, 54–69. [CrossRef]

21. Faverio, P.; Aliberti, S.; Bellelli, G.; Suigo, G.; Lonni, S.; Pesci, A.; Restrepo, M.I. The management of community-acquired pneumonia in the elderly. *Eur. J. Intern. Med.* **2014**, *25*, 312–319. [CrossRef]

22. Imai, K.; Kamio, N.; Cueno, M.E.; Saito, Y.; Inoue, H.; Saito, I.; Ochiai, K. Role of the histone H3 lysine 9 methyltransferase Suv39 h1 in maintaining Epsteinn-Barr virus latency in B95-8 cells. *FEBS J.* **2014**, *281*, 2148–2158. [CrossRef]

23. Liu, X.; Wetzler, L.M.; Nascimento, L.E.O.; Massari, P. Human Airway Epithelial Cell Responses to Neisseria lactamica and Purified Porin via Toll-Like Receptor 2-Dependent Signaling. *Infect. Immun.* **2010**, *78*, 5314–5323. [CrossRef]

24. Rydberg, C.; Månsson, A.; Uddman, R.; Riesbeck, K.; Cardell, L.-O. Toll-like receptor agonists induce inflammation and cell death in a model of head and neck squamous cell carcinomas. *Immunology* **2009**, *128*, e600–e611. [CrossRef]

25. Dogan, S.; Zhang, Q.; Pridmore, A.C.; Mitchell, T.J.; Finn, A.; Murdoch, C. Pneumolysin-induced CXCL8 production by nasopharyngeal epithelial cells is dependent on calcium flux and MAPK activation via Toll-like receptor 4. *Microbes Infect.* **2011**, *13*, 65–75. [CrossRef]

26. Hayden, M.S.; Ghosh, S. Signaling to NF-KappaB. *Genes Dev.* **2004**, *18*, 2195–2224. [CrossRef]

27. Hill, A.T.; Campbell, E.J.; Hill, S.L.; Bayley, D.L.; Stockley, R.A. Association between airway bacterial load and markers of airway inflammation in patients with stable chronic bronchitis. *Am. J. Med.* **2000**, *109*, 288–295. [CrossRef]

28. Nocker, R.E.; Schoonbrood, D.F.; Graaf, E.A.v.d.G.; Hack, C.E.; Lutter, R.; Jansen, H.M.; Out, T.A. Interleukin-8 in airway inflammation in patients with asthma and chronic obstructive pulmonary disease. *Int. Arch. Allergy Immunol.* **1996**, *109*, 183–191. [CrossRef]

29. Boutaga, K.; Savelkoul, P.H.M.; Winkel, E.G.; Van Winkelhoff, A.J. Comparison of Subgingival Bacterial Sampling with Oral Lavage for Detection and Quantification of Periodontal Pathogens by Real-Time Polymerase Chain Reaction. *J. Periodontol.* **2007**, *78*, 79–86. [CrossRef]

30. Von Troil-Linden, B.; Torkko, H.; Alaluusua, S.; Jousimies-Somer, H.; Asikainen, S. Salivary Levels of Suspected Periodontal Pathogens in Relation to Periodontal Status and Treatment. *J. Dent. Res.* **1995**, *74*, 1789–1795. [CrossRef]

31. Marik, P.E.; Kaplan, D. Aspiration Pneumonia and Dysphagia in the Elderly. *Chest* **2003**, *124*, 328–336. [CrossRef] [PubMed]

32. Ebihara, S.; Ebihara, T.; Kohzuki, M. Effect of Aging on Cough and Swallowing Reflexes: Implications for Preventing Aspiration Pneumonia. *Lung* **2011**, *190*, 29–33. [CrossRef] [PubMed]

33. Hamedi, M.; Belibasakis, G.N.; Cruchley, A.T.; Rangarajan, M.; Curtis, M.A.; Bostanci, N. Porphyromonas gingivalis culture supernatants differentially regulate interleukin-1beta and interleukin-18 in human monocytic cells. *Cytokine* **2009**, *45*, 99–104. [CrossRef] [PubMed]

34. Takahashi, N.; Honda, T.; Domon, H.; Nakajima, T.; Tabeta, K.; Yamazaki, K. Interleukin-1 receptor-associated kinase-M in gingival epithelial cells attenuates the inflammatory response elicited byPorphyromonas gingivalis. *J. Periodontal Res.* **2010**, *45*, 512–519. [CrossRef] [PubMed]

35. Sun, Y.; Shu, R.; Li, C.-L.; Zhang, M.-Z. Gram-Negative Periodontal Bacteria Induce the Activation of Toll-Like Receptors 2 and 4, and Cytokine Production in Human Periodontal Ligament Cells. *J. Periodontol.* **2010**, *81*, 1488–1496. [CrossRef]

36. Tietze, K.; Dalpke, A.; Morath, S.; Mutters, R.; Heeg, K.; Nonnenmacher, C. Differences in innate immune responses upon stimulation with gram-positive and gram-negative bacteria. *J. Periodontal Res.* **2006**, *41*, 447–454. [CrossRef]

37. Andrukhov, O.; Ertlschweiger, S.; Moritz, A.; Bantleon, H.P.; Rausch-Fan, X. Different effects of P. gingivalis LPS and E. coli LPS on the expression of interleukin-6 in human gingival fibroblasts. *Acta Odontol. Scand.* **2014**, *72*, 337–345. [CrossRef]
38. Burns, E.; Bachrach, G.; Shapira, L.; Nussbaum, G. Cutting Edge: TLR2 Is Required for the Innate Response toPorphyromonas gingivalis: Activation Leads to Bacterial Persistence and TLR2 Deficiency Attenuates Induced Alveolar Bone Resorption. *J. Immunol.* **2006**, *177*, 8296–8300. [CrossRef]
39. Hayashi, C.; Madrigal, A.G.; Liu, X.; Ukai, T.; Goswami, S.; Gudino, C.V.; Gibson, F.C.; Genco, C.A.; Gibson, I.F.C. Pathogen-mediated inflammatory atherosclerosis is mediated in part via Toll-like receptor 2-induced inflammatory responses. *J. Innate Immun.* **2010**, *2*, 334–343. [CrossRef]
40. Hajishengallis, G.; Tapping, R.I.; Harokopakis, E.; Nishiyama, S.-I.; Ratti, P.; Schifferle, R.E.; Lyle, E.A.; Triantafilou, M.; Triantafilou, K.; Yoshimura, F. Differential interactions of fimbriae and lipopolysaccharide from Porphyromonas gingivalis with the Toll-like receptor 2-centred pattern recognition apparatus. *Cell Microbiol.* **2006**, *8*, 1557–1570. [CrossRef]
41. Jain, S.; Coats, S.R.; Chang, A.M.; Darveau, R.P. A Novel Class of Lipoprotein Lipase-Sensitive Molecules Mediates Toll-Like Receptor 2 Activation by Porphyromonas gingivalis. *Infect. Immun.* **2013**, *81*, 1277–1286. [CrossRef]
42. Nichols, F.C.; Bajrami, B.; Clark, R.B.; Housley, W.; Yao, X. Free Lipid A Isolated from Porphyromonas gingivalis Lipopolysaccharide Is Contaminated with Phosphorylated Dihydroceramide Lipids: Recovery in Diseased Dental Samples. *Infect. Immun.* **2011**, *80*, 860–874. [CrossRef]
43. Kawai, T.; Akira, S. The role of pattern-recognition receptors in innate immunity: Update on Toll-like receptors. *Nat. Immunol.* **2010**, *11*, 373–384. [CrossRef] [PubMed]

Publisher's Note: MDPI stays neutral with regard to jurisdictional claims in published maps and institutional affiliations.

Journal of
Clinical Medicine

Article

Porphyromonas gingivalis Mfa1 Induces Chemokine and Cell Adhesion Molecules in Mouse Gingival Fibroblasts via Toll-Like Receptors

Yuhei Takayanagi [1], Takeshi Kikuchi [1,*], Yoshiaki Hasegawa [2], Yoshikazu Naiki [2], Hisashi Goto [1], Kousuke Okada [1], Iichiro Okabe [1], Yosuke Kamiya [1], Yuki Suzuki [1], Noritaka Sawada [1], Teppei Okabe [1], Yuki Suzuki [1], Shun Kondo [1], Tasuku Ohno [1], Jun-Ichiro Hayashi [1] and Akio Mitani [1]

1   Department of Periodontology, School of Dentistry, Aichi Gakuin University, 2-11 Suemoridori, Chikusaku, Nagoya, Aichi 464-8651, Japan; ag173d14@dpc.agu.ac.jp (Y.T.); hisashi@dpc.agu.ac.jp (H.G.); okakou@dpc.agu.ac.jp (K.O.); iichiro@dpc.agu.ac.jp (I.O.); k-yosuke@dpc.agu.ac.jp (Y.K.); yukip@dpc.agu.ac.jp (Y.S); ag173d11@dpc.agu.ac.jp (N.S.); ag183d03@dpc.agu.ac.jp (T.O.); ag193d11@dpc.agu.ac.jp (Y.S.); ag193d07@dpc.agu.ac.jp (S.K.); tasuku@dpc.agu.ac.jp (T.O.); jun1row@dpc.agu.ac.jp (J.-I.H.); minita@dpc.agu.ac.jp (A.M.)
2   Department of Microbiology, School of Dentistry, Aichi Gakuin University, 1-100 Kusumoto-cho, Chikusaku, Nagoya, Aichi 464-8651, Japan; yhase@dpc.agu.ac.jp (Y.H.); naikiy@dpc.agu.ac.jp (Y.N.)
*   Correspondence: tkikuchi@dpc.agu.ac.jp; Tel./Fax: +81-52-759-2150

Received: 6 November 2020; Accepted: 7 December 2020; Published: 10 December 2020

Abstract: *Porphyromonas gingivalis* Mfa1 fimbriae are thought to act as adhesion factors and to direct periodontal tissue destruction but their immunomodulatory actions are poorly understood. Here, we investigated the effect of Mfa1 stimulation on the immune and metabolic mechanisms of gingival fibroblasts from periodontal connective tissue. We also determined the role of Toll-like receptor (TLR) 2 and TLR4 in Mfa1 recognition. Mfa1 increased the expression of genes encoding chemokine (C-X-C motif) ligand (CXCL) 1, CXCL3, intercellular adhesion molecule (ICAM) 1 and Selectin endothelium (E) in gingival fibroblasts, but did not have a significant effect on genes that regulate metabolism. Mfa1-stimulated up-regulation of genes was significantly suppressed in *Tlr4* siRNA-transfected cells compared with that in control siRNA-transfected cells, which indicates that recognition by TLR4 is essential for immunomodulation by Mfa1. Additionally, suppression of *Tlr2* expression partially attenuated the stimulatory effect of Mfa1. Overall, these results help explain the involvement of *P. gingivalis* Mfa1 fimbriae in the progression of periodontal disease.

Keywords: *Porphyromonas gingivalis*; Mfa1; Toll-like receptors; gingival fibroblast

---

1. Introduction

Periodontitis is a chronic inflammatory disease caused by multiple periodontal pathogenic bacterial species [1]. Socransky et al. indicated that three bacteria, *Porphyromonas gingivalis*, *Tannerella forsythia*, and *Treponema denticola*, are the main cause of the onset and progression of periodontitis and they suggested that these three species be called "red complexes" [2].

Among the three species, *P. gingivalis* is the major pathogen associated with periodontitis. Riviere et al. found that *P. gingivalis* was more prevalent at diseased sites than at healthy sites in diseased subjects [3]. The number of *P. gingivalis* in plaque samples was associated with the plaque score and clinical attachment level [4]. From the perspective of oral flora, *P. gingivalis* is considered to act as a keystone pathogen that creates a dysbiosis between the host and the dental biofilm. This altered oral commensal microbiota is responsible for initiating pathological bone loss [5]. *P. gingivalis* expresses a number of potential virulence factors, including lipopolysaccharides, gingipines, and fimbriae [6].

Fimbriae are filamentous proteinaceous appendages on the surface of *P. gingivalis* bacteria that play a pivotal role in colonization through association with other bacteria and host tissues [7,8]. *P. gingivalis* ATCC strain 33,277 has fimbriae consisting of either FimA (*fimA* gene product) or Mfa1 (*mfa1* gene product), with apparent molecular masses of approximately 38 and 75 kDa, respectively [9].

Mfa1, encoded by the *mfa1* gene, is a structural subunit of a protein complex whose length varies from 60 to 500 nm [10]. In addition to the primary Mfa1 protein, mature fimbriae also have affiliated Mfa2–5 proteins. Mfa2 plays an anchor role, while Mfa3 can bind with Mfa1/2/4/5 in vitro to connect with other fimbrial subunits [11]. Recent data indicate that the C-terminal domain of Mfa1, rather than Mfa3, affects the aggregation and maturation of downstream fimbrial proteins [12].

Several studies have demonstrated different roles for FimA and Mfa1 fimbriae. FimA fimbriae act as an adhesive that mediates periodontal tissue colonization and host cell invasion [8,13,14]. They also induce inflammatory processes in periodontal tissues through several mechanisms [15–17]. FimA fimbriae promote early biofilm formation in a single species of *P. gingivalis*, whereas Mfa1 plays an inhibitory role in the formation of a homotypic biofilm in *P. gingivalis* [18]. However, although there are few reports on the host immune response to Mfa1 fimbriae, an Mfa1-deficient strain causes almost no alveolar bone resorption in a mouse model of oral infection [19].

The present study aimed to examine the effect of *P. gingivalis* Mfa1 stimulation on the immune and metabolic mechanisms of mouse gingival fibroblasts (MGFs) and to examine the effects of Toll-like receptor (TLR) 2 and TLR4 knock down on *P. gingivalis* Mfa1-stimulated MGFs. Our results show that Mfa1 fimbriae had a large effect on immunomodulation exerted by gingival fibroblasts, but did not have a significant effect on metabolic regulation. Our results also indicate that recognition of Mfa1 by TLR4 on MGFs is essential for the expression of genes related to cell migration and cell adhesion. Overall, these results help to explain how *P. gingivalis* Mfa1 fimbriae are involved in the progression of periodontal disease.

2. Materials and Methods

2.1. Cell Culture

MGFs were isolated from healthy gingival tissue from the palate of BALB/c mice which were purchased from CLEA Japan, Inc. (Tokyo, Japan). The MGFs were cultured in Minimum Essential Medium α (Thermo Fisher Scientific, Wilmington, DE, USA) containing 10% fetal bovine serum (Hyclone Laboratories Inc, Logan, UT, USA), 100 U/mL penicillin, and 100 µg/mL streptomycin at 37 °C in a CO_2 incubator. When the cells reached confluency, they were separated by treatment with 0.25% trypsin-EDTA (Thermo Fisher Scientific) and collected by centrifugation. This study was approved by the Institutional Animal Care and Use Committees of Aichi Gakuin University (AGUD438; approved date 31 March 2019) and all animal experiments were conducted following the national guidelines and the relevant national laws on the protection of animals. Ultrapure lipopolysaccharide from *P. gingivalis* was purchased for experiments (InvivoGen, San Diego, CA, USA).

2.2. Purification of Mfa1 Fimbriae

In this study, we used *P. gingivalis* mutant strains derived from ATCC 33,277. Mfa1 fimbriae were purified from JI-1, in which *fimA* was deleted, as described previously [20,21]. Mfa1 fimbriae were also purified from *mfa5* mutant FMFA5, in which *mfa5* was disrupted by *ermF-B*, and genetic complementation strain FMFA5C, as described previously [21,22]. Briefly, *P. gingivalis* cells were collected from 2 L of culture, suspended in 40 mL of 20 mM Tris/HCl buffer at pH 8.0, supplemented with protease inhibitors (1 mM phenylmethylsulfonyl fluoride and 1 mM Nα-p-tosyl-L-lysine chloromethyl ketone). The cells were disrupted by a French press. Unbroken cells were removed by centrifugation. The supernatant was subjected to precipitation at 50% ammonium sulfate saturation.

The precipitate was dialyzed with 20 mM Tris/HCl buffer at pH 8.0, and then applied to DEAE Sepharose Fast Flow chromatography (GE Healthcare Bio-Sciences AB, Uppsala, Sweden) with 50 mL of bed volume. After washing thoroughly with the buffer, sample was fractionated by a linear gradient elution with 400 mL of NaCl (0 to 0.3 M) in the buffer. Purity and identity were verified by sodium dodecyl sulfate-polyacrylamide gel electrophoresis (Bio-Rad Laboratories, Hercules, CA, USA) and transmission electron microscopy. Details are described in the Supplementary Material.

2.3. RT2 Profiler PCR Array Analysis

MGF (1×10^6 cells/dish) were seeded onto 60-mm dishes. When the cells reached confluency, they were incubated for 2 hours in the presence or absence of 1 µg/mL Mfa1, FMFA5, FMFA5C, FimA, or LPS of *P. gingivalis*. Total RNA was then collected. Complementary deoxyribonucleic acid (cDNA) was synthesized using the RT^2 First Strand Kit (Qiagen, Germantown, MD, USA) according to the manufacturer's instructions and then applied to the Mouse Antibacterial Response RT^2 Profile PCR Array (Qiagen) and the Mouse Extracellular Matrix & Adhesion Molecules RT^2 Profile PCR Array (Qiagen). Amplification was performed using the RT^2 SYBR Green/ROX qPCR master mix (Qiagen) and the StepOnePlus^{TM} Real-time PCR system (Thermo Fisher Scientific) with associated software (version 2.3; Thermo Fisher Scientific). CT values were transferred to an Excel file to build a table of CT values, which was then uploaded onto the data analysis web portal at http://www.qiagen.com/geneglobe.

2.4. Real-Time PCR

MGFs (1×10^6 cells/dish) were seeded onto 60-mm dishes. When the cells reached confluency, they were incubated for 2 hours in the presence or absence of 1 µg/mL JI-1, FMFA5, FMFA5C, FimA or LPS of *P. gingivalis*. Total RNA was then extracted with a Nucleospin RNA kit (Macherey-Nagel Inc., Bethlehem, PA, USA) according to the manufacturer's instructions, and purity and concentration were assessed by calculating the A230/A260 and A260/A280 ratios using a NanoDrop Lite (Thermo Fisher Scientific). To quantify mRNA, quantitative PCR was performed using the TaqMan gene expression assay (Thermo Fisher Scientific) for mouse *Cxcl1* (Mm04207460_m1), *Cxcl3* (Mm01701838), *Icam1* (Mm00516023_m1), *Sele* (Mm00441278_m1), *Tlr2* (Mm00442346_m1) and *Tlr4* (Mm00445273_m1) with the TaqMan Universal PCR Master Mix (Thermo Fisher Scientific). mRNA levels were normalized to the level of eukaryotic 18S rRNA (Hs99999901_s1). Quantitative PCR was performed using the StepOnePlus Real-Time System. PCR conditions were 10 minutes at 95 °C, followed by 40 cycles of 15 seconds at 95 °C and 1 minute at 60 °C. The relative amounts of target mRNAs were determined by subtracting the cycle threshold (CT) value for 18S rRNA from that for the gene (ΔCT). Then, the ΔCT value for the control group was subtracted from that for the experimental group (ΔΔCT). The results are expressed as the fold change (2-ΔΔCT) between the mRNA levels of control and experimental groups, where ΔΔCT was calculated as follows: ((CT for the target mRNA − CT for 18S rRNA) for the experimental group)–((CT for the target mRNA − CT for 18S rRNA) for the control group).

2.5. siRNA Transfection

MGFs were transfected with siRNA targeting TLR2 and TLR4 (Silencer Select Pre-designed siRNAs, Ambion, Austin, TX, USA) or non-targeting control siRNA (Ambion) using Lipofectamine RNAiMAX (Thermo Fisher Scientific) according to the manufacturer's protocol. Twenty-four hours after transfection, cells were stimulated with 1 µg/mL Mfa1, FMFA5, FMFA5C, FimA or LPS of *P. gingivalis* for 2 hours. The cells were then collected and TLR2 and TLR4 protein levels were determined by flow cytometry. Similarly, gene expression levels were determined by Real-Time PCR.

2.6. Flow Cytometry

MGFs (1×10^6 cells in 100 µL) were incubated with anti-mouse CD282 (TLR2) phycoerythrin (PE) (BioLegend, San Diego, CA, USA, Cat: 148604), anti-mouse CD284 (TLR4) phycoerythrin (PE) (BioLegend, Cat: 117605), or isotype control antibody phycoerythrin (PE) (BioLegend, Cat: 400508) and analyzed by flow cytometry using a MACSQuant analyzer and MACSQuantify software version 2.4 (Miltenyi Biotec, Tokyo, Japan).

2.7. Statistical Analysis

Data were analyzed using PASW Statistics software (version 18.0; SPSS Japan, Tokyo, Japan). Differences among groups were examined by one-factor analysis of variance (ANOVA) and Bonferroni's multiple comparison test. Comparisons of two independent groups were performed using Student's t-test. Data are expressed as the mean ± standard deviation (SD). Significance was accepted at $p < 0.05$.

3. Results

3.1. Analysis of Antibacterial Response-Associated Genes in Gingival Fibroblasts to Various Fimbriae

A mouse antibacterial response PCR array was used to investigate differences in the expression of 84 genes involved in bacteria-cell interactions. Figure 1 shows the fold changes in expression between control and 1 μg/mL various fimbriae or *P. gingivalis* LPS after stimulation for 2 hours. Among the 84 genes, *Cxcl1* and *Cxcl3* were upregulated in common by 4-fold in *JI-1*, *FMFA5* or *FMFA5C*-stimulated cells compared with non-stimulated cells (Figure 1A–C). Other genes that were elevated include *Nfkbia*, *Jun*, *Ccl5*, *Tlr6*, *Nod2* in JI-1 stimulated cells, Nfkbia, Jun, Irf5, Birc3 in FMFA5-stimulated cells, and Hsp90aa1, Nfkb1, Nfkbia, Jun, Nod1, Slpi, Tirap, Rela, Ccl5, Tlr2, Irf5, Irf7, Nod2, Birc3, Lcn2 in FMFA5C-stimulated cells. Many genes (*Tnfrsf1a*, *Mapk1*, *Tollip*, *Irak1*, *Fadd*, *Map3k7*, *Map2k4*, *Slpi*, *Ripk2*, *Il18*, *Rela*, *Irak3*, *Card6*, *Camp*, *Irf5*, *Birc3*, *Slc11a1*, *Tlr9*, *Casp1*, *Crp*), including *Cxcl3* were up-regulated in FimA-stimulated cells compared with non-stimulated cells (Figure 1D). *P. gingivalis* LPS stimulation changed several genes expression (*Ccl5*, *Tlr6*, *Irf5*, *Nod2*, *Casp1*) compared with non-stimulated cells (Figure 1E).

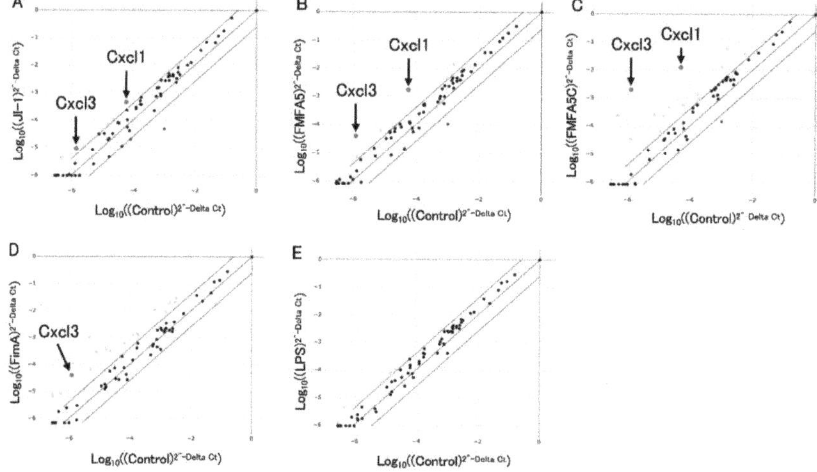

Figure 1. Analysis of antibacterial response-associated genes in mouse gingival fibroblasts in response to Mfa1, FimA, or *Porphyromonas gingivalis* LPS. Graphs show the fold changes of gene expression in cells stimulated with JI-1 (**A**), FMFA5 (**B**), FMFA5C (**C**), FimA (**D**), and LPS (**E**) compared with non-stimulated cells. *Cxcl1* and *Cxcl3* were up-regulated over 4-fold in JI-1, FMFA5 or FMFA5C-stimulated cells.

3.2. Confirmation of PCR Array Data for Selected Antibacterial Response-Associated Genes by Quantitative Real Time-PCR

To validate the PCR array data, Real Time-PCR showed that JI-1 stimulation increased the gene expression of the cell migration factors, *Cxcl1* and *Cxcl3*, compared with FimA stimulation (Figure 2). Furthermore, the highest increase in gene expression was observed with FMFA5 stimulation (Figure 2).

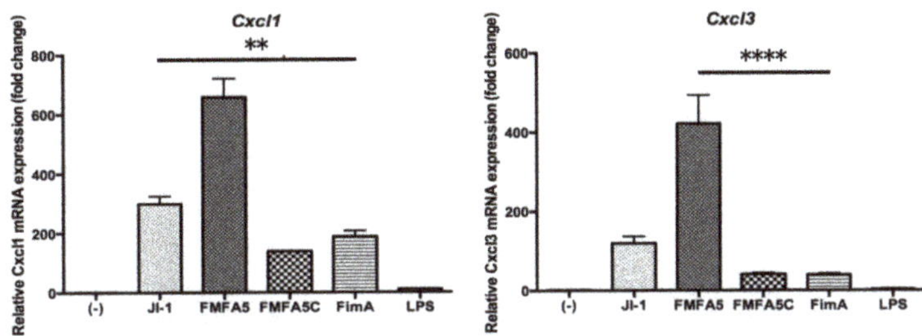

Figure 2. Mfa1 induces *Cxcl1* and *Cxcl3* in mouse gingival fibroblasts. Mouse gingival fibroblasts (MGFs) were cultured for 2 hours in the presence or absence of 1 µg/mL JI-1, FMFA5, FMFA5C, FimA, or LPS of *P. gingivalis* and then mRNA levels were examined using real-time PCR. Values are expressed as fold changes. Differences among groups were analyzed by one-way ANOVA. Data represent the mean + SD (n = 3). ** p < 0.01, **** p < 0.0001.

3.3. Analysis of Extracellular Matrix and Adhesion Molecule-Associated Genes in Gingival Fibroblasts in Response to Various Fimbriae

A mouse extracellular matrix and adhesion molecules PCR array was used to investigate differences in the expression of 84 genes involved in cell–cell and cell–matrix interactions. Figure 3 shows the fold changes in expression between control and 1 µg/mL various fimbriae or *P. gingivalis* LPS after stimulation for 2 hours. Among the 84 genes, *Icam1* expression was upregulated 4-fold in FMFA5-stimulated cells compared with non-stimulated cells (Figure 3B). FMFA5C stimulation similarly induced *Icam1* and also *Selectin E (Sele)* (Figure 3C). No obvious changes in gene expression were observed with JI-1, FimA or *P. gingivalis* LPS stimulation compared with unstimulated cells (Figure 3A,D,E).

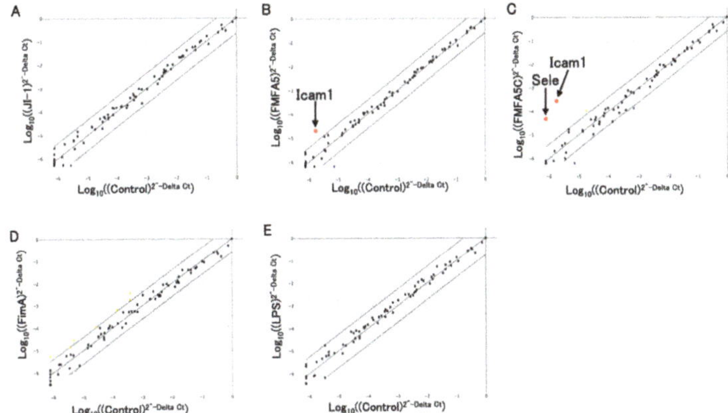

Figure 3. Analysis of extracellular matrix and adhesion molecule-associated genes in mouse gingival fibroblasts in response to Mfa1, FimA, or *P. gingivalis* LPS. Graphs show fold changes of gene expression in cells stimulated with JI-1 (**A**), FMFA5 (**B**), FMFA5C (**C**), FimA (**D**), and LPS (**E**) compared with non-stimulated cells. *Icam1* and *Sele* were up-regulated over 4-fold in FMFA5C-stimulated cells.

3.4. Confirmation of PCR Array Data for Selected Extracellular Matrix and Adhesion Molecule-Associated Genes by Quantitative RT-PCR

Expression of cell adhesion factors, *Icam1* and *Sele*, was increased to a greater extent by JI-1 stimulation compared with FimA stimulation (Figure 4). The highest increase in gene expression was observed with FMFA5 stimulation (Figure 4).

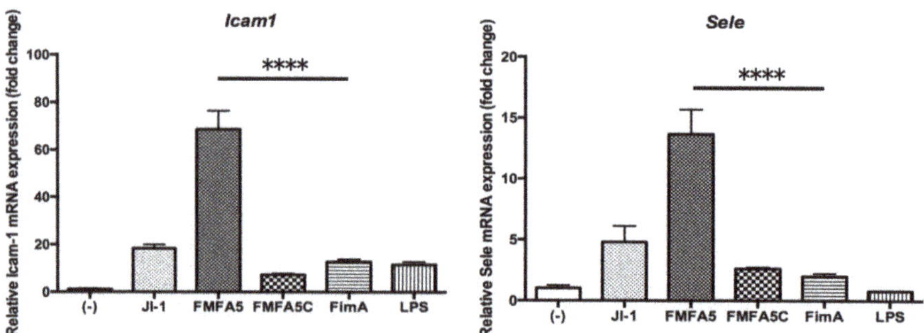

Figure 4. Mfa1 induces *Icam1* and *Sele* in mouse gingival fibroblasts. MGFs were cultured for 2 hours in the presence or absence of 1 µg/mL JI-1, FMFA5, FMFA5C, FimA, or LPS of *P. gingivalis* and then mRNA levels were examined using real-time PCR. Values are expressed as fold changes. Differences among groups were analyzed by one-way ANOVA. Data represent the mean + SD ($n = 3$). **** $p < 0.0001$.

3.5. Induction of Tlr2 and Tlr4 Gene Expression by Various Fimbriae and LPS Stimulation in Gingival Fibroblasts

Tlr2 showed the same change in expression as the other factors, such as *Cxcl1*, but *Tlr4*, the receptor for LPS, did not show any significant variation in gene expression in response to various stimulants (Figure 5).

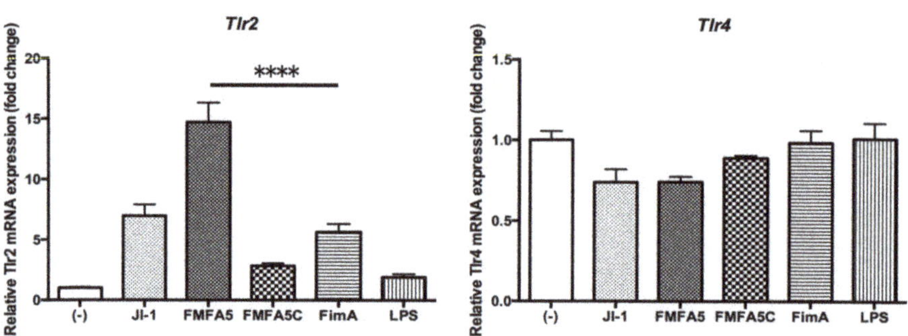

Figure 5. Mfa1 induces TLR2 but not TLR4 in mouse gingival fibroblasts. MGFs were cultured for 2 hours in the presence or absence of 1 µg/mL JI-1, FMFA5, FMFA5C, FimA or LPS of *P. gingivalis* and then mRNA levels were examined using real-time PCR. Values are expressed as fold changes. Differences among groups were analyzed by one-way ANOVA. Data represent the mean + SD ($n = 3$). **** $p < 0.0001$.

3.6. Transfection of Tlr2 and Tlr4 siRNA into Gingival Fibroblasts

Tlr2 and *Tlr4*siRNA-transfected gingival fibroblast cells showed obvious knockdown of *Tlr2* and *Tlr4*mRNA, respectively, compared with control siRNA-transfected gingival fibroblast cells (Figure 6A). FACS analysis confirmed a decrease in the surface expression of *Tlr2* and *Tlr4* in the respective siRNA-transfected gingival fibroblast cells compared with control cells (Figure 6B).

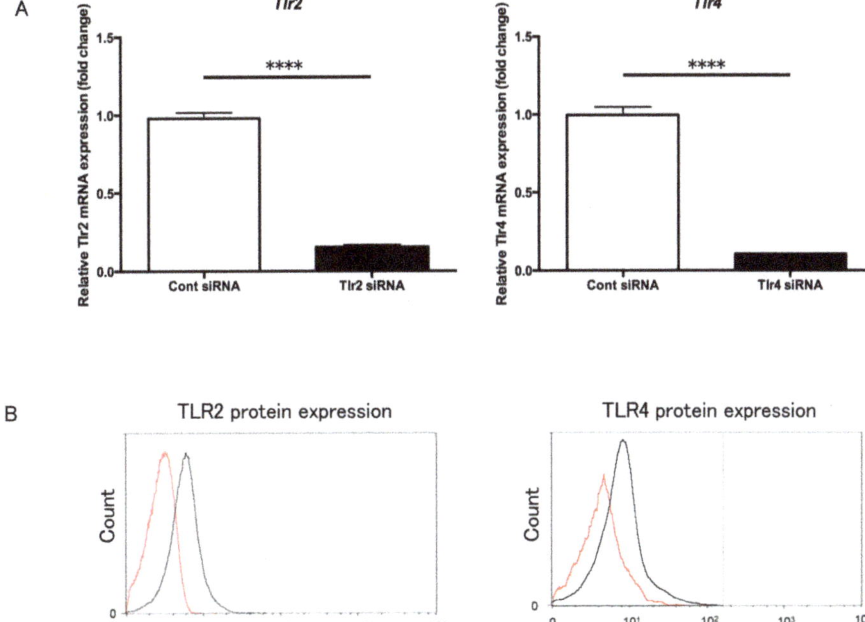

Figure 6. Suppression of *Tlr*2 and *Tlr*4 in mouse gingival fibroblasts using siRNA. (**A**) *Tlr*2 (left) and *Tlr*4 (right) mRNA levels were examined using real-time PCR. *Tlr*2 and *Tlr*4 siRNA-transfected cells showed clear knockdown of *Tlr*2 and *Tlr*4 mRNAs, respectively, compared with control siRNA-transfected cells. (**B**) TLR2 (left) and TLR4 (right) protein levels were examined using flow cytometry. *Tlr*2 or *Tlr*4 siRNA transfected cells (red line) showed decreased levels of the respective receptors compared with control siRNA-transfected cells (black line). **** $p < 0.0001$.

3.7. Expression of Selected Genes in Tlr2 and Tlr4 siRNA-Transfected Cells

Expression of the cell migration-related factor and cell adhesion factor genes in JI-1, FMFA5, and FMFA5C-stimulated *Tlr*4 siRNA-transfected cells were significantly suppressed compared with control siRNA-transfected cells (Figure 7). Also, the suppression of *Tlr*2 expression partially attenuated the stimulation effect (Figure 7).

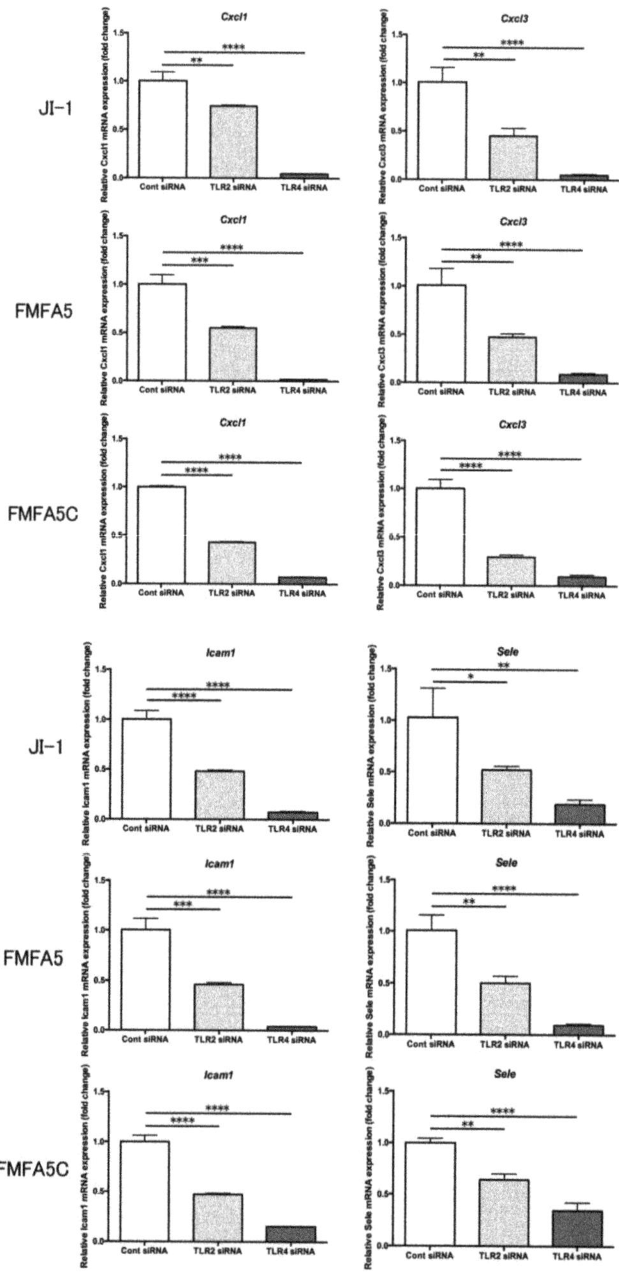

Figure 7. Involvement of TLR2 and TLR4 in the expression of Mfa1-induced cell migration and cell adhesion-related factors. *Tlr2* siRNA, *Tlr4* siRNA, or control siRNA-transfected mouse gingival fibroblasts were cultured for 2 hours in the presence or absence of 1 µg/mL JI-1, FMFA5, or FMFA5C and then mRNA levels were examined using real-time PCR. Values are expressed as fold changes. Differences among groups were analyzed by one-way ANOVA. Data represent the mean + SD ($n = 3$). * $p < 0.05$, ** $p < 0.01$, *** $p < 0.001$, **** $p < 0.0001$.

4. Discussion

In this study, we demonstrated that Mfa1 fimbriae stimulation markedly increased the expression of cell migration/cell adhesion-related genes in mouse gingival fibroblasts, and that the increase was more pronounced than with FimA stimulation. In addition, the ability of Mfa1 to regulate cell migration and cell adhesion was significantly attenuated in mouse gingival fibroblasts in which *Tlr4* expression was suppressed.

When the effect of JI-1 on gingival fibroblasts, which constitute the gingival connective tissue, was examined, a remarkable increase in the expression of *Cxcl1* and *Cxcl3*, which are involved in cell migration, was observed. Higher levels of CXCL1 were found in human and rat gingiva from periodontitis sites compared with periodontally healthy sites [23]. There was also a significant difference in the level of CXCL1 in gingival crevicular fluid between healthy and periodontitis subjects [24]. CXCL1 stimulates gingiva fibroblast migration [25] and it may be related to periodontal tissue healing. There is no literature reporting the relationship between CXCL3 and periodontitis, but its activity is similar to that of CXCL2, which may play an important role in the initiation of inflammation and subsequent periodontal tissue destruction [26], which suggests that it is involved in the progression of periodontitis. JI-1 stimulation produced a higher increase in *Cxcl1* and *Cxcl3* expression compared with FimA stimulation. FimA induces *Cxcl1* expression in mouse peritoneal macrophages [27]. It is possible that the immunomodulatory capacity of Mfa1 outperforms that of FimA for some factors in certain cell types.

Next, we investigated which part of the fimbriae structure is important for the immunomodulatory ability of Mfa1. FMFA5, with the Mfa3–5 tip structure of JI-1 removed, and FMFA5C, with the tip structure was complemented in FMFA5, were compared for their ability to induce immunomodulation. Compared with JI-1, a markedly higher increase in gene expression was observed with FMFA5 stimulation, while that of FMFA5C was significantly reduced. From this result, it is indicated that the incomplete fimbriae deficient in tip structure increase the stimulation ability.

When we confirmed the effects of Mfa1 on extracellular matrix and adhesion molecules, we found a marked increase in the expression of the cell adhesion factor genes, *Icam1* and *Sele*. ICAM1 deficiencies increase susceptibility to and severity of alveolar bone loss after *P. gingivalis* infection [28]. Selectin Platelet/Selectin E-deficient mice exhibit spontaneous early onset alveolar bone loss [29]. FimA and Mfa1 induce ICAM1 and Selectin E in human aortic endothelial cells [30]. These fimbriae appear to have similar effects on gingival fibroblasts. Similar to the results for cell migration factors, a markedly higher increase in cell adhesion factor gene expression was observed with FMFA5 stimulation and the stimulation ability of FMFA5C was significantly reduced compared with JI-1 stimulation. These results indicate that the regulation of cell adhesion factors by Mfa1 is also greatly influenced by the Mfa1 molecule in the shaft portion.

Cells use pattern recognition receptors, such as TLRs, to recognize pathogen-associated molecular patterns (PAMPs). Ten TLRs have been identified in humans and 12 in mice and the host innate immune response to pathogens is largely mediated via TLR signaling [31]. TLR2 and TLR4 are the most widely studied extracellular innate receptors that recognize various PAMPs and are likely to play a role in the pathogenesis of periodontitis [32]. TLR2 has been shown to be important for *P. gingivalis* to produce inflammatory cytokines [33,34]. TLR4 recognizes LPS (from *Escherichia coli*), which is a bacterial cell wall component [31]. Uniquely, *P. gingivalis* LPS is recognized by both TLR2 and TLR4 [35]. The $LPS_{1435/1449}$ and LPS_{1690} isoforms can produce opposite effects on TLR2 and TLR4 activation [36,37]. *P. gingivalis* LPS and *E. coli* LPS regulate cytokine production differently in human gingival fibroblasts [38]. The heterogeneity of *P. gingivalis* LPS might contribute to one of the strategies used by *P. gingivalis* to evade the innate host defense in gingival tissues [39]. FimA activates human peripheral blood monocytes via TLR2 and CD14 [40]. TLR2-dependent signaling leads to CD11b-CD18 activation in human monocytes upon recognition of FimA through CD14 [41]. However, the receptor for Mfa1 is still unclear. The gingival fibroblasts constitutively express TLR2 and TLR4 [42]. When we examined the expression of *Tlr2* and *Tlr4* genes in MGFs stimulated by various fimbriae and LPS,

Tlr2 showed the same changes in expression as the other factors, but *Tlr4* did not show any significant changes in expression in response to various fimbriae or LPS. Therefore, to clarify the receptor of Mfa1, MGFs in which *Tlr2* and *Tlr4* were knocked down were stimulated with Mfa1 and their reactivity was assessed. Control siRNA, *Tlr2* siRNA and *Tlr4* siRNA were introduced into MGFs, and cells were stimulated with JI-1, FMFA5, and FMFA5C. Gene expression of cell migration-related genes, *Cxcl1* and *Cxcl3*, and cell adhesion genes, *Icam1* and *Sele*, were analyzed. The suppression of TLR4 was significantly reduced by Mfa1 fimbriae stimulation. Also, suppression of *Tlr2* partially attenuated this stimulation. Recognition of Mfa1 by TLR4 is suggested to be essential for the expression of genes related to cell migration and cell adhesion. This result is different from previous reports in which anti-TLR2 antibody pretreatment significantly inhibited pro-inflammatory cytokines in mouse macrophages stimulated with Mfa1 fimbriae [43]. Conversely, Hajishengallis et al. reported that native Mfa1 induced proinflammatory cytokines in a CD14- and TLR2-dependent mode, which was likely due to a fimbriae-associated 12-kDa lipoprotein [44]. There is a similar report on FimA-like lipoproteins or lipopeptides associated with FimA that could, at least in part, account for signaling via TLR2 and subsequent TNF-α production in macrophages [45]. Furthermore, the following reports demonstrate the stimulation ability of LPS. A lipoprotein from *P. gingivalis* LPS was shown to be a principal component for TLR2-mediated cell activation [46]. A lipopolysaccharide preparation extracted from a *P. gingivalis* lipoprotein-deficient mutant showed a marked decrease in TLR2-mediated signaling [47]. Recombinant FimA stimulated cytokine release in THP-1 mononuclear cells via CD14 and TLR4 but not TLR2 [44], while recombinant FimA induced an inflammatory response via the TLR4/NF-kB signaling pathway in human peripheral blood mononuclear cells [48]. *P. gingivalis* lipid A and its synthetic counterpart activate cells through a TLR4-dependent pathway [46,49]. From our results, we speculate that wild-type purified Mfa1 is mainly recognized by TLR4, but TLR2 might recognize the lipoprotein of fimbriae and contributes to the overall action. In the future, it is necessary to confirm the reactivity of purified Mfa1 from which lipoprotein has been removed.

There are several reports on the importance of TLR4 in periodontitis. TLR4- but not TLR2-mediated stimulation, was positively associated with plaque score and bleeding on probing score of teeth from which the plaque samples were taken [4]. The ratio of TLR4/TLR2-mediated stimulation activity was also positively associated with probing depth and clinical attachment level [4]. TLR4- but not TLR2-stimulation of subgingival plaque is associated with plaque index [50]. Therefore, TLR4 may play an important role in the progression of periodontitis, in which stimulation by Mfa1 may play a role.

Recent reports suggest that intracellular DC-SIGN, an intracellularly expressed pattern recognition receptor, could be critical for recognition of Mfa1 by dendritic cells [51]. We assayed expression of DC-SIGN in MGFs after stimulation with various fimbriae, but we could not detect its expression. This may be because of differences between immunocompetent cells and periodontal tissue constituent cells.

In conclusion, Mfa1 fimbriae have a significant effect on immunomodulation in gingival fibroblasts of periodontal tissue. We also suggest that recognition of Mfa1 by TLR4 on MGFs is essential for the expression of genes related to cell migration and cell adhesion. More detailed analysis, such as using an animal infection model, is needed to assess the immunomodulatory capacity of Mfa1 fimbriae in the progression of periodontitis.

Supplementary Materials: The following are available online at http://www.mdpi.com/2077-0383/9/12/4004/s1, Figure S1: Purity of Mfa1 fimbriae.

Author Contributions: Conceptualization, Y.T., T.K., Y.H., and A.M.; methodology, Y.T., T.K., Y.H., and A.M.; validation, Y.T., T.K., and A.M.; formal analysis, Y.T., H.G., I.O., Y.S. (0000-0002-7536-5931), N.S., and T.O. (Tasuku Ohno); investigation, Y.T., Y.N., H.G., K.O., I.O., Y.K., Y.S. (0000-0002-7536-5931), N.S., T.O. (Teppei Okabe), and Y.S.(0000-0003-3660-6867); data curation, Y.K., Y.S. (0000-0003-3660-6867), and S.K.; writing—original draft preparation, Y.T., T.K., and A.M.; writing—review and editing, Y.T., T.K., Y.H., J.-I.H., and A.M.; visualization, Y.T. and T.K.; supervision, T.K., Y.H., J.H., and A.M.; project administration, A.M.; funding acquisition, T.K., Y.H. All authors have read and agreed to the published version of the manuscript.

Funding: This work was supported by a Grant-in-Aid for Scientific Research (No.20K09985, No. 20K09931 and No.17K11999) from the Ministry of Education, Culture, Sports, Science and Technology (MEXT) of Japan.

Acknowledgments: We thank Jeremy Allen, PhD, from Edanz Group (https://en-author-services.edanzgroup.com/) for editing a draft of this manuscript.

Conflicts of Interest: The authors declare no conflict of interest.

References

1. Pihlstrom, B.L.; Michalowicz, B.S.; Johnson, N.W. Periodontal diseases. *Lancet* **2005**, *366*, 1809–1820. [CrossRef]
2. Socransky, S.S.; Haffajee, A.D.; Cugini, M.A.; Smith, C.; Kent, R.L., Jr. Microbial complexes in subgingival plaque. *J. Clin. Periodontol.* **1998**, *25*, 134–144. [CrossRef] [PubMed]
3. Riviere, G.R.; Smith, K.S.; Tzagaroulaki, E.; Kay, S.L.; Zhu, X.; DeRouen, T.A.; Adams, D.F. Periodontal status and detection frequency of bacteria at sites of periodontal health and gingivitis. *J. Periodontol.* **1996**, *67*, 109–115. [CrossRef] [PubMed]
4. Yoshioka, H.; Yoshimura, A.; Kaneko, T.; Golenbock, D.T.; Hara, Y. Analysis of the activity to induce toll-like receptor (TLR)2- and TLR4-mediated stimulation of supragingival plaque. *J. Periodontol.* **2008**, *79*, 920–928. [CrossRef]
5. Darveau, R.P.; Hajishengallis, G.; Curtis, M.A. Porphyromonas gingivalis as a potential community activist for disease. *J. Dent. Res.* **2012**, *91*, 816–820. [CrossRef]
6. Lamont, R.J.; Jenkinson, H.F. Life below the gum line: Pathogenic mechanisms of Porphyromonas gingivalis. *Microbiol. Mol. Biol. Rev.* **1998**, *62*, 1244–1263. [CrossRef]
7. Hospenthal, M.K.; Costa, T.R.D.; Waksman, G. A comprehensive guide to pilus biogenesis in Gram-negative bacteria. *Nat. Rev. Microbiol.* **2017**, *15*, 365–379. [CrossRef]
8. Lamont, R.J.; Jenkinson, H.F. Subgingival colonization by Porphyromonas gingivalis. *Oral Microbiol. Immunol.* **2000**, *15*, 341–349. [CrossRef]
9. Yoshimura, F.; Murakami, Y.; Nishikawa, K.; Hasegawa, Y.; Kawaminami, S. Surface components of Porphyromonas gingivalis. *J. Periodontal. Res.* **2009**, *44*, 1–12. [CrossRef]
10. Enersen, M.; Nakano, K.; Amano, A. Porphyromonas gingivalis fimbriae. *J. Oral Microbiol.* **2013**, *5*, 20265. [CrossRef]
11. Ikai, R.; Hasegawa, Y.; Izumigawa, M.; Nagano, K.; Yoshida, Y.; Kitai, N.; Lamont, R.J.; Yoshimura, F.; Murakami, Y. Mfa4, an Accessory Protein of Mfa1 Fimbriae, Modulates Fimbrial Biogenesis, Cell Auto-Aggregation, and Biofilm Formation in Porphyromonas gingivalis. *PLoS ONE* **2015**, *10*, e0139454. [CrossRef] [PubMed]
12. Hall, M.; Hasegawa, Y.; Yoshimura, F.; Persson, K. Structural and functional characterization of shaft, anchor, and tip proteins of the Mfa1 fimbria from the periodontal pathogen Porphyromonas gingivalis. *Sci. Rep.* **2018**, *8*, 1793. [CrossRef] [PubMed]
13. Yilmaz, O.; Watanabe, K.; Lamont, R.J. Involvement of integrins in fimbriae-mediated binding and invasion by Porphyromonas gingivalis. *Cell Microbiol.* **2002**, *4*, 305–314. [CrossRef] [PubMed]
14. Amano, A.; Nakagawa, I.; Okahashi, N.; Hamada, N. Variations of Porphyromonas gingivalis fimbriae in relation to microbial pathogenesis. *J. Periodontol. Res.* **2004**, *39*, 136–142. [CrossRef] [PubMed]
15. Hajishengallis, G.; Wang, M.; Liang, S.; Shakhatreh, M.A.; James, D.; Nishiyama, S.; Yoshimura, F.; Demuth, D.R. Subversion of innate immunity by periodontopathic bacteria via exploitation of complement receptor-3. *Adv. Exp. Med. Biol.* **2008**, *632*, 203–219. [CrossRef] [PubMed]
16. Amano, A. Host-parasite interactions in periodontitis: Subgingival infection and host sensing. *Periodontology 2000* **2010**, *52*, 7–11. [CrossRef] [PubMed]
17. Amano, A. Host-parasite interactions in periodontitis: Microbial pathogenicity and innate immunity. *Periodontology 2000* **2010**, *54*, 9–14. [CrossRef]
18. Kuboniwa, M.; Amano, A.; Hashino, E.; Yamamoto, Y.; Inaba, H.; Hamada, N.; Nakayama, K.; Tribble, G.D.; Lamont, R.J.; Shizukuishi, S. Distinct roles of long/short fimbriae and gingipains in homotypic biofilm development by Porphyromonas gingivalis. *BMC Microbiol.* **2009**, *9*, 105. [CrossRef]

19. Umemoto, T.; Hamada, N. Characterization of biologically active cell surface components of a periodontal pathogen. The roles of major and minor fimbriae of Porphyromonas gingivalis. *J. Periodontol.* **2003**, *74*, 119–122. [CrossRef]

20. Hasegawa, Y.; Nagano, K.; Ikai, R.; Izumigawa, M.; Yoshida, Y.; Kitai, N.; Lamont, R.J.; Murakami, Y.; Yoshimura, F. Localization and function of the accessory protein Mfa3 in Porphyromonas gingivalis Mfa1 fimbriae. *Mol. Oral Microbiol.* **2013**, *28*, 467–480. [CrossRef]

21. Hasegawa, Y.; Nagano, K.; Murakami, Y.; Lamont, R.J. Purification of Native Mfa1 Fimbriae from Porphyromonas gingivalis. *Methods Mol. Biol.* **2021**, *2210*, 75–86. [CrossRef] [PubMed]

22. Hasegawa, Y.; Iijima, Y.; Persson, K.; Nagano, K.; Yoshida, Y.; Lamont, R.J.; Kikuchi, T.; Mitani, A.; Yoshimura, F. Role of Mfa5 in Expression of Mfa1 Fimbriae in Porphyromonas gingivalis. *J. Dent. Res.* **2016**, *95*, 1291–1297. [CrossRef] [PubMed]

23. Rath-Deschner, B.; Memmert, S.; Damanaki, A.; Nokhbehsaim, M.; Eick, S.; Cirelli, J.A.; Gotz, W.; Deschner, J.; Jager, A.; Nogueira, A.V.B. CXCL1, CCL2, and CCL5 modulation by microbial and biomechanical signals in periodontal cells and tissues-in vitro and in vivo studies. *Clin. Oral Investig.* **2020**, *24*, 3661–3670. [CrossRef] [PubMed]

24. Sakai, A.; Ohshima, M.; Sugano, N.; Otsuka, K.; Ito, K. Profiling the cytokines in gingival crevicular fluid using a cytokine antibody array. *J. Periodontol.* **2006**, *77*, 856–864. [CrossRef] [PubMed]

25. Buskermolen, J.K.; Roffel, S.; Gibbs, S. Stimulation of oral fibroblast chemokine receptors identifies CCR3 and CCR4 as potential wound healing targets. *J. Cell. Physiol.* **2017**, *232*, 2996–3005. [CrossRef] [PubMed]

26. Miyauchi, M.; Kitagawa, S.; Hiraoka, M.; Saito, A.; Sato, S.; Kudo, Y.; Ogawa, I.; Takata, T. Immunolocalization of CXC chemokine and recruitment of polymorphonuclear leukocytes in the rat molar periodontal tissue after topical application of lipopolysaccharide. *Histochem. Cell. Biol.* **2004**, *121*, 291–297. [CrossRef]

27. Hanazawa, S.; Murakami, Y.; Takeshita, A.; Kitami, H.; Ohta, K.; Amano, S.; Kitano, S. Porphyromonas gingivalis fimbriae induce expression of the neutrophil chemotactic factor KC gene of mouse peritoneal macrophages: Role of protein kinase C. *Infect. Immun.* **1992**, *60*, 1544–1549. [CrossRef]

28. Baker, P.J.; DuFour, L.; Dixon, M.; Roopenian, D.C. Adhesion molecule deficiencies increase Porphyromonas gingivalis-induced alveolar bone loss in mice. *Infect. Immun.* **2000**, *68*, 3103–3107. [CrossRef]

29. Niederman, R.; Westernoff, T.; Lee, C.; Mark, L.L.; Kawashima, N.; Ullman-Culler, M.; Dewhirst, F.E.; Paster, B.J.; Wagner, D.D.; Mayadas, T.; et al. Infection-mediated early-onset periodontal disease in P/E-selectin-deficient mice. *J. Clin. Periodontol.* **2001**, *28*, 569–575. [CrossRef]

30. Takahashi, Y.; Davey, M.; Yumoto, H.; Gibson, F.C., 3rd; Genco, C.A. Fimbria-dependent activation of pro-inflammatory molecules in Porphyromonas gingivalis infected human aortic endothelial cells. *Cell Microbiol.* **2006**, *8*, 738–757. [CrossRef]

31. Akira, S.; Takeda, K. Toll-like receptor signalling. *Nat. Rev. Immunol.* **2004**, *4*, 499–511. [CrossRef] [PubMed]

32. Hajishengallis, G.; Lambris, J.D. Microbial manipulation of receptor crosstalk in innate immunity. *Nat. Rev. Immunol.* **2011**, *11*, 187–200. [CrossRef] [PubMed]

33. Burns, E.; Bachrach, G.; Shapira, L.; Nussbaum, G. Cutting Edge: TLR2 is required for the innate response to Porphyromonas gingivalis: Activation leads to bacterial persistence and TLR2 deficiency attenuates induced alveolar bone resorption. *J. Immunol.* **2006**, *177*, 8296–8300. [CrossRef] [PubMed]

34. Watanabe, N.; Yokoe, S.; Ogata, Y.; Sato, S.; Imai, K. Exposure to Porphyromonas gingivalis Induces Production of Proinflammatory Cytokine via TLR2 from Human Respiratory Epithelial Cells. *J. Clin. Med.* **2020**, *9*, 3433. [CrossRef]

35. Darveau, R.P.; Pham, T.T.; Lemley, K.; Reife, R.A.; Bainbridge, B.W.; Coats, S.R.; Howald, W.N.; Way, S.S.; Hajjar, A.M. Porphyromonas gingivalis lipopolysaccharide contains multiple lipid A species that functionally interact with both toll-like receptors 2 and 4. *Infect. Immun.* **2004**, *72*, 5041–5051. [CrossRef]

36. Olsen, I.; Singhrao, S.K. Importance of heterogeneity in Porhyromonas gingivalis lipopolysaccharide lipid A in tissue specific inflammatory signalling. *J. Oral Microbiol.* **2018**, *10*, 1440128. [CrossRef]

37. Herath, T.D.K.; Darveau, R.P.; Seneviratne, C.J.; Wang, C.Y.; Wang, Y.; Jin, L. Heterogeneous Porphyromonas gingivalis LPS modulates immuno-inflammatory response, antioxidant defense and cytoskeletal dynamics in human gingival fibroblasts. *Sci Rep.* **2016**, *6*, 29829. [CrossRef]

38. Andrukhov, O.; Ertlschweiger, S.; Moritz, A.; Bantleon, H.P.; Rausch-Fan, X. Different effects of P. gingivalis LPS and E. coli LPS on the expression of interleukin-6 in human gingival fibroblasts. *Acta Odontol. Scand.* **2014**, *72*, 337–345. [CrossRef]

39. Herath, T.D.; Darveau, R.P.; Seneviratne, C.J.; Wang, C.Y.; Wang, Y.; Jin, L. Tetra- and penta-acylated lipid A structures of Porphyromonas gingivalis LPS differentially activate TLR4-mediated NF-kappaB signal transduction cascade and immuno-inflammatory response in human gingival fibroblasts. *PLoS ONE* **2013**, *8*, e58496. [CrossRef]

40. Ogawa, T.; Asai, Y.; Hashimoto, M.; Uchida, H. Bacterial fimbriae activate human peripheral blood monocytes utilizing TLR2, CD14 and CD11a/CD18 as cellular receptors. *Eur. J. Immunol.* **2002**, *32*, 2543–2550. [CrossRef]

41. Harokopakis, E.; Hajishengallis, G. Integrin activation by bacterial fimbriae through a pathway involving CD14, Toll-like receptor 2, and phosphatidylinositol-3-kinase. *Eur. J. Immunol.* **2005**, *35*, 1201–1210. [CrossRef]

42. Tabeta, K.; Yamazaki, K.; Akashi, S.; Miyake, K.; Kumada, H.; Umemoto, T.; Yoshie, H. Toll-like receptors confer responsiveness to lipopolysaccharide from Porphyromonas gingivalis in human gingival fibroblasts. *Infect. Immun.* **2000**, *68*, 3731–3735. [CrossRef] [PubMed]

43. Hiramine, H.; Watanabe, K.; Hamada, N.; Umemoto, T. Porphyromonas gingivalis 67-kDa fimbriae induced cytokine production and osteoclast differentiation utilizing TLR2. *FEMS Microbiol. Lett.* **2003**, *229*, 49–55. [CrossRef]

44. Hajishengallis, G.; Martin, M.; Sojar, H.T.; Sharma, A.; Schifferle, R.E.; DeNardin, E.; Russell, M.W.; Genco, R.J. Dependence of bacterial protein adhesins on toll-like receptors for proinflammatory cytokine induction. *Clin. Diagn. Lab. Immunol.* **2002**, *9*, 403–411. [CrossRef]

45. Aoki, Y.; Tabeta, K.; Murakami, Y.; Yoshimura, F.; Yamazaki, K. Analysis of immunostimulatory activity of Porphyromonas gingivalis fimbriae conferred by Toll-like receptor 2. *Biochem. Biophys. Res. Commun.* **2010**, *398*, 86–91. [CrossRef]

46. Hashimoto, M.; Asai, Y.; Ogawa, T. Separation and structural analysis of lipoprotein in a lipopolysaccharide preparation from Porphyromonas gingivalis. *Int. Immunol.* **2004**, *16*, 1431–1437. [CrossRef]

47. Asai, Y.; Hashimoto, M.; Fletcher, H.M.; Miyake, K.; Akira, S.; Ogawa, T. Lipopolysaccharide preparation extracted from Porphyromonas gingivalis lipoprotein-deficient mutant shows a marked decrease in toll-like receptor 2-mediated signaling. *Infect. Immun.* **2005**, *73*, 2157–2163. [CrossRef]

48. Cai, J.; Chen, J.; Guo, H.; Pan, Y.; Zhang, Y.; Zhao, W.; Li, X.; Li, Y. Recombinant fimbriae protein of Porphyromonas gingivalis induces an inflammatory response via the TLR4/NFkappaB signaling pathway in human peripheral blood mononuclear cells. *Int. J. Mol. Med.* **2019**, *43*, 1430–1440. [CrossRef] [PubMed]

49. Reife, R.A.; Coats, S.R.; Al-Qutub, M.; Dixon, D.M.; Braham, P.A.; Billharz, R.J.; Howald, W.N.; Darveau, R.P. Porphyromonas gingivalis lipopolysaccharide lipid A heterogeneity: Differential activities of tetra- and penta-acylated lipid A structures on E-selectin expression and TLR4 recognition. *Cell Microbiol.* **2006**, *8*, 857–868. [CrossRef] [PubMed]

50. Ziauddin, S.M.; Montenegro Raudales, J.L.; Sato, K.; Yoshioka, H.; Ozaki, Y.; Kaneko, T.; Yoshimura, A.; Hara, Y. Analysis of Subgingival Plaque Ability to Stimulate Toll-Like Receptor 2 and 4. *J. Periodontol.* **2016**, *87*, 1083–1091. [CrossRef]

51. El-Awady, A.R.; Miles, B.; Scisci, E.; Kurago, Z.B.; Palani, C.D.; Arce, R.M.; Waller, J.L.; Genco, C.A.; Slocum, C.; Manning, M.; et al. Porphyromonas gingivalis evasion of autophagy and intracellular killing by human myeloid dendritic cells involves DC-SIGN-TLR2 crosstalk. *PLoS Pathog.* **2015**, *10*, e1004647. [CrossRef] [PubMed]

Publisher's Note: MDPI stays neutral with regard to jurisdictional claims in published maps and institutional affiliations.

Journal of
Clinical Medicine

Article

Effect of Azithromycin on Proinflammatory Cytokine Production in Gingival Fibroblasts and the Remodeling of Periodontal Tissue

Takatoshi Nagano [1,*], Takao Yamaguchi [2], Sohtaro Kajiyama [1], Takuma Suzuki [1], Yuji Matsushima [1], Akihiro Yashima [1], Satoshi Shirakawa [1] and Kazuhiro Gomi [1]

[1] Department of Periodontology, Tsurumi University School of Dental Medicine, Yokohama 230-8501, Japan; kajiyama-sotaro@tsurumi-u.ac.jp (S.K.); suzuki-takuma@tsurumi-u.ac.jp (T.S.); matsushima-y@tsurumi-u.ac.jp (Y.M.); yashima-akihiro@tsurumi-u.ac.jp (A.Y.); shirakawa-satoshi@tsurumi-u.ac.jp (S.S.); gomi-k@tsurumi-u.ac.jp (K.G.)
[2] Aiko Yamaguchi Dental Clinic, Atsugi 243-0028, Japan; liberty@v006.vaio.ne.jp
* Correspondence: nagano-takatoshi@tsurumi-u.ac.jp; Tel.: +81-45-580-8434

Abstract: Previous reports have shown that azithromycin (AZM), a macrolide antibiotic, affects collagen synthesis and cytokine production in human gingival fibroblasts (hGFs). However, there are few reports on the effect of AZM on human periodontal ligament fibroblasts (hPLFs). In the present study, we comparatively examined the effects of AZM on hGFs and hPLFs. We monitored the reaction of AZM under lipopolysaccharide (LPS) stimulation or no stimulation in hGFs and hPLFs. Gene expression analyses of interleukin-6 (IL-6), interleukin-8 (IL-8), matrix metalloproteinase-1 (MMP-1), matrix metalloproteinase-2 (MMP-2), and Type 1 collagen were performed using reverse transcription-polymerase chain reaction (RT-PCR). Subsequently, we performed Western blotting for the analysis of the intracellular signal transduction pathway. In response to LPS stimulation, the gene expression levels of IL-6 and IL-8 in hGFs increased due to AZM in a concentration-dependent manner, and phosphorylation of nuclear factor kappa B (NF-κB) was also promoted. Additionally, AZM caused an increase in MMP-1 expression in hGFs, whereas it did not affect the expression of any of the analyzed genes in hPLFs. Our findings indicate that AZM does not affect hPLFs and acts specifically on hGFs. Thus, AZM may increase the expression of IL-6 and IL-8 under LPS stimulation to modify the inflammatory response and increase the expression of MMP-1 to promote connective tissue remodeling.

Keywords: azithromycin; human gingival fibroblast; human periodontal ligament fibroblast; IL-6; IL-8; MMP-1; MMP-2

Citation: Nagano, T.; Yamaguchi, T.; Kajiyama, S.; Suzuki, T.; Matsushima, Y.; Yashima, A.; Shirakawa, S.; Gomi, K. Effect of Azithromycin on Proinflammatory Cytokine Production in Gingival Fibroblasts and the Remodeling of Periodontal Tissue. *J. Clin. Med.* **2021**, *10*, 99. https://doi.org/10.3390/jcm10010099

Received: 10 December 2020
Accepted: 27 December 2020
Published: 30 December 2020

Publisher's Note: MDPI stays neutral with regard to jurisdictional claims in published maps and institutional affiliations.

1. Introduction

We previously demonstrated the clinical utility of full-mouth as well as partial-mouth scaling and root planing (SRP) with the administration of the macrolide antibiotic azithromycin (AZM) [1–4]. Studies have indicated that while performing full-mouth SRP (FM-SRP) with AZM administration, inflammation was reduced and periodontal pockets improved rapidly within an extremely short period of time. This appears to be the result of the synergistic effect of the control of periodontopathic bacteria due to the antibacterial action of AZM and mechanical plaque control by SRP.

However, when other antibiotics were used in the same manner, results equivalent to or better than those of AZM could not be obtained [5], especially with regard to the reduction in inflammation of periodontal pockets during the early stage after treatment [6]. Thus, it appears that the effect of AZM is not limited to its antibacterial action, suggesting the possible involvement of another mechanism.

Some reports have indicated that AZM is effective in treating cyclosporine-induced gingival overgrowth [7–9]. In 2008, Kim et al. reported that AZM inhibits the proliferation

61

of gingival fibroblasts and activations of matrix metalloproteinase-1 (MMP-1) and matrix metalloproteinase-2 (MMP-2). These metalloproteinases play a central role in collagen degradation. This may thereby reduce the volume of connective tissue constituents within the gingival tissue, resulting in amelioration of the gingival overgrowth associated with drug-induced fibrosis [10].

AZM and clarithromycin are known to regulate immune function and have been indicated to be particularly useful in treating pulmonary diseases, such as cystic fibrosis and chronic panacinar bronchitis [11,12]. A study using human bronchial epithelial cells reported that AZM's efficacy against long-term chronic inflammation is due to increased interleukin-8 (IL-8) production [13]. Furthermore, it has been suggested that AZM increases IL-8 production in human gingival fibroblasts more than other macrolide antibiotics (i.e., erythromycin, josamycin), resulting in further promotion of neutrophil migration during the inflammatory response triggered by periodontal pathogens [14]. However, many points of these theories remain unclear.

The purpose of this study was to investigate the reaction of human gingival fibroblasts (hGFs) and human periodontal ligament fibroblasts (hPLFs) when FM-SRP is performed with AZM administration. Therefore, in the present study, we selected interleukin-6 (IL-6) and IL-8, cytokines involved in the regulation of the inflammatory response, to conduct experiments to examine the effects of AZM and cell behavior with hGFs and hPLFs stimulated by lipopolysaccharide (LPS). MMP-1, MMP-2, and Type I-Collagen were studied to investigate the effects of AZM on periodontal tissue remodeling. Further, we examined the effects of AZM on the intracellular signal transduction pathway of hGFs.

2. Materials and Methods

The present study was approved by the institutional review board of the School of Dental Medicine, Tsurumi University (approval number: 1035) and was conducted in accordance with their regulations.

HGFs and hPLFs were obtained from patients who were determined to require tooth extraction for orthodontic treatment or oral surgery upon visiting Tsurumi University Dental Hospital, and who also gave consent to participate in the experiment. HGFs and hPLFs were prepared as described previously [15,16].

Using cells in the third to sixth passages, we investigated the effects of AZM on hGFs and hPLFs.

The experiment protocol is demonstrated in Figure 1.

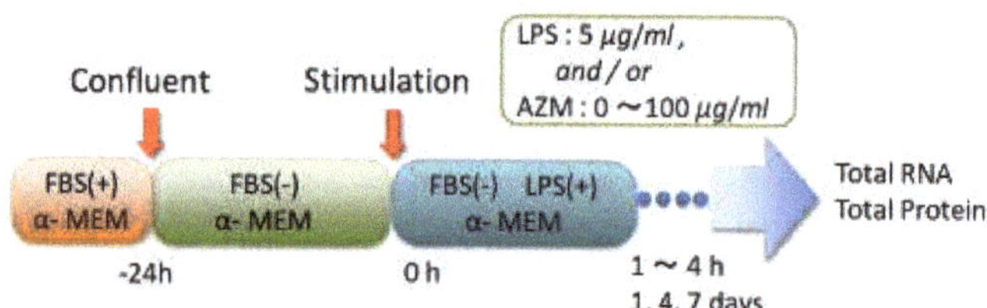

Figure 1. Experimental protocol.

2.1. Cell Culture

Gingival tissue and periodontal ligament tissue was collected, finely chopped, and cultured, and the outgrowth cells of each were used as hGFs and hPLFs, respectively. Alpha-minimum essential media (α-MEM) (SIGMA) containing 10% fetal bovine serum (FBS) (ICN Biomedicals Inc., Solon, OH, USA) and 2% penicillin–streptomycin (Pn-St) (SIGMA, St. Louis, MO, USA) were used as serum and cultured at 37 °C in the presence of 5% CO_2. A subculture was performed using 0.1% trypsin–0.04% EDTA (SIGMA).

2.2. MTS Assay (Cell Proliferation Assay)

To investigate the effects of AZM on the cell proliferation ability of hGFs and hPLFs, the cells were uniformly seeded on a 96-well plate with 1.0×10^4 cells/well, and cultured in α-MEM containing 10% FBS and 1% Pn-St at 37 °C in the presence of 5% CO_2.

At 1, 4, and 7 d after the addition of AZM (Pfizer Japan Inc., Tokyo, Japan), 20 µL of Cell Titer 96 Aqueous One Solution Reagent (Promega, Madison, WI, USA) was added and incubated for 60 min. Absorbance was subsequently measured at a wavelength of 490 nm.

2.3. Reverse Transcription-Polymerase Chain Reaction (RT-PCR)

Both hGFs and hPLFs were seeded separately in 6-well plates with a cell count of 1.0×10^4 cells/well, and the culture medium was replaced every two days.

After confluence was observed, the experiment was conducted by transferring the cells to a stimulating medium, to which a sample was added 24 h after exchanging with α-MEM without the addition of serum. The stimulation medium was created by adding an LPS (LPS; *Escherichia coli* 0111: B4, SIGMA) and AZM to α-MEM to a final concentration of 0 or 5 µg/mL and 0 to 100 µg/mL, respectively, at 37 °C after pre-incubation in a 5% CO_2 environment. The gene expression levels of the inflammatory cytokines IL-6 and IL-8 were examined after a specific period. The gene expression levels of MMP-1, MMP-2, and Type I-Collagen, which are involved in connective tissue remodeling, were examined 1, 4, and 7 days after AZM was added without LPS.

Total RNA was extracted from cultured cells using RNAqueous (Applied Biosystems, Austin, TX, USA), and cDNA was prepared using Ready-to-go You-Prime First-Strand Beads (GE Healthcare Life Sciences, Piscataway, NJ, USA). Glyceraldehyde 3-phosphate dehydrogenase (GAPDH) (Clontec, Palo Alto, CA, USA) was used as the internal standard.

RT-PCRs were performed using specific primers and Taq DNA polymerase (Takara Bio Inc., Shiga, Japan) (Applied Biosystems/Gene Amp PCR System 9700). PCR products were detected by performing 4.5% polyacrylamide gel electrophoresis (PAGE), staining with ethidium bromide, and subsequent exposure to ultraviolet light. The results were recorded.

The gene sequences of each primer are shown in Figure 2.

GAPDH	Sense	5'-ACC ACA GTC CAT GCC ATC AC-3'	452bp
	Antisense	5'-TCC ACC ACC CTG TTG CTG TA-3'	
IL-6	Sense	5'-ATG AAC TCC TTC TCC ACA AGC GC-3'	680bp
	Antisense	5'-GAA GAG CCC TCA GGC TGG ACT G-3'	
IL-8	Sense	5'-CTG CGC CAA CAC AGA AAT TAT-3'	238bp
	Antisense	5'-ATT GCA TCT GGC AAC CCT ACA-3'	
Type I Collagen	Sense	5'-CTG GCA AAG AAG GCG GCA AA-3'	500bp
	Antisense	5'-CTC ACC ACG ATC ACC ACT CT-3'	
MMP1	Sense	5'-AAG CCA TAT ATG GAC GTT CC-3'	586bp
	Antisense	5'-TCT GGA GAG TCA AAA TTC TC-3'	
MMP2	Sense	5'-CTG CTG AAG GAC ACA CTA AAG-3'	605bp
	Antisense	5'-CAT CCT TCT CAA AGT TGT AGG-3'	

Figure 2. PCR primers.

2.4. Western Blotting

Total protein was extracted from the cultured hGFs under the same conditions maintained for total RNA extraction, and an extraction buffer containing 2% sodium dodecyl sulfate (SDS) and 0.1 M dithiothreitol (DTT) was used. After pretreatment, 10% SDS-PAGE was performed to separate the proteins and transfer them to the polyvinylidene difluoride (PVDF) membrane (Bi-Rad Laboratories, Hercules, CA, USA). After blocking with bovine serum albumin (BSA), an antibody specific for each protein (nuclear factor kappa B (NF-κB), phospho NF-κB, p38 mitogen-activated protein kinase (MAPK), phospho p38 MAPK, Jun N-terminal kinase (JNK), and phospho JNK; Cell Signaling Technology, Danvers, MA, USA) was used as the primary antibody, and horseradish peroxidase (HRP)-labeled anti-rabbit immunoglobulin G (IgG) antibody (Cell Signaling Technology) was used as the secondary antibody. Immunological detection was performed using enhanced chemiluminescence (ECL) Western blotting detection reagents (GE Healthcare Life Sciences Corp, Verdesian, MA, USA), and the activation of intracellular signal transduction molecules associated with the expression of inflammatory cytokines was investigated. β-actin (Cell Signaling Technology) was used as the internal standard.

2.5. Data Analysis and Statistical Processing

Based on the results of our pilot study, we calculated the number of samples in this study using the formula $n = 16(SD/\Delta)^2$, where SD is the average standard deviation between the two groups and Δ is the difference between the average values of the two groups. It was calculated at a significance level α of 0.05 and power of 80% ($\beta = 0.2$). As a result, all experiments were performed with $n = 3$. All the experiments were performed at least three times.

The results of RT-PCR electrophoretic imaging were standardized using GAPDH with a Densitograph Lane & Spot Analyzer (ATTO, Tokyo, Japan), and semi-quantification was performed. Students t-tests were used for data analysis of the MTS assay, and statistical significance was defined as $p < 0.05$.

3. Results

3.1. MTS Assay (Cell Proliferation Assay)

MTS assay results showed that AZM did not statistically affect the proliferative activity of hGFs and hPLFs (Figures 3 and 4).

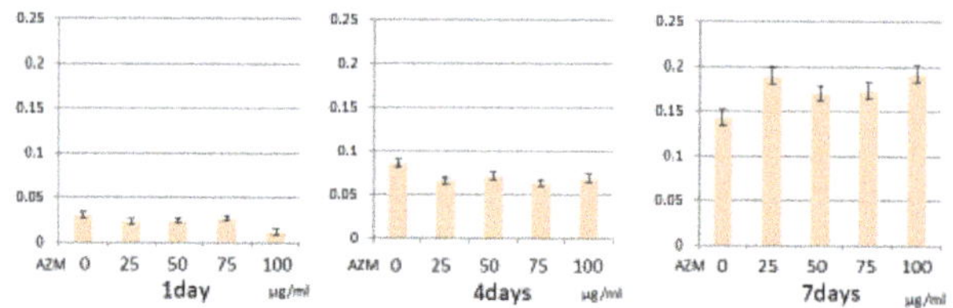

Figure 3. Cell proliferation ability of human gingival fibroblasts (hGFs) with azithromycin (AZM). These experiments were performed in triplicate, and data are presented as the mean ± SD.

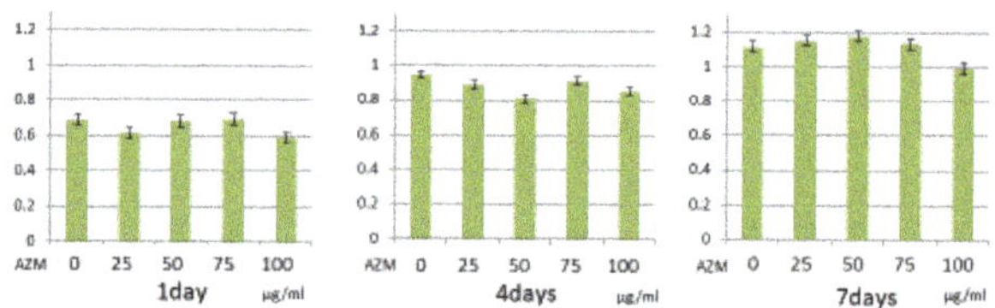

Figure 4. Cell proliferation ability of human periodontal ligament fibroblasts (hPLFs) with AZM. These experiments were performed in triplicate, and data are presented as the mean ± SD.

3.2. IL-6 and IL-8 Gene Expression

In hGFs, the gene expression levels of IL-6 and IL-8 were found to increase due to stimulation with LPS (Figure 5), whereas no changes in IL-6 and IL-8 gene expression levels were observed with the addition of AZM alone (Figure 6). Thus, it was demonstrated that AZM by itself does not amplify the gene expression levels of IL-6 and IL-8.

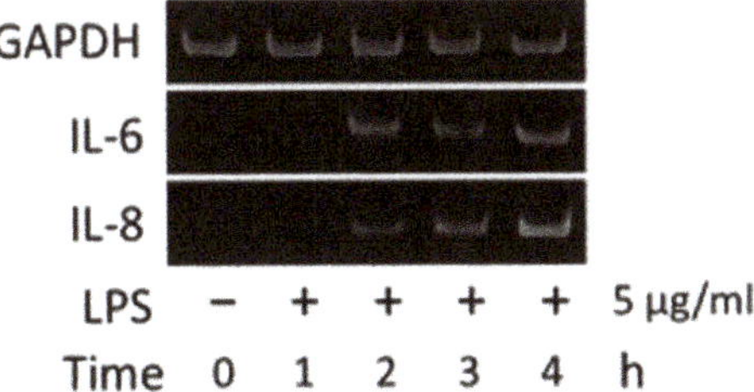

Figure 5. LPS enhances IL-6 and IL-8 gene expression in hGFs.

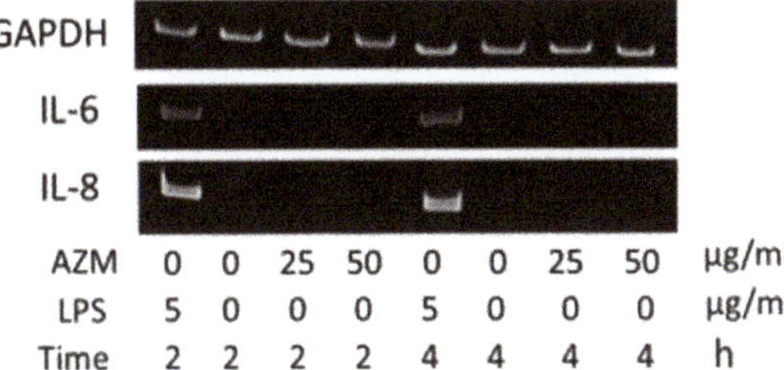

Figure 6. AZM alone does not enhance gene expression of IL-6 and IL-8 in hGFs.

Moreover, in the case of hGFs, the gene expression levels of IL-6 and IL-8 increased in a concentration-dependent manner due to AZM in the presence of LPS (Figure 7). However, hPLFs did not show any changes with regard to IL-6 and IL-8 gene expression (Figure 8).

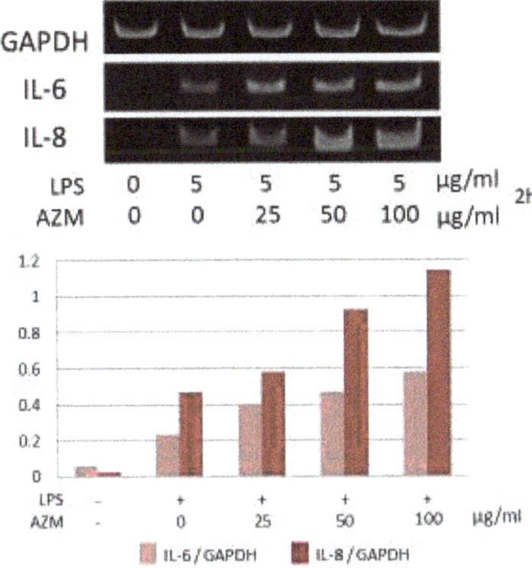

Figure 7. Gene expression of IL-6 and IL-8 in hGFs in the presence of LPS and AZM. The most typical data were shown.

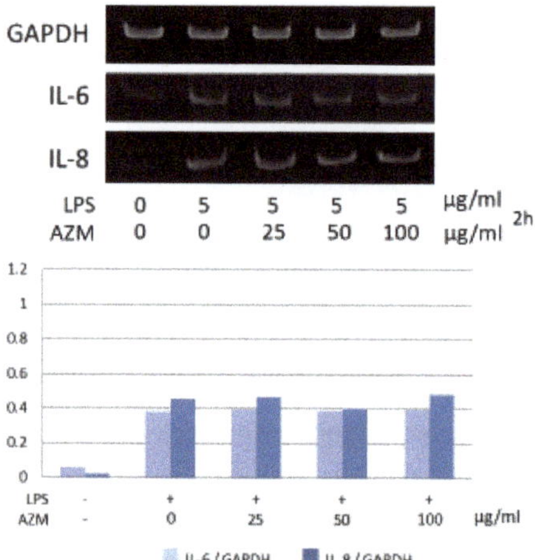

Figure 8. Gene expression of IL-6 and IL-8 in hPLFs in the presence of LPS and AZM. The most typical data were shown.

3.3. Western Blotting

Western blot of hGFs showed an increase in the phosphorylation of NF-κB. However, no change was observed in the phosphorylation patterns of p38 MAPK and JNK in the presence of both LPS and AZM (Figure 9).

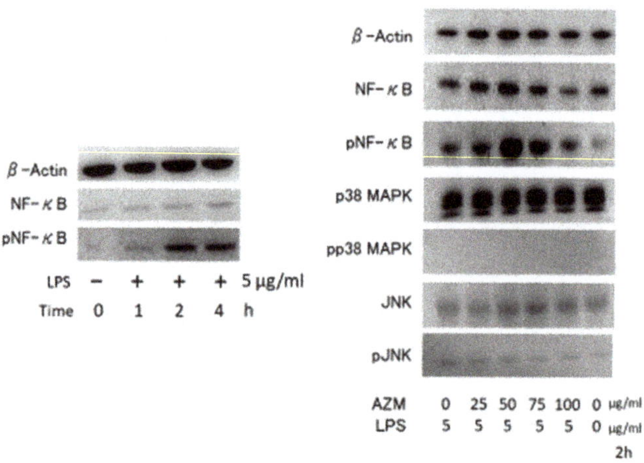

Figure 9. Effects on intracellular signaling molecules in hGFs in the presence of LPS and AZM.

3.4. Gene Expression Levels of MMP-1, MMP-2, and Type I-Collagen

Increased MMP-1 gene expression levels were observed in hGFs due to the effects of AZM (Figure 10), whereas no significant changes were observed in the gene expression levels of MMP-2 and Type I-Collagen. In hPLFs, no prominent change was observed in the gene expression levels of MMP-1, MMP-2, and Type I-Collagen (Figure 11).

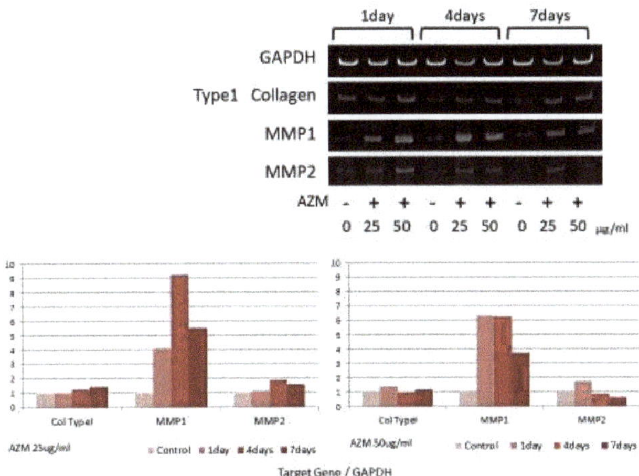

Figure 10. Gene expression of Type I-Collagen, MMP-1, and MMP-2 in hGFs due to the influence of AZM. The most representative data were shown.

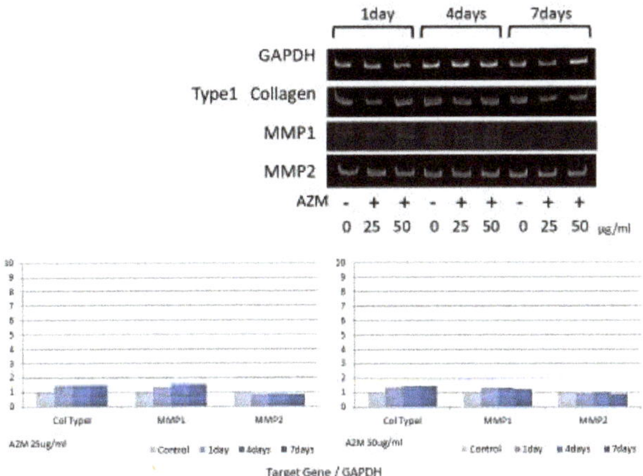

Figure 11. Gene expression of Type I-Collagen, MMP-1, and MMP-2 in hPLFs due to the influence of AZM. The most representative data were shown.

4. Discussion

AZM is a macrolide antibiotic with a long half-life and a wide antibacterial spectrum. It has been demonstrated to be useful in periodontal treatment due to its ability to inhibit biofilm formation [17]. AZM is characterized by phagocyte delivery, which allows for selective uptake by phagocytic cells, causing accumulation at the site of inflammation. AZM accumulated locally in the inflamed gingival tissue temporarily increased the expression of IL-6 and IL-8, which further promoted neutrophil migration. It is possible that there may be a mechanism by which high concentrations of AZM are selectively accumulated in the periodontal tissue at an early stage as a result of the aforementioned synergistic effects.

When performing periodontal treatment under AZM administration, the antibacterial action of AZM may facilitate the control of periodontopathic bacteria by creating a microenvironment in which a high concentration of AZM selectively gathers locally in the periodontal tissue at an early stage. As a result, the inhibitory or destructive action of AZM against biofilm formation may ameliorate microbial accumulation in the periodontal pocket. It has also been reported that AZM does not significantly affect cell adhesion ability and human epithelial cell death [18]. The results of the present study indicate that AZM does not affect the cell proliferation activity of either hGFs or hPLFs, confirming that AZM can be safely administered in periodontal treatment.

In the present study, the gene expression of IL-6 and IL-8 of hGFs increased in a concentration-dependent manner due to the combined effects of AZM and LPS. Furthermore, the use of AZM alone did not affect the gene expression of IL-6 and IL-8. Moreover, hPLFs showed no significant changes in IL-6 and IL-8 gene expression. The ability of AZM to modify the inflammatory response is a characteristic phenomenon in hGFs, and the use of AZM alone does not appear to be able to modify or regulate the response. Thus, these results suggest that the action of AZM at the site of inflammation might temporarily increase the inflammatory response in the presence of LPS. However, in a study using KB cells, a human oral epithelial cell line, it was reported that AZM inhibited the expression of IL-8 induced by LPS [19]. Based on these findings, the effects of AZM could differ, depending on the target cell.

In 2015, Doyle et al. reported that the expression of the inflammatory cytokines IL-6, IL-8, monocyte chemoattractant protein-1 (MCP-1), and growth-regulated oncogene (GRO), previously induced by LPS derived from *Porphyromonas gingivalis* against hGFs, was reduced due to the effects of AZM [20]. In that study, 10 μg/mL of AZM was used, and the results were recorded 24 h after the addition of AZM. In the present study, however, the maximum concentration of AZM used was 100 μg/mL, the stimulation time was particularly short (2 h), and the experimental conditions differed. Based on these findings, it appears that AZM may play a role in strengthening the inflammatory response during the very early stage in the presence of LPS, and subsequently reducing the inflammatory response in latter stages. Thus, its effects may be present for a long period. However, since hGFs may behave and react differently depending on the intracellular concentration of AZM and the environment of action, further research in this area is needed.

Moreover, in the present study, phosphorylation of NF-κB was also promoted when LPS and AZM acted simultaneously, compared with when stimulation with LPS alone was performed, whereas there were no significant changes in the phosphorylation of p38 MAPK and JNK. These findings indicate that an increased production of inflammatory cytokines in hGFs may be caused by the effects of AZM due to LPS stimulation. This may be mediated by the activation of the signal transduction pathway of NF-κB. A previous report showed that the simultaneous action of AZM and tumor necrosis factor-alpha (TNF-α) in infant tracheal aspirate cells inhibited NF-κB signal activation [21]. It has also been reported that when the effects of AZM were investigated using murine macrophage cell lines, the NF-κB signal was inhibited [22]. A study using human peripheral blood mononuclear cells reported that AZM had a major effect on JNK expression [23]. Furthermore, in 2015 it was reported that AZM inhibited the phosphorylation of p38 MAPK and extracellular-signal-regulated kinase (ERK) signaling pathways in human gingival epithelial cells activated by TNF-α stimulation [24]. The reason for non-concurrence of the results of the present study with those of the previous one may be because the effects of AZM on the intracellular signal transduction pathway may differ, depending on factors such as the degree and timing of inflammation, tissue and cell type, patient age, and study designs. Extracellular vesicles are currently known to be the main source of proinflammatory cytokines. Especially, microvesicles may function as strong regulators of the immune system [25]. Given that the microvesicle function and network of signal transduction pathways in cells are complicated, further studies are necessary to clarify these findings.

In 1995, it was reported that AZM improved cyclosporine-induced gingival overgrowth [26], and continued research indicated that said effect was due to the promotion of phagocytosis of fibroblasts by AZM [27]. The results of the present study indicate that AZM acted specifically on hGFs and increased the expression of MMP-1. This suggests that the action of MMP-1 produced by gingival fibroblasts on collagen promoted its decomposition, thereby inducing periodontal tissue remodeling. This mechanism might explain the clinical findings indicating that gingival retraction results in the early and dramatic reduction of periodontal pockets when performing FM-SRP during AZM administration. However, since a significant reaction to the expression of MMP-1 was not observed in hPLFs, it appears that this unique effect of AZM was specific to the gingival tissue.

Some researchers are of the view that the clinical results of FM-SRP without AZM administration are comparable to those in the case of conventional SRP [28]. However, it has been previously reported that the formation of human osteoclasts is inhibited due to the effects of AZM [29]. Furthermore, performing FM-SRP with AZM administration can inhibit the production of inflammatory mediators and increase total body temperature [30,31]. Hence, applying AZM to periodontal treatment may produce various secondary benefits and demonstrate efficacy in the clinical setting.

These findings suggest that when FM-SRP is performed with AZM administration, AZM amplifies the gene expression of IL-6 and IL-8, which are involved in neutrophil migration in the inflamed gingival region. This contributes to the amelioration of inflammation during the early stage. This suggests the existence of a mechanism by which the periodontal pockets are reduced as a result of the remodeling of the gingival connective tissue due to the increased gene expression of MMP-1.

5. Conclusions

The results of the present study suggest that in the presence of LPS, AZM may activate the NF-κB signaling dental pathway in hGFs, resulting in the increased production of inflammatory cytokines. AZM may also increase the expression of MMP-1 in hGFs, thereby promoting gingival connective tissue remodeling.

Author Contributions: Conceptualization, T.N., T.Y. and K.G.; methodology, T.N., T.Y., S.K., T.S., Y.M., A.Y. and S.S.; manuscript preparation, T.N. and K.G. All authors have read and agreed to the published version of the manuscript.

Funding: This work was supported by JSPS KAKENHI, Grant-in-Aid for Young Scientists (B), 22792101 to T.N.

Informed Consent Statement: Informed consent was obtained from all subjects involved in the study.

Data Availability Statement: All relevant data are available upon request. Please address requests to Takatoshi Nagano.

Acknowledgments: The authors thank the study participants and the staff of Tsurumi University Dental Hospital.

Conflicts of Interest: All authors declare that there are no conflicting interests.

References

1. Gomi, K.; Yashima, A.; Nagano, T.; Kanazashi, M.; Maeda, N.; Arai, T. Effects of full-mouth scaling and root planing in conjunction with systemically administrated azithromycin. *J. Periodontol.* **2007**, *78*, 422–429. [CrossRef] [PubMed]
2. Gomi, K.; Yashima, A.; Iino, F.; Kanazashi, M.; Nagano, T.; Shibukawa, N.; Ohshima, T.; Maeda, N.; Arai, T. Drug concentration in inflamed periodontal tissues with systemically administrated azithromycin. *J. Periodontol.* **2007**, *78*, 918–923. [CrossRef] [PubMed]
3. Gomi, K.; Matsushima, Y.; Ujiie, Y.; Shirakawa, S.; Nagano, T.; Kanazashi, M.; Yashima, A. Full-mouth scaling and root planning combined with azithromycin to treat peri-implantitis. *Aust. Dent. J.* **2015**, *60*, 503–510. [CrossRef]
4. Yashima, A.; Gomi, K.; Maeda, N.; Arai, T. One-Stage Full-Mouth Versus Partial-Mouth Scaling and Root Planing During the Effective Half-Life of Systemically Administered Azithromycin. *J. Periodontol.* **2009**, *80*, 1406–1413. [CrossRef] [PubMed]
5. Herrera, D.; Alonso, B.; León, R.; Roldán, S.; Sanz, M. Antimicrobial therapy in periodontitis: The use of systemic antimicrobials against the subgingival biofilm. *J. Clin. Periodontol.* **2008**, *35*, 45–66. [CrossRef] [PubMed]
6. Lang, N.P.; Tan, W.C.; Krähenmann, M.A.; Zwahlen, M. A systematic review of the effects of full-mouth debridement with and without antiseptics in patients with chronic periodontitis. *J. Clin. Periodontol.* **2008**, *35*, 8–21. [CrossRef] [PubMed]
7. Gomez, E.; Sanchez-Nunez, M.; Sánchez, J.E.; Corte, C.; Aguado, S.; Portal, C.; Baltar, J.; Alarez-Grande, J. Treatment of cyclosporin-induced gingival hyperplasia with azithromycin. *Nephrol. Dial. Transplant.* **1997**, *12*, 2694–2697. [CrossRef]
8. Citterio, F.; Di Pinto, A.; Borzi, M.T.; Scatà, M.C.; Foco, M.; Pozzetto, U.; Castagneto, M. Azithromycin treatment of gingival hyperplasia in kidney transplant recipients is effective and safe. *Transplant. Proc.* **2001**, *33*, 2134–2135. [CrossRef]
9. Tokgoz, B.; Sari, H.I.; Yildiz, O.; Aslan, S.; Sipahioglu, M.; Okten, T.; Oymak, O.; Utas, C. Effects of azithromycin on cyclocporine-induced gingival hyperplasia in renal transplant patients. *Transplant. Proc.* **2004**, *36*, 2699–2702. [CrossRef]

10. Kim, J.-Y.; Park, S.-H.; Cho, K.-S.; Kim, H.-J.; Lee, C.-K.; Park, K.-K.; Choi, S.-H.; Chung, W.-Y. Mechanism of azithromycin treatment on gingival overgrowth. *J. Dent. Res.* **2008**, *87*, 1075–1079. [CrossRef]

11. Yousef, A.A.; Jaffe, A. The Role of Azithromycin in Patients with Cystic Fibrosis. *Paediatr. Respir. Rev.* **2010**, *11*, 108–114. [CrossRef] [PubMed]

12. Schultz, M.J. Macrolide activities beyond their antimicrobial effects: Macrolides in diffuse panbronchiolitis and cystic fibrosis. *J. Antimicrob. Chemother.* **2004**, *54*, 21–28. [CrossRef] [PubMed]

13. Shinkai, M.; Foster, G.H.; Rubin, B.K. Macrolide antibiotics modulate ERK phosphorylation and IL-8 and GM-CSF production by human bronchial epithelial cells. *Am. J. Physiol. Lung Cell. Mol. Physiol.* **2006**, *290*, L75–L85. [CrossRef] [PubMed]

14. Kamemoto, A.; Ara, T.; Hattori, T.; Fujinami, Y.; Imamura, Y.; Wang, P.-L. Macrolide antibiotics like azithromycin increase lipopolysaccharide-induced IL-8 production by human gingival fibroblasts. *Eur. J. Med. Res.* **2009**, *14*, 309–314. [CrossRef]

15. Takada, H.; Mihara, I.; Morisaki, S.; Hamada, S. Induction of interleukin-1 and -6 in human gingival fibroblast cultures stimulated with Bacteroides lipopolysaccharides. *Infect. Immun.* **1991**, *59*, 295–301. [CrossRef]

16. Somerman, M.; Archer, S.; Imm, G.; Foster, R. A Comparative Study of Human Periodontal Ligament Cells and Gingival Fibroblasts in vitro. *J. Dent. Res.* **1988**, *67*, 66–70. [CrossRef]

17. Hirsch, R.; Deng, H.; Laohachai, M.N. Azithromycin in periodontal treatment: More than an antibiotic. *J. Periodontal Res.* **2012**, *47*, 137–148. [CrossRef]

18. Inoue, K.; Kumakura, S.-I.; Uchida, M.; Tsutsui, T. Effects of eight antibacterial agents on cell survival and expression of epithelial-cell- or cell-adhesion-related genes in human gingival epithelial cells. *J. Periodontal Res.* **2004**, *39*, 50–58. [CrossRef]

19. Matsumura, Y.; Mitani, A.; Suga, T.; Kamiya, Y.; Kikuchi, T.; Tanaka, S.; Aino, M.; Noguchi, T. Azitromycin may inhibit interleukin-8 through suppression of Rac1 and a Nuclear Factor-Kappa B pathway in KB cells stimulated with lipopolysaccharide. *J. Periodontol.* **2011**, *82*, 1623–1631. [CrossRef]

20. Doyle, C.J.; Fitzsimmons, T.R.; Marchant, C.; Dharmapatni, A.A.S.S.K.; Hirsch, R.; Bartold, P.M. Azithromycin suppresses P. gingivalis LPS-induced pro-inflammatory cytokine and chemokine production by human gingival fibroblasts in vitro. *Clin. Oral Investig.* **2015**, *19*, 221–227. [CrossRef]

21. Aghai, Z.H.; Kode, A.; Saslow, J.G.; Nakhla, T.; Farhath, S.; Stahl, G.E.; Eydelman, R.; Strande, L.; Leone, P.; Rahman, I. Azithromycin Suppresses Activation of Nuclear Factor-kappa B and Synthesis of Pro-inflammatory Cytokines in Tracheal Aspirate Cells From Premature Infants. *Pediatr. Res.* **2007**, *62*, 483–488. [CrossRef] [PubMed]

22. Choi, E.Y.; Jin, J.Y.; Choi, J.I.; Choi, I.S.; Kim, S.J. Effect of azithromycin on Prevotella intermedia lipopolysaccharide-induced production of interleukin-6 in murine macropahges. *Eur. J. Pharmacol.* **2014**, *729*, 10–16. [CrossRef] [PubMed]

23. Hiwatashi, Y.; Maeda, M.; Fukushima, H.; Onda, K.; Takana, S.; Utsumi, H.; Hirano, T. Azithromycin suppresses proliferation, interleukin production and mitogen-activates protein kinases in human peripheral-blood mononuclear cells stimulated with bacterial super antigen. *J. Pharm. Pharmacol.* **2011**, *63*, 1320–1326. [CrossRef] [PubMed]

24. Miyagawa, T.; Fujita, T.; Yumoto, H.; Yoshimoto, T.; Kajiya, M.; Ouhara, K.; Matsuda, S.; Shiba, H.; Matsuo, T.; Kurihara, H. Azithromycin recovers reductions in barrier function in human gingival epithelial cells stimulated with tumor necrosis factor-α. *Arch. Oral Biol.* **2016**, *62*, 64–69. [CrossRef] [PubMed]

25. Słomka, A.; Urban, S.K.; Lukacs-Kornek, V.; Żekanowska, E.; Kornek, M. Large Extracellular Vesicles: Have We Found the Holy Grail of Inflammation? *Front. Immunol.* **2018**, *9*, 2723. [CrossRef] [PubMed]

26. Wahlstrom, E.; Zamora, J.U.; Teichman, S. Improvement in Cyclosporine-Associated Gingival Hyperplasia with Azithromycin Therapy. *N. Engl. J. Med.* **1995**, *332*, 753–754. [CrossRef] [PubMed]

27. Paik, J.-W.; Kim, C.-S.; Cho, K.-S.; Chai, J.-K.; Kim, C.-K.; Choi, S.-H. Inhibition of Cyclosporin A-Induced Gingival Overgrowth by Azithromycin through Phagocytosis: An In Vivo and In Vitro Study. *J. Periodontol.* **2004**, *75*, 380–387. [CrossRef]

28. Wennstrom, J.L.; Tomasi, C.; Bertelle, A.; Dellasega, E. Full-mouth ultrasonic debridement versus quadrant scaling and root planing as an initial approach in the treatment of chronic periodontitis. *J. Clin. Periodontol.* **2005**, *32*, 851–859. [CrossRef]

29. Gannon, S.C.; Cantley, M.; Haynes, D.R.; Hirsch, R.; Bartold, P.M. Azithromycin suppresses human osteoclast formation and activity in vitro. *J. Cell. Physiol.* **2013**, *228*, 1098–1107. [CrossRef]

30. Morozumi, T.; Yashima, A.; Gomi, K.; Ujiie, Y.; Izumi, K.; Akizuki, T.; Mitanni, K.; Takamatsu, H.; Minabe, M.; Miyauchi, S.; et al. Increased systemic levels of inflammatory mediators following one-stage full-mouth scaling and root planning. *J. Periodontal Res.* **2018**, *53*, 536–544. [CrossRef]

31. Yashima, A.; Morozumi, T.; Yoshie, H.; Hokari, T.; Izumi, Y.; Akizuki, T.; Mizutani, K.; Takamatsu, H.; Minabe, M.; Miyauchi, S.; et al. Biological responses following one-stage full-mouth scaling and root planing with and without azithromycin: Multicenter randomized trial. *J. Periodontal Res.* **2019**, *54*, 709–719. [CrossRef] [PubMed]

Journal of
Clinical Medicine

Article

Association of Periodontal Status, Number of Teeth, and Obesity: A Cross-Sectional Study in Japan

Norio Aoyama [1,*], Toshiya Fujii [1], Sayuri Kida [1], Ichirota Nozawa [2], Kentaro Taniguchi [1], Motoki Fujiwara [2], Taizo Iwane [3], Katsushi Tamaki [2] and Masato Minabe [1]

[1] Division of Periodontology, Department of Oral Interdisciplinary Medicine, Graduate School of Dentistry, Kanagawa Dental University, 82 Inaoka-cho, Yokosuka, Kanagawa 238-8580, Japan; t.fujii@kdu.ac.jp (T.F.); kdsyr61@gmail.com (S.K.); k.taniguchi@kdu.ac.jp (K.T.); minabe@kdu.ac.jp (M.M.)

[2] Division of Prosthodontic Dentistry for Function of TMJ and Occlusion, Department of Oral Interdisciplinary Medicine, Graduate School of Dentistry, Kanagawa Dental University, 82 Inaoka-cho, Yokosuka, Kanagawa 238-8580, Japan; nozawa@kdu.ac.jp (I.N.); fuji12422003@yahoo.co.jp (M.F.); tamaki@kdu.ac.jp (K.T.)

[3] School of Nutrition and Dietetics, Faculty of Health and Social Services, Kanagawa University of Human Services, 1-10-1 Heisei-cho, Yokosuka, Kanagawa 238-8522, Japan; iwane-d2k@kuhs.ac.jp

* Correspondence: aoyama@kdu.ac.jp; Tel.: +81-46-845-3160

Abstract: Recent reports have shown an association between obesity and periodontitis, but the precise relationship between these conditions has yet to be clarified. The purpose of this study was to compare the status of periodontitis, tooth loss, and obesity. Participants comprised 235 patients at the Center for Medical and Dental Collaboration in Kanagawa Dental University Hospital between 2018 and 2020. Clinical examinations such as blood testing, body composition analysis, periodontal measurement, assessment of chewing ability, salivary testing, and oral malodor analysis were performed. Periodontal inflamed surface area (PISA) was significantly associated with the number of teeth and body mass index (BMI). The number of teeth was negatively associated with age, but positively with chewing ability. Chewing ability was associated negatively with age, and positively with high-sensitivity C-reactive protein (hsCRP). The level of methyl-mercaptan in breath and protein and leukocyte scores from salivary testing were positively associated with PISA. The rate of insufficient chewing ability was increased in subjects with hemoglobin (Hb)A1c $\geq$ 7%. The high PISA group showed increased hsCRP. BMI as an obesity marker was positively associated with PISA, indicating periodontal inflammation. Chewing ability was related to serum markers such as HbA1c and hsCRP.

Keywords: chewing ability; infection; inflammation; periodontal medicine

Citation: Aoyama, N.; Fujii, T.; Kida, S.; Nozawa, I.; Taniguchi, K.; Fujiwara, M.; Iwane, T.; Tamaki, K.; Minabe, M. Association of Periodontal Status, Number of Teeth, and Obesity: A Cross-Sectional Study in Japan. *J. Clin. Med.* **2021**, *10*, 208. https://doi.org/10.3390/jcm10020208

Received: 14 December 2020
Accepted: 7 January 2021
Published: 8 January 2021

Publisher's Note: MDPI stays neutral with regard to jurisdictional claims in published maps and institutional affiliations.

1. Introduction

Obesity is becoming a serious problem for public health and is associated with many kinds of health concerns, such as diabetes mellitus, hypertension, dyslipidemia, and cardiovascular diseases [1,2]. Periodontitis is a common chronic inflammatory disease, characterized by gingival inflammation and destruction of tissue around the teeth, resulting in tooth loss [3]. Recent reports showed a strong association between periodontitis and general health. Periodontitis and tooth loss may be risk factors for cardiovascular disease [4]. Ischemic heart disease remains the single largest cause of death in countries of all income groups with a rise in obesity and diabetes mellitus [5]. Patients with periodontitis should be advised that there is a higher risk for cardiovascular diseases, such as myocardial infarction or stroke [6].

An association between obesity and periodontitis has recently been recognized [7]. The available evidence suggests a significantly positive association between periodontal disease and obesity not only in adults but also in children [8]. Obesity may influence host susceptibility to periodontopathic bacteria via inflammatory reactions [9]. Specific

71

periodontal pathogens may thus play roles in obese patients [10]. Although the underlying pathophysiological mechanism remains unclear, the development of insulin resistance as a consequence of a chronic inflammatory state and oxidative stress could be implicated in the association between obesity and periodontitis [11]. A recent report showed that obesity is also associated with a higher risk of tooth loss [12]. Periodontal diseases and chewing disorders are associated with poor nutrition [13]. Periodontitis patients exhibited a higher body mass index (BMI) and altered diet practices [14]. A change in food intake might be a factor connecting oral health and obesity.

The world is aging, and Japan is the most aging country in the world [15]. Although the percentage of overweight/obese people in Japan is lower than that in other countries [16], many Japanese patients suffer from diabetes and periodontitis. The associations among periodontal disease, tooth loss, and obesity have been reported; however, data on the Japanese population is limited. The aim of the present cross-sectional study was thus to clarify relationships between periodontitis, tooth loss, and obesity within the Japanese population.

2. Materials and Methods

2.1. Study Population

Subjects were recruited from patients at the Center for Medical and Dental Collaboration in Kanagawa Dental University Hospital between 2018 and 2020. Data were obtained from 235 subjects in this study. Inclusion criteria were as follows: age $\geq$ 20 years; and consent to participate in this study. The criterion of 20 years old was used according to the definition of adult in Japan. Exclusion criteria were as follows: antibiotic intake within the past 2 months; severe systemic infection; or pregnancy or lactating status. The ethics committees of the School of Dentistry at Kanagawa Dental University approved the present study (approval no. 665), and the protocol conformed to the 1975 Declaration of Helsinki, as revised in 2013. Each participant was informed of the aims and procedures of the study. Written informed consent was provided by all participants prior to participation.

2.2. Clinical Examinations

All examinations were performed at the Center for Medical and Dental Collaboration in Kanagawa Dental University Hospital. General information on subjects such as age and sex were collected from medical records. Body composition analysis was performed using an analyzer (InBody 460; InBody Japan, Tokyo, Japan), and body indices such as height, weight, and BMI were recorded. Peripheral blood samples were obtained and serum levels of high-sensitivity C-reactive protein (hsCRP) and hemoglobin (Hb)A1c were measured.

Two trained periodontists (N.A. and M.M.) counted the number of residual teeth, excluding wisdom teeth. Probing pocket depth and bleeding on probing at six points per tooth for all teeth were measured using a manual probe (PCP-UNC 15; Hu-Friedy, Chicago, IL, USA). Using these periodontal parameters, periodontal inflamed surface area (PISA) was calculated as previously described [17].

Chewing ability was calculated using a Gluco Sensor GS-II (GC, Tokyo, Japan) in accordance with the instructions from the manufacturer. Oral malodor was objectively evaluated using OralChroma (FIS, Itami, Japan) as previously described [18]. The salivary test was performed using Sill-Ha (Arkray Inc., Kyoto, Japan) in accordance with the instructions from the manufacturer. Briefly, subjects rinsed their mouth with a dedicated solution for 10 s, then saliva samples were obtained with the solution. The oral rinse solution was applied to a test strip and placed in the instrument for testing. Protein score and leukocyte score were calculated from this salivary test, which indicated the level of inflammation in the gingiva.

2.3. Statistical Analysis

The Shapiro-Wilk test was performed to test the normality of data distributions. Numerical data are presented as mean $\pm$ standard deviation for parameters showing normal distributions and as median and interquartile range for skewed distributions.

Spearman's correlation coefficient was used to calculate correlations between values. The chi-square test was performed to compare the subject rate with insufficient chewing ability. The Wilcoxon test was used to compare differences in hsCRP between groups. JMP version 14.2.0 software (SAS Institute Inc., Cary, NC, USA) was used for all statistical analyses. Values of $p < 0.05$ were considered statistically significant.

3. Results

The characteristics of the subjects in this study are shown in Table 1. BMI ranges of 25–29.9 for overweight and ≥ 30 for obesity were used, because most of the epidemiologic data on obesity are based on this classification [19].

Table 1. Subject characteristics.

Variables	
N	235
Female (n, (%))	155 (66%)
Age (years)	67.2 ± 12.6 [1]
Height (cm)	157.5 ± 8.6 [1]
Weight (kg)	57.6 ± 11.7 [1]
BMI (kg/m^2)	23.1 ± 3.9 [1]
Overweight (BMI 25–29.9 kg/m^2) (n, (%))	55 (23%)
Obese (BMI $\geq$ 30 kg/m^2) (n, (%))	8 (3.4%)
Number of teeth	25 (21, 27) [2]
hsCRP (mg/dL)	0.078 ± 0.112 [1]
HbA1c (%)	5.85 ± 0.75 [1]
Chewing ability (mg/dL)	192.0 ± 69.5 [1]
Methyl-mercaptan	9 (4, 28) [2]
Protein score	61.3 ± 25.0 [1]
Leukocyte score	47.1 ± 23.4 [1]

[1] Data are shown as mean $\pm$ standard deviation in age, height, weight, body mass index (BMI), high-sensitivity C-reactive protein (hsCRP), hemoglobin (Hb) A1c, chewing ability, protein score, and leukocyte score. [2] Number of teeth and methyl-mercaptan are shown as median (interquartile range).

The relationships between PISA and parameters such as number of teeth, age, BMI, hsCRP, HbA1c, and chewing ability are shown in Figure 1.

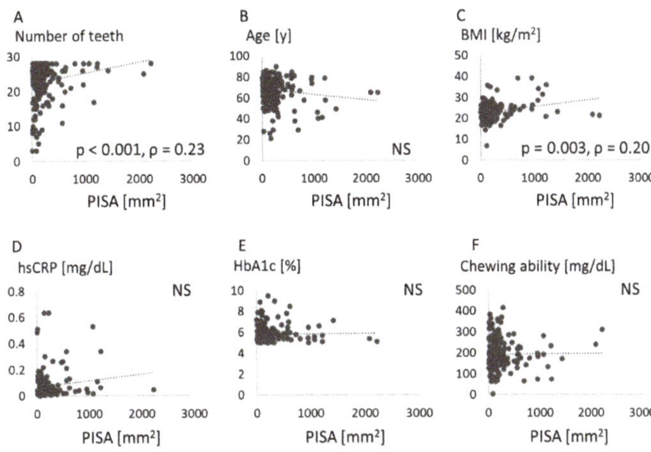

Figure 1. Association between each score and periodontal inflamed surface area. Number of teeth (**A**), age (**B**), body mass index (BMI) (**C**), high-sensitivity C-reactive protein (hsCRP) (**D**), hemoglobin (Hb)A1c (**E**), and chewing ability (**F**) are compared with periodontal inflamed surface area (PISA). NS: not significant.

PISA was significantly associated with the number of teeth and BMI. Associations between number of teeth and parameters such as age, BMI, hsCRP, HbA1c, and chewing ability are shown in Figure 2. The number of teeth was negatively associated with age, but positively with chewing ability.

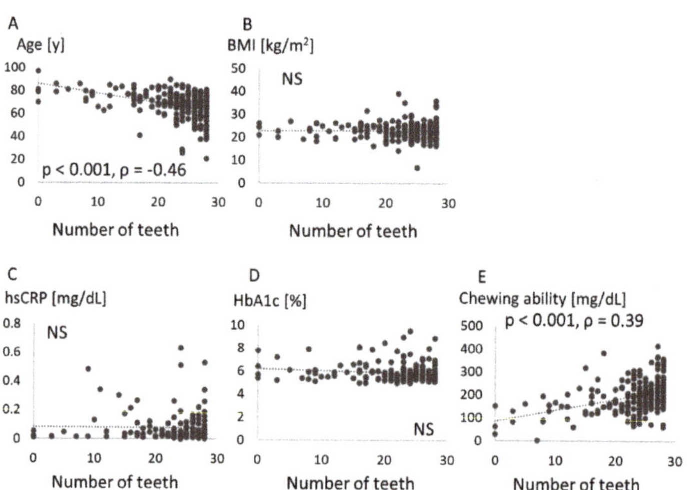

Figure 2. Association between each score and number of teeth. Age (**A**), body mass index (BMI) (**B**), high-sensitivity C-reactive protein (hsCRP) (**C**), hemoglobin (Hb)A1c (**D**), and chewing ability (**E**) are compared with number of teeth. NS: not significant.

Figure 3 shows the relationship between chewing ability and values like age, BMI, hsCRP, and HbA1c. Chewing ability was negatively associated with age, and positively with hsCRP.

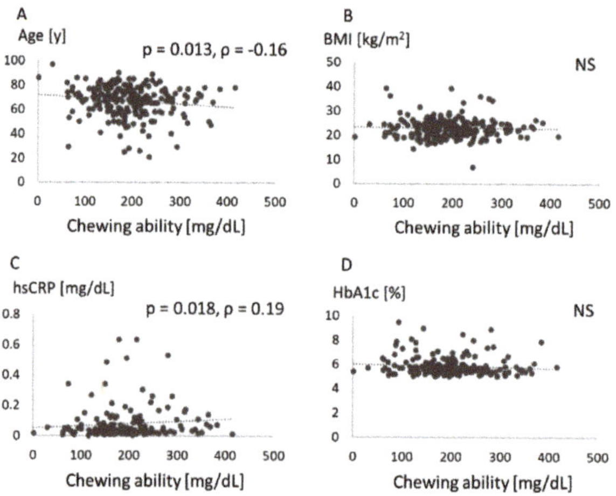

Figure 3. Association between each score and chewing ability. Age (**A**), body mass index (BMI) (**B**), high-sensitivity C-reactive protein (hsCRP) (**C**), and hemoglobin (Hb)A1c (**D**) are compared with number of teeth. NS: not significant.

Figure 4 shows the associations between BMI and values such as age, hsCRP, and HbA1c. BMI was associated with serum levels of hsCRP and HbA1c.

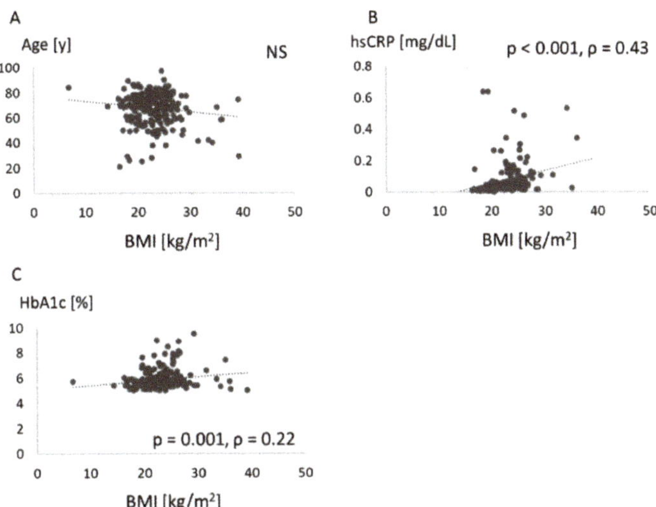

Figure 4. Association between each score and body mass index. Age (**A**), high-sensitivity C-reactive protein (hsCRP) (**B**), and hemoglobin (Hb)A1c (**C**) are compared with body mass index (BMI). NS: not significant.

Figure 5 shows the associations between PISA and scores for halitosis and salivary test results. The level of methyl-mercaptan in breath was positively associated with PISA, as were protein score and leukocyte score from salivary testing.

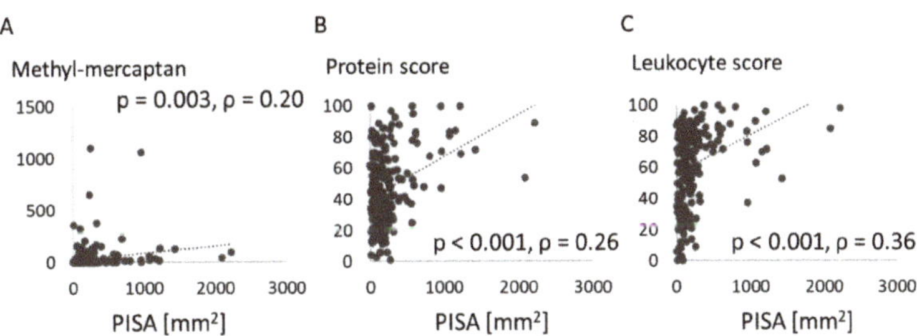

Figure 5. Association between oral parameters and periodontal inflamed surface area. Methyl-mercaptan in breath (**A**) and protein (**B**) and leukocyte (**C**) scores in saliva are compared with periodontal inflamed surface area (PISA).

Table 2 shows an association between chewing ability and HbA1c level. The rate of insufficient chewing ability (<150 mg/dL) was 22% in subjects with HbA1c < 7%, compared to 44% in subjects with HbA1c $\geq$ 7% (p = 0.047, chi-square test).

Table 2. Chewing ability and HbA1c.

	HbA1c < 7%	HbA1c $\geq$ 7%	*p*
Insufficient chewing ability [1]	22%	44%	0.047 [2]

[1] Cut-off point for chewing ability was set as 150 mg/dL. Hb, hemoglobin. [2] The chi-square test was used to compare groups.

Figure 6 compares serum hsCRP levels between subjects with low and high PISA. A cut-off of 300 mm^2 was used for PISA. The high-PISA group displayed increased levels of hsCRP (p = 0.046, Wilcoxon test).

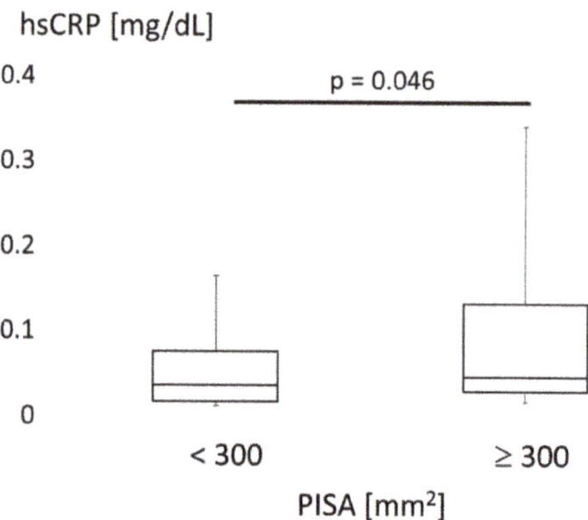

Figure 6. Comparison of high-sensitivity C-reactive protein. High-sensitivity C-reactive protein (hsCRP) is compared between high ($\geq$300 mm^2) and low (<300 mm^2) periodontal inflamed surface area (PISA) groups. Box plots show medians, 25th and 75th percentiles as boxes, and 10th and 90th percentiles as whiskers. The Wilcoxon test was used.

4. Discussion

In this study, PISA was positively associated with BMI, while the number of teeth was not associated with BMI. Moreover, we found that chewing ability was related to systemic markers such as HbA1c and hsCRP. Insufficient chewing ability was found in those with HbA1c $\geq$ 7.

A relationship between periodontitis and obesity has been widely recognized [7–11], and the association was confirmed in the Japanese population [20]. The present study confirmed that periodontal destruction was increased in obese patients (Figure 1). Vallim et al. reported that obesity could represent a risk factor for tooth loss over 5 years [21], but no relationship between BMI and number of teeth was identified in our study (Figure 2). Other measures of obesity such as percent body fat and body fat volume might be associated with tooth loss.

Links between periodontal disease or tooth loss and diabetes are also known. Tooth loss and periodontal attachment were increased by hyperglycemia in individuals with diabetes [21]. In our data, however, the HbA1c level was not associated with PISA or the number of teeth (Figures 1 and 2). The periodontal treatment reportedly improved HbA1c by reducing CRP [22]. An association between periodontal disease and hyperglycemia might be seen in specific patients with increased systemic inflammatory markers via periodontal inflammation. In our study, elevated levels of hsCRP were found in patients with increased PISA (Figure 6), while the hsCRP level was not associated with PISA (Figure 1). An association between glycemic control and systemic inflammation in people with established diabetes has been suggested [23]. Periodontitis may impair blood glucose regulation in healthy subjects in conjunction with elevated CRP levels [24]. More precise classifications might be effective in clarifying these relationships.

The present study also measured chewing ability. Masticatory dysfunction may be an important risk factor for mortality [25]. After controlling for possible confounding factors, the number of functional teeth and periodontal status were common factors associated with malocclusion [26]. The importance of mastication assessment in the diagnosis of periodontitis patients has been proposed [27]. As a result, chewing ability was related to systemic markers such as HbA1c and hsCRP (Figure 3, Table 2). To the best of our knowledge, this is the first report to investigate the relationship between chewing ability and serum glycemic and inflammatory markers. A lower number of teeth was associated with lower masticatory ability, which in turn was associated with lower plasma albumin levels and lower BMI [28]. A high level of HbA1c was suggested to be associated with poor masticatory function and severe periodontitis [29]. In the present study, chewing ability was associated with the number of teeth (Figure 2), but appeared unrelated to PISA (Figure 1) and BMI (Figure 3). Because chewing ability was related to oral health-related quality of life and general health [30], we should consider mastication in periodontal medicine.

Nutrition could be a crucial link between the oral condition and systemic health. Proper food intake is essential to maintain a healthy life. Obesity, diabetes, and other chronic non-communicable diseases (NCDs) are increasing globally, and malnutrition with inappropriate food intake leads to health issues [31]. The inclusion of nutrition is strongly recommended as a key focal point for all health professionals [32]. Chewing ability can influence food intake, and malnutrition induces insufficient general conditions. Nutrition might play a specific role in periodontal medicine. Further investigations regarding food intake are needed.

A strong association is well known to exist between periodontal and systemic diseases. A score reflecting the total inflammatory burden of periodontitis has been considered necessary, particularly in the field of periodontal medicine. Periodontal epithelial surface area (PESA) and PISA were thus proposed by Nesse et al. [17]. PESA indicates the surface area of the entire periodontal pocket epithelium, while PISA reflects the surface area of the bleeding pocket epithelium and the inflammatory burden posed by periodontitis. This conversion of periodontal inflammation to an individual value using PISA facilitates communication with healthcare workers outside of dental specialties. As a result, PISA has gained use in many studies in the field of periodontal medicine [33–35]. On this basis, we used PISA in this study to indicate periodontal inflammation.

Salivary testing was performed in this study and revealed that protein score and leukocyte score were positively associated with PISA. The utility of salivary testing has been confirmed in various reports [36–38]. Kim et al. [36] conducted a study of 10 male and female adults and indicated the measurement principles of salivary testing. Swiatkowska et al. [37] assessed 25 patients and showed that the parameters estimated by the analyzer correlated with the results from common salivary kits and oral health indices. Irie et al. [38] investigated 125 pediatric patients, between 3 and 18 years old. They found that leukocyte and protein scores changed according to gingival inflammation. Although those researchers investigated relatively small numbers of subjects, the utility of salivary analyzers was confirmed in the present study of 235 participants. Prediction of risks for periodontal disease using simple evaluation methods is an important issue, so the use of this salivary test is effective in screening for periodontal diseases.

Recently, a new classification scheme for periodontitis has been adopted, using a multidimensional staging and grading system [39]. Staging is dependent on disease severity as well as the complexity of disease management, while grading provides supplemental information about the biological features of the disease. However, this new system still shows some difficulties with classification and its authors have mentioned the need for the development of methodologies to accurately assess periodontal tissues. To better understand periodontal disease and its effects on systemic health, issues such as inflammation, infection, level of destruction and oral function should be considered [40,41]. Both inflammation and oral function are particularly important to evaluate the effects of periodontal disease on systemic health. The new classification also pays attention to the number of teeth and occlusal function, and PISA received attention as an inflammatory marker in the grade classification.

Some limitations of this study must be recognized and considered. We did not account for the medical conditions of participants in this study, because a wide variety of diseases and medications were involved. However, information on systemic diseases and medication is important when considering different patient backgrounds. Next, participants in this study were not limited to those on their first visit to the dental hospital. Therefore, the phases of dental treatment differed between subjects. These issues should be kept in mind when interpreting the data. Moreover, the sample size calculation was not performed before the research.

In conclusion, BMI as an obesity marker was positively associated with PISA, a marker of periodontal inflammation. Relationships between chewing ability and serum markers such as HbA1c and hsCRP were also found. Associations of periodontal status, number of teeth, and systemic condition should be considered in clinical settings.

Author Contributions: Conceptualization, N.A., K.T. (Katsushi Tamaki) and M.M.; methodology, N.A., K.T. (Katsushi Tamaki) and M.M.; validation, N.A., K.T. (Katsushi Tamaki) and M.M.; formal analysis, N.A.; investigation, N.A., T.F., S.K., I.N., K.T. (Kentaro Taniguchi), M.F. and T.I.; resources, N.A., K.T. (Katsushi Tamaki) and M.M.; data curation, N.A., K.T. (Katsushi Tamaki) and M.M.; writing—original draft preparation, N.A.; writing—review and editing, T.F., S.K., I.N., K.T. (Kentaro Taniguchi), M.F., T.I., K.T. (Katsushi Tamaki) and M.M.; visualization, N.A.; supervision, M.M.; project administration, N.A.; funding acquisition, N.A., K.T. (Katsushi Tamaki) and M.M. All authors have read and agreed to the published version of the manuscript.

Funding: This research was funded by JSPS KAKENHI (grant numbers 18K09925 and 19K10119), Ministry of Education, Culture, Sports, Science and Technology of Japan, Grants for Project Research and Young Investigator Project Research in Kanagawa Dental University and Uehara Memorial Foundation.

Institutional Review Board Statement: The study was conducted according to the guidelines of the Declaration of Helsinki, and approved by the Institutional Review Board of Kanagawa Dental University (protocol code 665, 13/May/2020).

Informed Consent Statement: Informed consent was obtained from all subjects involved in the study.

Data Availability Statement: The data presented in this study are available on request from the corresponding author.

References

1. Kopelman, P.G. Obesity as a medical problem. *Nature* **2000**, *404*, 635–643.
2. Bouldin, M.J.; Ross, L.A.; Sumrall, C.D.; Loustalot, F.V.; Low, A.K.; Land, K.K. The effect of obesity surgery on obesity comorbidity. *Am. J. Med. Sci.* **2006**, *331*, 183–193. [CrossRef]
3. Williams, R.C. Periodontal disease. *N. Engl. J. Med.* **1990**, *322*, 373–382. [CrossRef]
4. Dietrich, T.; Jimenez, M.; Krall Kaye, E.A.; Vokonas, P.S.; Garcia, R.I. Age-dependent associations between chronic periodontitis/edentulism and risk of coronary heart disease. *Circulation* **2008**, *117*, 1668–1674.
5. Nowbar, A.N.; Gitto, M.; Howard, J.P.; Francis, D.P.; Al-Lamee, R. Mortality from ischemic heart disease. *Circ. Cardiovasc. Qual. Outcomes* **2019**, *12*, e005375. [CrossRef]
6. Sanz, M.; Del Castillo, A.M.; Jepsen, S.; Gonzalez-Juanatey, J.R.; D'Aiuto, F.; Bouchard, P.; Chapple, I.; Dietrich, T.; Gotsman, I.; Graziani, F.; et al. Periodontitis and cardiovascular diseases: Consensus report. *J. Clin. Periodontol.* **2020**, *47*, 268–288.
7. Jepsen, S.; Suvan, J.; Deschner, J. The association of periodontal diseases with metabolic syndrome and obesity. *Periodontology 2000* **2020**, *83*, 125–153.
8. Martens, L.; De Smet, S.; Yusof, M.Y.P.M.; Rajasekharan, S. Association between overweight/obesity and periodontal disease in children and adolescents: A systematic review and meta-analysis. *Eur. Arch. Paediatr. Dent.* **2017**, *18*, 69–82. [CrossRef]
9. Genco, R.J.; Grossi, S.G.; Ho, A.; Nishimura, F.; Murayama, Y. A proposed model linking inflammation to obesity, diabetes, and periodontal infections. *J. Periodontol.* **2005**, *76*, 2075–2084.
10. Thanakun, S.; Izumi, Y. Effect of periodontitis on adiponectin, C-reactive protein, and immunoglobulin G against Porphyromonas gingivalis in Thai people with overweight or obese status. *J. Periodontol.* **2016**, *87*, 566–576. [CrossRef]
11. Martinez-Herrera, M.; Silvestre-Rangil, J.; Silvestre, F.J. Association between obesity and periodontal disease. A systematic review of epidemiological studies and controlled clinical trials. *Med. Oral Patol. Oral Cir. Bucal* **2017**, *22*, e708–e715. [CrossRef]
12. Vallim, A.C.; Gaio, E.J.; Oppermann, R.V.; Rösing, C.K.; Albandar, J.M.; Susin, C.; Haas, A.N. Obesity as a risk factor for tooth loss over 5 years: A population-based cohort study. *J. Clin. Periodontol.* **2021**, *48*, 14–23. [CrossRef]
13. Isola, G. Current evidence of natural agents in oral and periodontal health. *Nutrients* **2020**, *12*, 585. [CrossRef]
14. Almoznino, G.; Gal, N.; Levin, L.; Mijiritsky, E.; Weinberg, G.; Lev, R.; Zini, A.; Touger-Decker, R.; Chebath-Taub, D.; Shay, B. Diet practices, body mass index, and oral health-related quality of life in adults with periodontitis—A case-control study. *Int. J. Environ. Res. Public Health* **2020**, *17*, 2340. [CrossRef]
15. Kanasi, E.; Ayilavarapu, S.; Jones, J. The aging population: Demographics and the biology of aging. *Periodontology 2000* **2016**, *72*, 13–18. [CrossRef]
16. Nishi, N. Monitoring obesity trends in health Japan 21. *J. Nutr. Sci. Vitaminol.* **2015**, *61*, S17–S19. [CrossRef]

17. Nesse, W.; Abbas, F.; van der Ploeg, I.; Spijkervet, F.K.; Dijkstra, P.U.; Vissink, A. Periodontal inflamed surface area: Quantifying inflammatory burden. *J. Clin. Periodontol.* **2008**, *35*, 668–673. [CrossRef]
18. Kameyama, A.; Ishii, K.; Tomita, S.; Tatsuta, C.; Sugiyama, T.; Ishizuka, Y.; Takahashi, T.; Tsunoda, M. Correlations between Perceived Oral Malodor Levels and Self-Reported Oral Complaints. *Int. J. Dent.* **2015**, *2015*, 343527.
19. Caballero, B. Humans against obesity: Who will win? *Adv. Nutr.* **2019**, *10*, S4–S9. [CrossRef]
20. Aoyama, N.; Suzuki, J.I.; Thanakun, S.; Izumi, Y.; Minabe, M.; Isobe, M. Elevated concentrations of specific periodontopathic pathogens associated with severe periodontitis in Japanese patients with cardiovascular disease and concomitant obesity. *J. Oral Biosci.* **2018**, *60*, 54–58.
21. Botero, J.E.; Yepes, F.L.; Roldán, N.; Castrillón, C.A.; Hincapie, J.P.; Ochoa, S.; Ospina, C.A.; Becerra, M.A.; Jaramillo, A.; Gutierrez, S.J.; et al. Tooth and periodontal clinical attachment loss are associated with hyperglycemia in patients with diabetes. *J. Periodontol.* **2012**, *83*, 1245–1250. [CrossRef] [PubMed]
22. Katagiri, S.; Nitta, H.; Nagasawa, T.; Uchimura, I.; Izumiyama, H.; Inagaki, K.; Kikuchi, T.; Noguchi, T.; Kanazawa, M.; Matsuo, A.; et al. Multi-center intervention study on glycohemoglobin (HbA1c) and serum, high-sensitivity CRP (hs-CRP) after local anti-infectious periodontal treatment in type 2 diabetic patients with periodontal disease. *Diabetes Res. Clin. Pract.* **2009**, *83*, 308–315. [CrossRef] [PubMed]
23. King, D.E.; Mainous, A.G., 3rd; Buchanan, T.A.; Pearson, W.S. C-reactive protein and glycemic control in adults with diabetes. *Diabetes Care* **2003**, *26*, 1535–1539. [CrossRef] [PubMed]
24. Susanto, H.; Nesse, W.; Dijkstra, P.U.; Hoedemaker, E.; van Reenen, Y.H.; Agustina, D.; Vissink, A.; Abbas, F. Periodontal inflamed surface area and C-reactive protein as predictors of HbA1c: A study in Indonesia. *Clin. Oral Investig.* **2012**, *16*, 1237–1242. [CrossRef]
25. Nomura, Y.; Kakuta, E.; Okada, A.; Otsuka, R.; Shimada, M.; Tomizawa, Y.; Taguchi, C.; Arikawa, K.; Daikoku, H.; Sato, T.; et al. Effects of self-assessed chewing ability, tooth loss and serum albumin on mortality in 80-year-old individuals: A 20-year follow-up study. *BMC Oral Health* **2020**, *20*, 122. [CrossRef]
26. Kosaka, T.; Ono, T.; Kida, M.; Kikui, M.; Yamamoto, M.; Yasui, S.; Nokubi, T.; Maeda, Y.; Kokubo, Y.; Watanabe, M.; et al. A multifactorial model of masticatory performance: The Suita study. *J. Oral Rehabil.* **2016**, *43*, 340–347. [CrossRef]
27. Barbe, A.G.; Javadian, S.; Rott, T.; Scharfenberg, I.; Deutscher, H.C.D.; Noack, M.J.; Derman, S.H.M. Objective masticatory efficiency and subjective quality of masticatory function among patients with periodontal disease. *J. Clin. Periodontol.* **2020**, *47*, 1344–1353. [CrossRef]
28. Okamoto, N.; Amano, N.; Nakamura, T.; Yanagi, M. Relationship between tooth loss, low masticatory ability, and nutritional indices in the elderly: A cross-sectional study. *BMC Oral Health* **2019**, *19*, 110. [CrossRef]
29. Nakazawa, M.; Yamamoto, T.; Muto, E.; Seino, H.; Sasaki, S.; Kakugawa, T.; Shirai, Y.; Mochida, Y.; Fuchida, S.; Minabe, M. Association of glycosylated hemoglobin a1c with the masticatory function and periodontitis in type 2 diabetes patients hospitalized for an education program: A cross-sectional study. *J. Jpn. Soc. Periodontol.* **2020**, *62*, 74–81. (In Japanese) [CrossRef]
30. Brennan, D.S.; Spencer, A.J.; Roberts-Thomson, K.F. Tooth loss, chewing ability and quality of life. *Qual. Life Res.* **2008**, *17*, 227–235. [CrossRef]
31. Darnton-Hill, I.; Nishida, C.; James, W.P. A life course approach to diet, nutrition and the prevention of chronic diseases. *Public Health Nutr.* **2004**, *7*, 101–121. [CrossRef] [PubMed]
32. Herman, D.R.; Baer, M.T.; Adams, E.; Cunningham-Sabo, L.; Duran, N.; Johnson, D.B.; Yakes, E. Life Course Perspective: Evidence for the role of nutrition. *Matern. Child Health J.* **2014**, *18*, 450–461. [CrossRef] [PubMed]
33. Nesse, W.; Linde, A.; Abbas, F.; Spijkervet, F.; Dijkstra, P.U.; de Brabander, E.C.; Gerstenbluth, I.; Vissink, A. Dose-response relationship between periodontal inflamed surface area and HbA1c in type 2 diabetics. *J. Clin. Periodontol.* **2009**, *36*, 295–300. [CrossRef] [PubMed]
34. Temelli, B.; Yetkin Ay, Z.; Aksoy, F.; Büyükbayram, H.İ.; Kumbul Doğuç, D.; Uskun, E.; Varol, E. Platelet indices (mean platelet volume and platelet distribution width) have correlations with periodontal inflamed surface area in coronary artery disease patients: A pilot study. *J. Periodontol.* **2018**, *89*, 1203–1212. [CrossRef] [PubMed]
35. Leira, Y.; Rodríguez-Yáñez, M.; Arias, S.; López-Dequidt, I.; Campos, F.; Sobrino, T.; D'Aiuto, F.; Castillo, J.; Blanco, J. Periodontitis is associated with systemic inflammation and vascular endothelial dysfunction in patients with lacunar infarct. *J. Periodontol.* **2019**, *90*, 465–474. [CrossRef] [PubMed]
36. Kim, H.; Kim, J. Usefulness of salivary testing machine on oral care management: Pilot study. *Int. J. Clin. Prev. Dent.* **2018**, *14*, 89–94. [CrossRef]
37. Świątkowska, M.; Piekoszewska-Ziętek, P.; Gozdowski, D.; Olczak-Kowalczyk, D. The usefulness of Spotchem® Analyser (Arkray) in determining the risk in oral diseases in adolescents—A pilot study. *Nowa Stomatol.* **2018**, *23*, 9–17.
38. Irie, Y.; Tatsukawa, N.; Iwamoto, Y.; Nakano, M.; Ogasawara, T.; Sakurai, K.; Mitsuhata, C.; Kozai, K. Investigation of pediatric specifications for the Salivary Multi Test® saliva test system. *Pediatric. Dent. J.* **2018**, *28*, 148–153. [CrossRef]
39. Papapanou, P.N.; Sanz, M.; Buduneli, N.; Dietrich, T.; Feres, M.; Fine, D.; Flemmig, T.F.; Garcia, R.; Giannobile, W.V.; Graziani, F.; et al. Periodontitis: Consensus report of workgroup 2 of the 2017 World Workshop on the Classification of Periodontal and Peri-Implant Diseases and Conditions. *J. Clin. Periodontol.* **2018**, *45* (Suppl. 20), S162–S170.

40. Offenbacher, S.; Barros, S.P.; Beck, J.D. Rethinking periodontal inflammation. *J. Periodontol.* **2008**, *79*, 1577–1584. [CrossRef]
41. Loos, B.G.; Van Dyke, T.E. The role of inflammation and genetics in periodontal disease. *Periodontology 2000* **2020**, *83*, 26–39. [CrossRef] [PubMed]

Journal of *Clinical Medicine*

Article

Gan-Lu-Yin (Kanroin), Traditional Chinese Herbal Extracts, Reduces Osteoclast Formation In Vitro and Prevents Alveolar Bone Resorption in Rat Experimental Periodontitis

Yuji Inagaki *, Jun-ichi Kido, Yasufumi Nishikawa, Rie Kido, Eijiro Sakamoto, Mika Bando, Koji Naruishi [iD], Toshihiko Nagata and Hiromichi Yumoto [iD]

Department of Periodontology and Endodontology, Institute of Biomedical Sciences, Tokushima University Graduate School, Tokushima 770-8504, Japan; kido.jun-ichi@tokushima-u.ac.jp (J.K.); nishikawa.yasufumi@tokushima-u.ac.jp (Y.N.); rie.kido@tokushima-u.ac.jp (R.K.); sakamoji@tokushima-u.ac.jp (E.S.); banchi@tokushima-u.ac.jp (M.B.); naruishi@tokushima-u.ac.jp (K.N.); nagata.toshihiko@tokushima-u.ac.jp (T.N.); yumoto@tokushima-u.ac.jp (H.Y.)
* Correspondence: yinazo@tokushima-u.ac.jp; Tel.: +81-88-633-7344

Abstract: Gan-Lu-Yin (GLY), a traditional Chinese herbal medicine, shows therapeutic effects on periodontitis, but that mechanism is not well known. This study aims to clarify the precise mechanism by investigating the inhibitory effects of GLY extracts on osteoclastogenesis in vitro and on bone resorption in periodontitis in vivo. RAW264.7 cells are cultured with soluble receptor activator of nuclear factor-kappa B (sRANKL) and GLY extracts (0.01–1.0 mg/mL), and stained for tartrate-resistant acid phosphatase (TRAP) to evaluate osteoclast differentiation. Experimental periodontitis is induced by placing a nylon ligature around the second maxillary molar in rats, and rats are administered GLY extracts (60 mg/kg) daily for 20 days. Their maxillae are collected on day 4 and 20, and the levels of alveolar bone resorption and osteoclast differentiation are estimated using micro-computed tomography (CT) and histological analysis, respectively. In RAW264.7 cells, GLY extracts significantly inhibit sRANKL-induced osteoclast differentiation at a concentration of more than 0.05 mg/mL. In experimental periodontitis, administering GLY extracts significantly decreases the number of TRAP-positive osteoclasts in the alveolar bone on day 4, and significantly inhibits the ligature-induced bone resorption on day 20. These results show that GLY extracts suppress bone resorption by inhibiting osteoclast differentiation in experimental periodontitis, suggesting that GLY extracts are potentially useful for oral care in periodontitis.

Keywords: Gan-Lu-Yin; periodontitis; herbal medicine; osteoclastogenesis

Citation: Inagaki, Y.; Kido, J.; Nishikawa, Y.; Kido, R.; Sakamoto, E.; Bando, M.; Naruishi, K.; Nagata, T.; Yumoto, H. Gan-Lu-Yin (Kanroin), Traditional Chinese Herbal Extracts, Reduces Osteoclast Formation In Vitro and Prevents Alveolar Bone Resorption in Rat Experimental Periodontitis. *J. Clin. Med.* **2021**, *10*, 386. https://doi.org/10.3390/jcm10030386

Received: 1 December 2020
Accepted: 15 January 2021
Published: 20 January 2021

1. Introduction

Periodontal diseases are caused by periodontopathic bacteria in dental plaque, occlusal trauma, and several systemic diseases, and induced inflammation in periodontal tissues and resorption of alveolar bone. Periodontal diseases are strongly associated with various systemic inflammatory diseases, such as rheumatoid arthritis (RA), diabetes, and cardiovascular diseases, including coronary heart disease (CHD) and stroke. Recently, it has been reported that periodontal diseases stimulate the immune systems not only in periodontal lesions, but also in various systemic tissues [1,2]. Because periodontal disease affects the expression of biomarkers, including cytokines, oxidative stress markers, receptors, and immunoglobulins, not only in periodontal tissues, but also in saliva, serum, and systemic tissues, many biomarkers for periodontitis has been systemically researched to date and recently expected to be a valuable prognostic biomarker of periodontitis and systematically inflammatory diseases.

Pro-inflammatory cytokines, such as tumor necrosis factor (TNF-α) and interlukin (IL-6) directly, and indirectly activated osteoclast differentiation through receptor activator of nuclear factor-κB ligand (RANKL)/RANK pathway and induced alveolar bone

81

resorption in periodontitis [3]. The pathological bone resorption is caused by stimulating osteoclast differentiation and formation, and this process is regulated by some enzymes and transcription factors, including cathepsin K, nuclear factor of activated T-cells cytoplasmic 1 (NFATc1), and dendritic cell-specific transmembrane protein (DC-STAMP) [3–8]. TNF-α antagonists, inhibitors of the IL-1 receptor, IL-6 receptor and RANKL/RNAK interaction, and cathepsin K inhibitor are thought to inhibit alveolar bone resorption in periodontitis [3]; however, these agents have not been used as an internal medication to prevent alveolar bone resorption in periodontitis. Seto et al. [9] and Tokunaga et al. [10] reported that simvastatin and parathyroid hormone-stimulated differentiation of rat calvarial cells and recovered the ligature-induced alveolar bone resorption in rat experimental periodontitis. Furthermore, calcitonin significantly decreased the numbers of inflammatory cells and tartrate-resistant acid phosphatase (TRAP)-positive cells (osteoclasts) and inhibited the ligature-induced alveolar bone resorption in experimental periodontitis [11]. These agents and hormones showed the suppression of alveolar bone resorption when they were locally injected into rat periodontal tissues.

Some medical plant extracts and herbal medicines show therapeutic effects against periodontal diseases with inflammation and alveolar bone resorption [12–14]. Harmine, a β-carboline alkaloid from *Peganum harmala*, inhibited NFATc1 expression and RANKL-induced osteoclast formation [15], and alisol-B, a phyto-steroid from *Alisma orientale* Juzepczuk, inhibited RANKL/RANK signaling, NFATc1 expression and osteoclast formation [16], and glymnasterkoreayne F from *Gymnaster koraiensis* suppressed NFATc1 expression, decreased the levels of cathepsin K and TRAP and inhibited osteoclast differentiation from bone marrow-derived macrophages [17]. On the other hand, the traditional Chinese and Japanese medicines contain multiple components derived from herbs and natural plant extracts, have pharmacological effects with anti-inflammatory, anti-oxidant, and anti-microbial characteristics, etc., for treatment of periodontal diseases [18,19]. Shi-Quan-Da-Bu-Tang (Juzentaihoto), a medicine that contains 10 herbs, showed antibacterial activity against *Porphyromonas gingivalis* and reduced *P. gingivalis*-induced alveolar bone resorption by inhibiting osteoclast differentiation in rat experimental periodontitis [20]. Da-Huang-Gan-Cao-Tang (Daiokanzoto), a crude extracts of Rhubarb rhizomes and Glycyrrhiza roots, was thought to suppress bone resorption in periodontitis, since this medicine inhibited *P. gingivalis*-induced nuclear factor (NF-κB) activity, IL-6 expression and matrix metalloproteinase (MMP)-1 activity in human gingival epithelial cells and fibroblasts [19,21].

In the clinical stage, Gan-Lu-Yin (GLY) formula (Kanroin) is approved as a medicine to treat oral inflammations, such as periodontitis, stomatitis, and glossodynia, by the Japanese Ministry, Labor and Welfare, and contains nine herbs, including Artemisia Capillaris flower (*Artemisia capillaris* Thunberg), Scutellaria root (*Scutellaria baicalensis* Georgi), Loquat leaf (*Eriobotrya japonica* Lindley), Immature orange (*Citrus aurantium* Linné var. *daidai* Makino), Rehmannia root (*Rehmannia glutinosa* Liboschitz var. *purpurea* Makino), Asparagus root (*Asparagus cochinchinensis* Merrill), Ophiopogon root (*Ophiopogon japonicus* Ker-Gawler), Dendrobium (*Dendrobium nobile* Lindley), and Glycyrrhiza (*Glycyrrhiza uralensis* Fisher) [22–24]. GLY extracts suppressed the migration of vascular smooth muscle cells (VSMCs) by inhibiting MMP-2 and -9 [24] and inhibited vascular endothelial growth factor (VEGF) expression and tube formation in human umbilical vein endothelial cells (HUVEC) [25], and further suppressed TNF-α expression in human oral cancer cells through ERK and NF-κB pathway [22], suggesting that GLY extracts have an anti-inflammatory reaction and inhibitory action of tissue destruction. Although GLY extracts have been used for medical treatments of periodontitis, the effect of GLY extracts on alveolar bone metabolism, and their detailed mechanism are not well known. In the present study, to investigate the possibility of GLY extracts as the therapeutic agent for periodontitis, we examined the inhibitory effects of GLY extracts on the differentiation of pre-osteoclastic cells using a murine osteoclast precursor cells treated with sRANKL in vitro. Further, we assessed the preventive effect of GLY extracts on alveolar bone resorption in rat experimental periodontitis using micro-computed tomography (CT) and histological sections in vivo.

2. Experimental Section

2.1. Reagents

The powder of GLY extracts was supplied by Sunstar (Osaka, Japan). Briefly, the powder was prepared by extracting and drying a mixture of Artemisia Capillaris flower, Scutellaria root, Loquat leaf, Immature orange, Rehmannia root, Asparagus root, Ophiopogon root, Dendrobium, and Glycyrrhiza. Each herb were equally weighted (12.5 g) and extracted with boiling in 125 mL of distilled water; then, the extract was filtered and dried by heating under reduced pressure. Alpha-modified Eagle's minimal essential medium (α-MEM) and TRAP staining kit were purchased from Wako (catalog number 135-15175 and 294-67001, Osaka, Japan). Recombinant murine sRANKL was from Peprotech (catalog number 315-11, Rocky Hill, NJ, USA), and rabbit polyclonal antibodies against cathepsin K, NFATc1, ephrin B2, and mouse monoclonal antibody against β-actin were from Abcam (catalog number ab19027, ab25916, ab150411 and ab49900, Cambridge, UK). Mouse monoclonal antibody against DC-STAMP was from EMD Millipore (catalog number MABF39-I, Temecula, CA, USA). Horseradish peroxidase (HRP)-conjugated goat anti-rabbit IgG antibody, HRP-conjugated goat anti-rat IgG antibody, and HRP-conjugated horse anti-mouse IgG antibody were from Cell Signaling Technology (catalog number 7074, 7077, and 7076, Beverly, MA, USA).

2.2. Cell Culture and Osteoclastogenesis

The murine macrophage cell line, RAW264.7, was obtained from the American Type Culture Collection (ATCC, Rockville, MD, USA), and were seeded in 6-wells or 12-wells plates (IWAKI, Chiba, Japan) at 10,000 cells per cm^2 and cultured in α-MEM supplemented with 10% fetal bovine serum (FBS) at 37 °C in a humidified atmosphere of 5% CO_2 and 95% air. RAW264.7 cells were differentiated to osteoclasts by sRANKL (50 ng/mL) according to the method of Zhang et al. [26]. To examine the effect of GLY extracts on osteoclast differentiation, the cells were cultured with sRANKL and GLY extracts (0.01–1.0 mg/mL) for five days and stained with the TRAP staining kit according to the manufacturer's instructions. TRAP-positive multinucleated cells (MNCs) with more than three nuclei were counted under a phase contrast microscope with x100 magnification, and its number was counted by an independent examiner in a blind manner. The significantly differences were calculated with the number of TRAP-positive MNCs of sRANKL alone treatment as a control. These procedures were repeated three times by using triplicate RAW264.7 cell cultures.

2.3. Cell Viability Assay

Cell viability was examined using a Cell Counting Kit-8 (CCK-8; catalog number CK04, Dojindo, Kumamoto, Japan) according to the manufacturer's instructions. Briefly, RAW264.7 cells were seeded at 5000 cells/well in a 96-well plate (SUMITOMO BAKELITE, Tokyo, Japan) and cultured in a growth medium (α-MEM–10% FBS). After 48 h, cells were exposed with medium alone or 0.01–1.0 mg/mL GLY extracts for 48 h, and then incubated with CCK-8 solution. After the incubation of cells with CCK-8 solution for 1 h, the absorbance of each well was measured at 450 nm using a microplate reader (iMark^{TM}; Bio-Rad, Hercules, CA, USA). The percentages of cell viability were calculated according to the values of the medium alone as 100%. These procedures were repeated two times by using triplicate RAW264.7 cell cultures.

2.4. Western Blot Analysis

RAW264.7 cells differentiated to osteoclasts by sRANKL were cultured with GLY extracts (0.05–0.5 mg/mL) in 6-wells plates for five days. The cells were lysed in 50 µL of RIPA lysis buffer (catalog number sc-24948, Santa Cruz Biotechnology Inc., Santa Cruz, CA, USA). Equal amounts of proteins (30 µg) were denatured and separated by 10% sodium dodecyl sulfate polyacrylamide gel electrophoresis (SDS-PAGE) and transferred to nitrocellulose membranes (Bio-Rad). After blocking the membranes with 5% non-fat dry milk in 50 mM Tris-buffered saline (pH 7.6) containing 0.05% Tween 20, proteins on the membrane were reacted with optimally diluted primary antibodies of β-actin, cathepsin K, DC-STAMP, ephrin B2, and NFATc1 overnight at 4 °C, and then incubated with each HRP-conjugated secondary antibodies (1/2000 dilution) at room temperature for 1 h. After washing a membrane, the blots were developed by Amercham ECL Western Blotting Detection Reagents and visualized using Image Quant LAS 500 (GE Healthcare, Little Ckalfont, UK). These procedures were repeated 2–3 times.

2.5. Animals

Experiments using animal was approved by the Animal Research Control Committee of the Tokushima University Graduate School (T28-90). The study complied with the Animal Research: Reporting in vivo Experiments guidelines developed by the National Centre for the Replacement, Refinement, and Reduction of Animals in Research (NC3Rs). Forty-two male 8-week-old Wistar rats (220–230 g, Charles River Laboratories Japan, Inc., Yokohama, Japan) were housed in a temperature- and humidity-controlled room (23 ± 1 °C and 60 ± 5% relative humidity) with 12 h light/dark cycle and fed with standard rodent chow and water ad libitum.

2.6. Experimental Periodontitis

Twenty-eight rats were randomly divided into two groups as follows; (1) a GLY administration group in which rats were orally administered with GLY extracts (60 mg/kg) for 20 days from the day of ligaturing (n = 18) and (2) a non-administration group in which rats were orally-administered with the same volume of distilled water for 20 days as control (n = 10). The number of rats was justified using size calculation based on the previous report [27]. However, considering the possibility that the rats could be lost by administering GLY extracts during the experimental period, a sample size of 18 rats was planned in the GLY administration group compared to 10 rats in the control group. Experimental periodontitis of rats was induced according to the previous method [28–31]. Briefly, the cervical area of the right second molar of the rat maxilla was ligatured with nylon thread (No. 5-0; Natsume Corporation, Tokyo, Japan) under anesthesia with sodium pentobarbital (day 0). The ligature was knotted on the buccomesial to confirm the ligature remained during the experimental period, and the ligature was checked every day to ensure subgingival placement. The left second molar of rat maxilla was used as the non-ligatured control.

2.7. Microcomputed Tomography (μCT) Analysis of Alveolar Bone

Eighteen rats, which consist of twelve rats of the GLY administration group and six rats of control, were sacrificed after a measurement of body weight on day 20, and their maxillae and peripheral blood were collected. The specimens of alveolar bone were immediately fixed in 10% neutral-buffered formalin, and alveolar bone resorption levels were analyzed using micro-CT system (SkyScan 1176, Bruker, Billerica, MA, US) according to the previous method [28,29]. The distance from the buccal cement-enamel junction (CEJ) to the alveolar bone crest (ABC) of the second molar was measured in the frontal section as a marker of bone height. To ensure reproducibility of the alignment of the micro-CT image, the buccal cusp tip of the second molar was placed such that they superimposed on the corresponding palatal cusp tip. The distance between CEJ and ABC was measured at four points of the buccal side of the second molar, including mesial, distal, and middle (furcation) sites, and its average was calculated.

2.8. Histological Analysis

Ten rats, which consist of six rats of the GLY administration group and four rats of control, were sacrificed on day 4 after ligature placement, and their maxillae were dissected. The specimens of alveolar bone were immediately fixed in 10% neutral-buffered formalin and decalcified with 10% EDTA for 21 days, and embedded in paraffin (Paraplust

Plus, Sigma, St Louis, MO). Histological sections with alveolar bone were longitudinally prepared at 6 μm-thick widths and stained with the TRAP staining kit according to the manufacturer's instructions. The TRAP-positive MNCs in the stained section were observed as osteoclasts using an optical microscope (Microphoto V series, Nikon, Tokyo, Japan), and the TRAP-positive MNCs were defined by cells with more than three nuclei under a light microscope and counted according to a previous report [11]. Briefly, the number of TRAP-positive MNCs in the alveolar bone was measured in the square of 450 × 600 μm microscopic visual fields at a magnification of x200, using 16 sections from four control rats (non-ligatured left side, $n = 8$; ligatured right side, $n = 8$), and 24 sections from six GLY administration rats (non-ligatured left side, $n = 12$; ligatured right side, $n = 12$). The same visual fields of every second molar were analyzed by an independent examiner in a blind manner using a light microscope and digital camera and averaged the number of osteoclasts in buccal and palatal sides.

2.9. Analysis of Bone Metabolism Marker in Rat Serum

Rats were administered GLY extracts or distilled water for 20 days, and their 5mL of peripheral blood was collected to determine levels of bone metabolism marker. Sera were prepared from the blood samples, and the amounts of NTx-1 (cross-linked N-telopeptide of collagen type I) and osteocalcin in sera were determined using each ELISA kit (catalog number LS-F21857 and LS-F22801, Life Span BioSciences, Inc., Seattle, WA, US) according to the manufacturer's instructions.

2.10. Analysis of Systemic Bone Volume Fraction and Bone Mineral Density

Fourteen rats were randomly divided into the GLY administration group and the non-administration group as control ($n = 7$/group). The administration group was orally-administered with GLY extracts (60 mg/kg) every day, and the non-administration group was administered with the same volume of distilled water every day. On day 28, the rats' bodyweights were measured, and the rats were sacrificed, followed by the collection of femurs. The femurs were immediately fixed in 70 % ethanol, and then bone volume fraction (BV/TV: bone volume/tissue volume), bone mineral contents (BMC), and bone mineral density (BMD) were analyzed by Kureha Special Laboratory (Fukushima, Japan).

2.11. Statistical Analysis

Statistical analyses were performed with SPSS Statistics version 20 (Armonk, NY, USA). After a Anderson–Darling test, comparisons between two experimental groups were performed by Student's t-test, and those among three groups or more were analyzed using one-way ANOVA followed by a post-hoc Tukey-Kramer test. In all statistical analyses, p values of less than 0.05 were considered significant.

3. Results

3.1. Effect of GLY Extracts on Cell Proliferation of RAW264.7 Cells

The inhibitory effect of GLY extracts on the proliferation of RAW264.7 cells was investigated at 48 h-culture. 0.01–0.1 mg/mL of GLY extracts did not significantly influence the cell viability of RAW264.7 cells (Figure 1). However, GLY extracts at a high concentration of 0.2–1.0 mg/mL significantly inhibited the cell viability of RAW264.7 cells and decreased its viability to approximately 20–30% compared to control (GLY: 0 mg/mL).

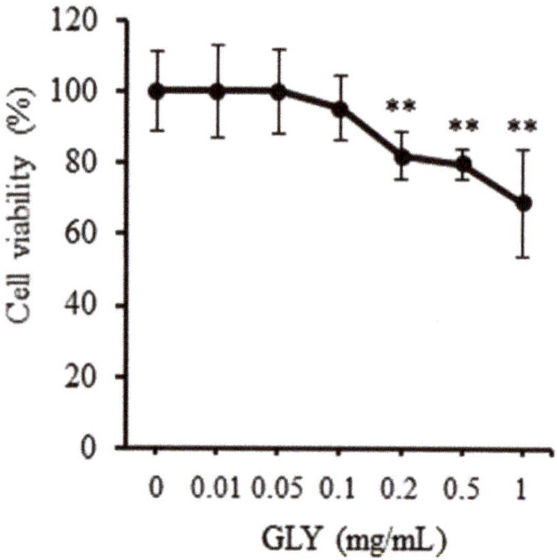

Figure 1. Effects of Gan-Lu-Yin (GLY) extracts on cell viability in RAW264.7 cells. The cells were seeded at 5000 cells/well in a 96-well plate. After sub-confluency, the cells were incubated in the absence or presence of various concentrations (0.01, 0.05, 0.1, 0.2, 0.5, and 1.0 mg/mL) of GLY extracts. The cell viability was assessed for the cultivation of 48 h using Cell Counting Kit-8. The percentages of cell viability were calculated according to the values of the absence of GLY extracts as 100%. Data are presented as the mean ± SD. ** $p < 0.01$ compared with the cultivation with 10% fetal bovine serum (FBS) alone. These procedures were repeated two times by using triplicate RAW264.7 cell cultures.

3.2. GLY Extracts Inhibit Osteoclast Differentiation

To investigate the effects of GLY extracts on osteoclast differentiation, RAW264.7 cells were cultured with sRANKL and GLY extracts. When RAW264.7 cells were stimulated with sRANKL (50 ng/mL), the cells were differentiated to TRAP-positive MNCs, osteoclasts. GLY extracts inhibited sRANKL-induced osteoclast differentiation and decreased the number and size of osteoclasts (Figure 2A), and significantly decreased the number of osteoclasts in a dose-dependent manner (0.05–0.2 mg/mL) (Figure 2B). GLY extracts at high concentration (0.5 and 1.0 mg/mL) completely suppressed sRANKL-induced osteoclast differentiation (Figure S1). These results showed that GLY extracts at a concentration of more than 0.05 mg/mL inhibited osteoclast differentiation in vitro.

3.3. GLY Extracts Inhibited the Expressions of Osteoclast Markers in RAW264.7 Cells

sRANKL (50 ng/mL) increased the expression of osteoclast markers, including NFATc1, cathepsin K, ephrin B2 and DC-STAMP in RAW264.7 cells (Figure 3). When sRANKL-induced osteoclastic cells were cultured with GLY extracts, GLY extracts inhibited the expression of cathepsin K, ephrin B2 and DC-STAMP in a dose-dependent manner (Figure 3A,B; 0.05–0.5 mg/mL). In contrast, NFATc1 protein level in sRANKL-induced osteoclastic cells was not influenced by GLY extracts at 0.05–0.5 mg/mL concentration.

(A)

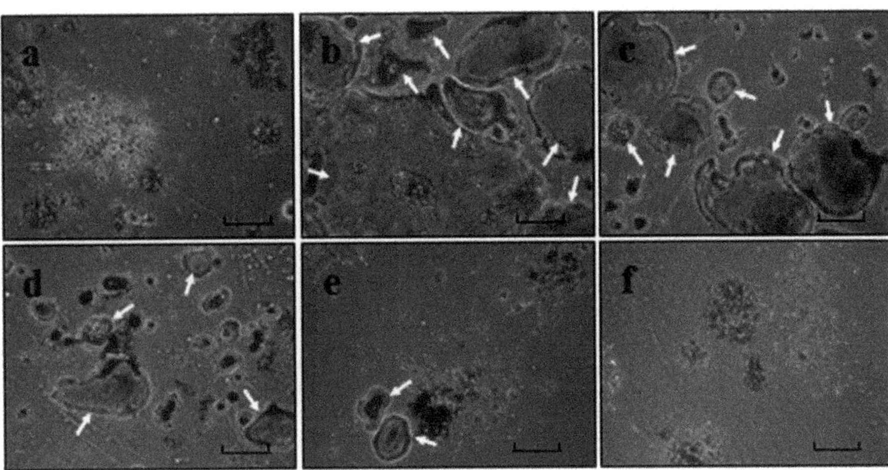

(B)

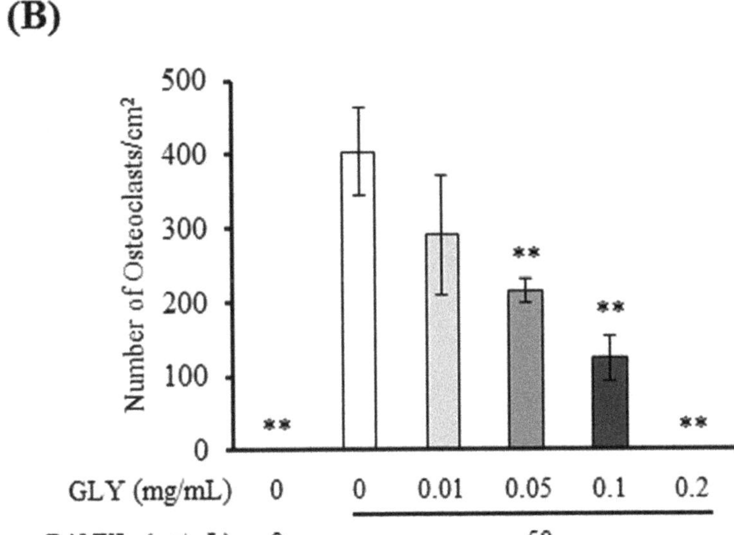

Figure 2. Effects of GLY extracts on the receptor activator of nuclear factor-κB ligand (RANKL)-induced osteoclast differentiation. (**A**) Microscopic observations for tartrate-resistant acid phosphatase (TRAP) staining of osteoclast-like positive multinucleated cells (MNCs) derived from soluble RANKL (sRANKL)-treated RAW264.7 cells in the absence or presence of various concentrations of the GLY extracts for five days. (**a**) no-treated with sRANKL and absences of GLY extracts; (**b**) absence of GLY extracts; (**c**) 0.01 mg/mL of GLY extracts; (**d**) 0.05 mg/mL of GLY extracts; (**e**) 0.1 mg/mL of GLY extracts; (**f**) 0.2 mg/mL of GLY extracts. Magnification x100, the arrowheads indicate TRAP-positive osteoclasts. Scale bars represent 300 μm. (**B**) Quantitative analysis of TRAP-positive MNCs with three or more nuclei in each well. Data are presented as the mean ± SD. ** $p < 0.01$ compared with the sRANKL treatment in the absence of GLY extracts. These procedures were repeated three times by using triplicate RAW264.7 cell cultures.

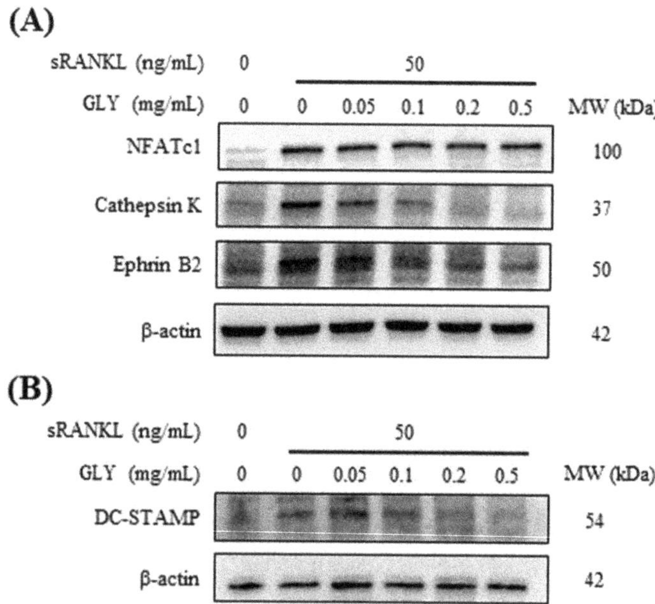

Figure 3. Effects of GLY extracts on the expression of osteoclast-specific markers in RAW264.7 cells. (**A**) sRANKL-treated RAW264.7 cells were incubated in the absence or presence of various concentrations (0.05, 0.1, 0.2 and 0.5 mg/mL) of GLY extracts for five days. The expression of NFATc1, cathepsin K and ephrin B2 in the cells were evaluated using Western blot analysis. B-actin served as a loading control. (**B**) The expression of DC-STAMP and β-actin in the cells were evaluated using Western blot analysis.

3.4. GLY Extracts Inhibited Alveolar Bone Resorption in Rat Experimental Periodontitis

The effect of GLY extracts on alveolar bone resorption was evaluated using rats with ligature-induced periodontitis (Figure 4A). There were no significant differences in body weight of rats between the administration group and non-administration group on day 20 of ligature placement, and the mean of rat body weight was 365 ± 16 g and 380 ± 20 g, respectively (Figure 4B). In μCT images of alveolar bone of rats with periodontitis, bone resorption was observed around the ligatured second molar (right side) as compared with the non-ligatured side (left side) (Figure 4C). Although the length between cement-enamel junction (CEJ) and alveolar bone crest (ABC) was 116.7 ± 31.3 μm in the non-ligatured control group, bone resorption in the ligatured control group remarkably increased, and its length was 1134.7 ± 231.2 μm. When GLY extracts were administered to rats, the length in the ligatured GLY administration group was 829.6 ± 173.6 μm in spite of the length in the non-ligatured GLY administration group was 134.8 ± 28.5 μm (Figure 4D). The administration of GLY extracts decreased the level of bone resorption of the ligatured control without GLY extracts by approximately 27%, suggesting that GLY extracts significantly inhibited the ligature-induced bone resorption.

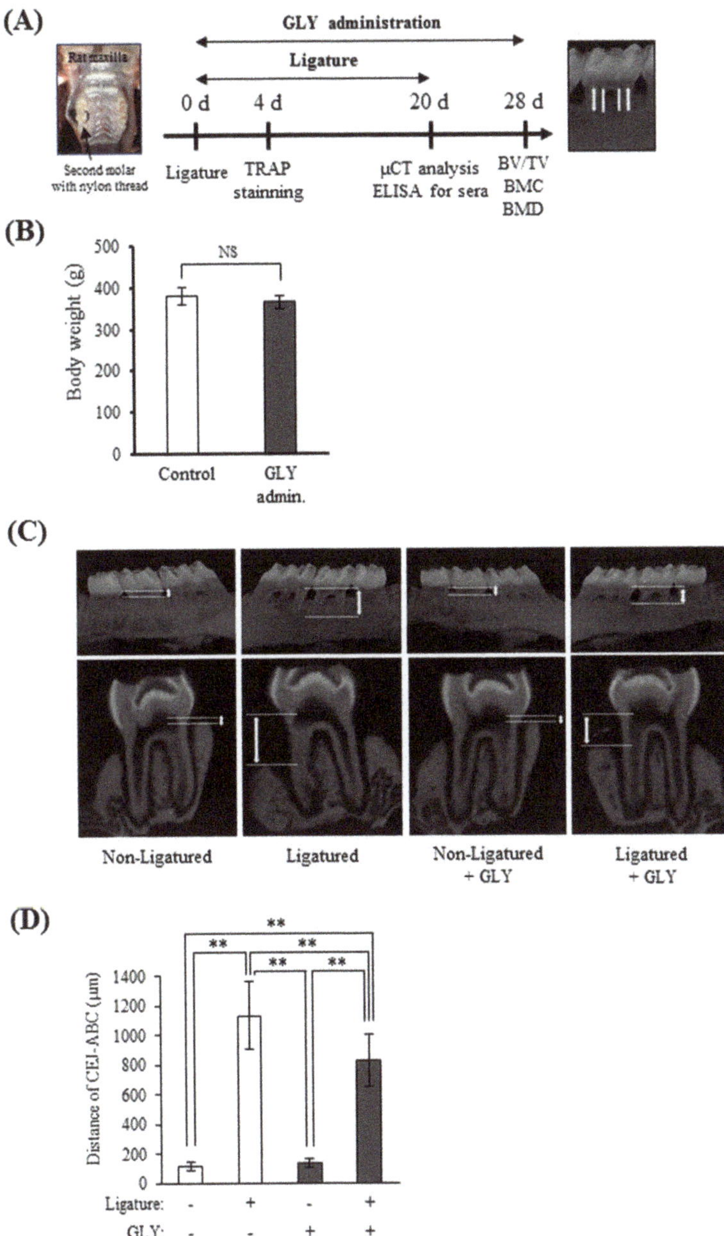

Figure 4. The effects of GLY extracts on bone resorption in ligature-induced periodontitis. (**A**) Experimental design and the points of linear measurement of the cement-enamel junction (CEJ) to the alveolar bone crest (ABC) by micro-CT analysis. (**B**) Body weights of rats on day 20 from the placement of ligature. Control; group of oral administration of distilled water for 20 days (*n* = 6), GLY admin.; group of oral administration with 60 mg/kg of GLY extracts for 20 days (*n* = 12). (**C**) Representative micro-CT images of the frontal and sagittal sections of maxillary secondary molars from rats. (**D**) Distance from the CEJ to the ABC as a marker of alveolar bone resorption in micro-CT. Data are presented as the mean ± SD from six or twelve rats per group. Double asterisk indicates *p* < 0.01. NS indicates no significant differences.

3.5. GLY Extracts Inhibited Osteoclast Differentiation in Rat Experimental Periodontitis

The effect of GLY extracts on osteoclast differentiation was histologically investigated at ligatured buccal and palatal sides around rat maxillary secondary molar. The TRAP-positive MNCs, osteoclasts, were not observed in the non-ligatured control group and non-ligatured GLY administration group, but the numerous osteoclasts were observed in the ligatured control group (Figure 5A and Figure S2). The number of osteoclasts were suppressed by GLY extracts, and the inhibitory rates of osteoclast differentiation were approximately 55% at ligatured control group (Figure 5B).

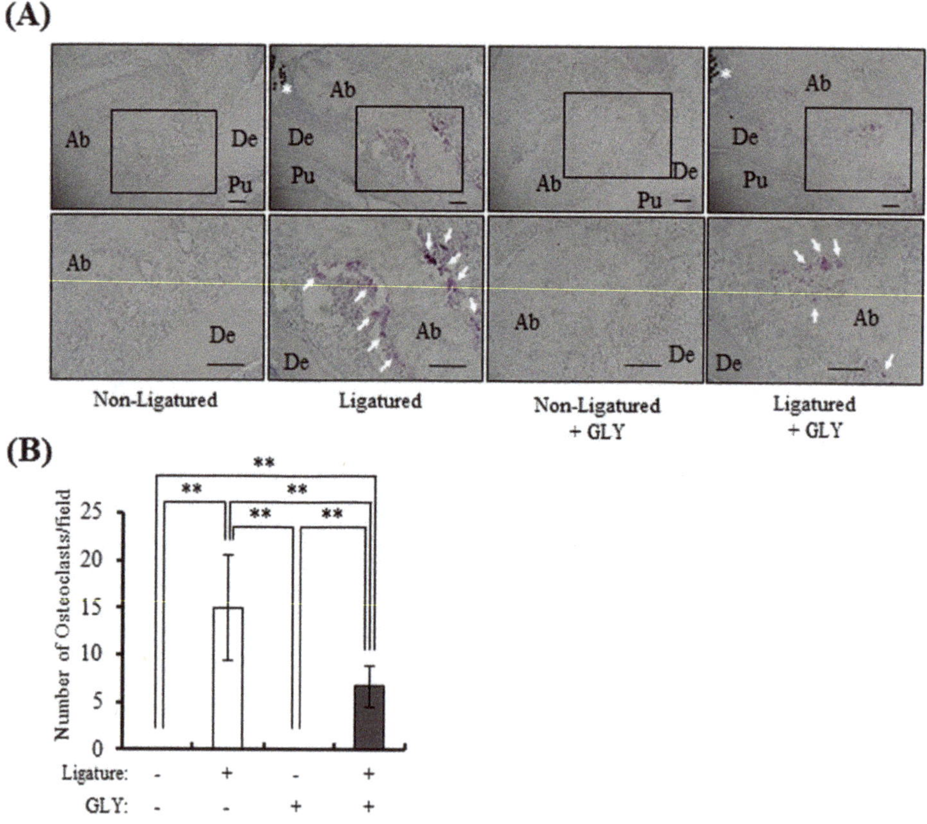

Figure 5. The effect of GLY extracts on osteoclast differentiation in ligature-induced periodontitis. Microscopic observations were performed on the decalcified paraffin sections with TRAP staining on day 4 from the placement of ligature. (**A**) Representative images have shown TRAP-positive osteoclasts in the alveolar bone around maxillary secondary molar in the non-ligatured and ligatured buccal side; upper panels: x100 magnification, lower panels: x200 magnification. Ab, alveolar bone; De, dentin; Pu, pulp; the asterisk indicates the nylon ligature, and the arrowheads indicate TRAP-positive osteoclasts, Scale bars represent 100 μm. (**B**) Quantitative analysis of TRAP-positive osteoclasts on day 4 in the ligatured buccal and palatal side. TRAP-positive cells per visual field, counted 16–24 visual fields. Data are presented as the mean ± SD from four or six rats per group. Double asterisk indicates $p < 0.01$. Control; group of oral administration of distilled water for four days ($n = 4$), GLY admin.; group of oral administration with 60 mg/kg of GLY extracts for four days ($n = 6$).

3.6. GLY Extracts Decreased Serum NTx-1 Level in Rat Experimental Periodontitis

The level of NTx-1, bone resorption marker, in sera from the GLY administration group (4.4 ± 1.4 ng/mL) were significantly lower than that of the non-administration group (7.2 ± 1.7 ng/mL) (Figure 6A), whereas there was no significant difference in the level of osteocalcin, osteoblastic cell marker, between two groups (Figure 6B).

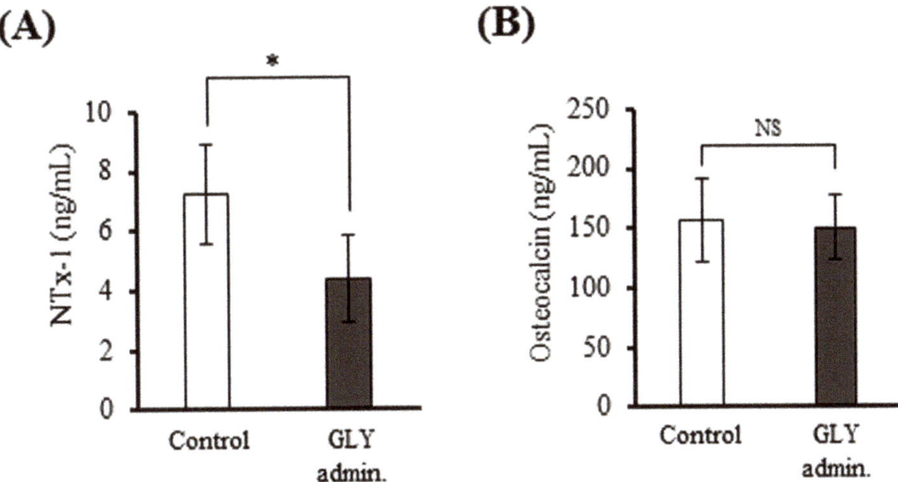

Figure 6. Comparison of the number of bone biomarkers in sera of ligature-induced periodontitis. (**A**) Serum concentrations of NTx-1 as a bone resorption marker. (**B**) Serum concentrations of osteocalcin as a bone formation marker. Data are presented as the mean ± SD from six or twelve rats per group. * $p < 0.05$ compared with control. NS indicates no significant differences. Control; group of administration of distilled water for 20 days ($n = 6$), GLY admin.; group of oral administration with 60 mg/kg of GLY extracts for 20 days ($n = 12$).

3.7. GLY Extracts Not Affect Bone Structure and Bone Mineral Density in Rat with Experimental Periodontitis

To investigate the effects of GLY extracts on systemic bone metabolism, BV/TV, BMC, and BMD were assessed using the rat femurs. Figure 7A shows no differences in the body weight between the GLY administration group (384 ± 26 g) and control (382 ± 12 g). BV/TV of the GLY administration group and control were 12.1 ± 2.5 and 13.2 ± 2.1 %, respectively (Figure 7B). BMC of the GLY administration group and control were 322.9 ± 12.2 mg and 318.9 ± 24.3 mg, respectively (Figure 7C). Further, BMD of the GLY administration group and control were 113.1 ± 2.2 mg/cm^2 and 114.9 ± 2.5 mg/cm^2, respectively (Figure 7D). These μCT and DXA (Dual energy X-ray Absorptiometry) assessments indicated that no differences in BV/TV, BMC, and BMD between the GLY administration group and the non-administration control group. Thus, in this experimental period, GLY extracts did not systemically affect bone metabolism in vivo.

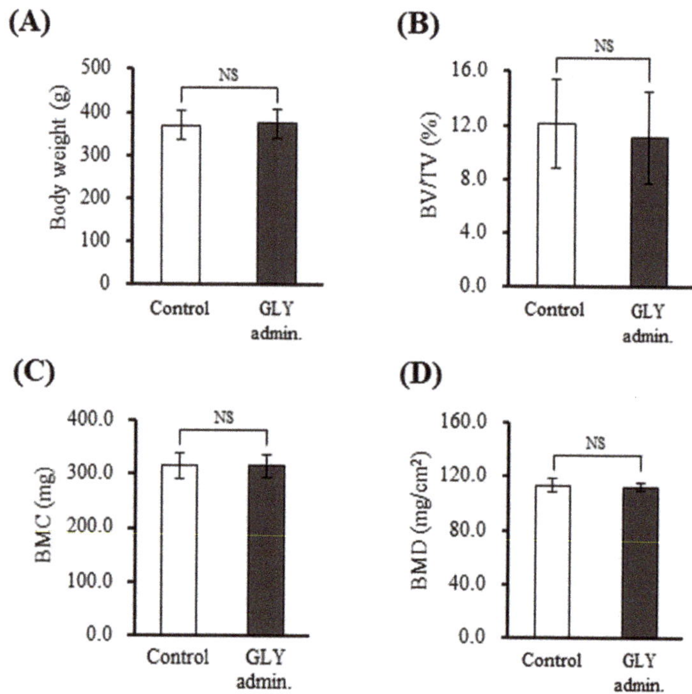

Figure 7. Effects of GLY extracts on systemic bone metabolism. (**A**) Body weights of rats, (**B**) Bone volume fraction (BV/TV, %), (**C**) Bone mineral contents (BMC, mg), (**D**) Bone mineral density (BMD, mg/cm^2). Data are presented as the mean ± SD from seven rats per group. NS indicates no significant differences. Control; group of administration of distilled water for 28 days (n = 7), GLY admin.; group of oral administration with 60 mg/kg GLY extracts for 28 days (n = 7).

4. Discussion

The present in vitro study using RAW264.7 cells showed that GLY extracts significantly inhibited sRANKL-induced osteoclast differentiation and decreased the expressions of osteoclast differentiation markers, such as cathepsin K, ephrin B2, and DC-STAMP. Further, the present in vivo study showed that administering GLY extracts for rat experimental periodontitis inhibited the ligature-induced alveolar bone resorption and significantly decreased the number of TRAP-positive osteoclasts around a ligature in the alveolar bone. GLY formula consists of nine herbs and contains 14 main components with pharmacological effects, including baicalin, baicalein, oroxylin A-7-O-glucuronide, wogonin, oroxylin A, naringin, neohesperidin, liquiritigenin, liquiritin and glycyrrhizic acid [32]. GLY extracts have anti-inflammatory and anti-tumor effects, which suppress TNF-α level in human oral carcinoma cells [22], angiogenesis in HUVEC [25], and migration of VSMCs [24]. Regarding the effective concentration, GLY extracts showed an inhibition of TNF-α level in oral carcinoma cells at more than 0.2 mg/mL, cell migration, and MMP-9 production in VSMCs at more than 0.125 mg/mL, and angiogenesis of HUVEC at 1.0–1.5 mg/mL [22,24,25]. In contrast, in osteoclast precursor (RAW264.7 cells), GLY extracts exhibited significant inhibitory effects on osteoclast differentiation, the production of cathepsin K, and ephrin B2 at comparatively low concentration, such as 0.05 and 0.1 mg/mL (Figures 2 and 3). Moreover, in the present study, GLY extracts at more than 0.2 mg/mL significantly attenuated the cell proliferation (Figure 1). However, several studies indicated that GLY extracts inhibit cell viability at 0.25 mg/mL on murine leukemia cells, 0.5 mg/mL on VSMCs and 1.5 mg/mL on HUVEC, respectively [23–25]. Our results suggested that osteoclasts may have a high sensitivity for GLY extracts.

On the other hand, Baicalin, at a high concentration of 200 mg/kg, suppressed the ligature-induced alveolar bone resorption, and slightly increased the area of collagen fibers in rat gingival connective tissues when that area of gingival tissues in the ligatured group was compared [33]. Glycyrrhizic acid and Naringin inhibited RANKL-induced osteoclast differentiation in murine bone marrow macrophages (BMMs), and preserved bone mass and trabecular structure in proximal tibial trabecular of ovariectomized mice, and suppressed bone resorption in pit assay [34,35]. Furthermore, Wogonin inhibited LPS-induced osteoclast formation in co-cultures with mouse calvaria osteoblasts and BMMs and suppressed lipopolysaccharide (LPS)-induced RANKL expression and recovered LPS-decreased osteoprotegerin production in osteoblasts [36]. In the present study, GLY extracts suppressed osteoclast differentiation in vitro and bone resorption in vivo, and these effects were similar to the effects of Bacicalin, Glycyrrhizic acid, Naringin, and Wogonin on bone metabolism, suggesting that multiple components in GLY extracts influence the inhibition of alveolar bone resorption. Wogonin contained in GLY extracts influenced osteoblasts, as well as osteoclasts [36]. We did not investigate the direct effect of GLY extracts on osteoblasts in the present study, but an administration of GLY extracts to rats did not change osteocalcin level in rat serum. Since osteocalcin is an indicator of bone formation activity in osteoblasts [37], the reason for the different effect on osteoblasts between GLY extracts and wogonin is not known. However, there may be differences between in vitro and in vivo, or the effects derived from other components in GLY extracts.

Regarding the inhibitory effects of GLY extracts on osteoclast differentiation, GLY extracts suppressed RANKL-induced cathepsin K, DC-STAMP, and ephrin B2 productions in RAW264.7 cells (Figure 3). Cathepsin K, a member of the papain family of cysteine proteases, is highly expressed in osteoclasts and plays an important role in bone resorption by degrading type I collagen and osteopontin in bone tissues [4,38]. DC-STAMP is essential for cell-cell fusion of mononuclear pre-osteoclasts and to mature osteoclasts with multi-nuclear [5–7]. Cathepsin K and DC-STAMP are thought as osteoclastic specific markers to control pathological bone resorption in osteogenic diseases because they are closely related to osteoclast differentiation and function. Glycyrrhizin (glycyrrhizic acid) and Naringin contained in GLY extracts also suppressed the expressions of cathepsin K and DC-STAMP mRNAs, and then abrogated osteoclast formation in BMMs [35,39]. The inhibitory effect of GLY extracts on cathepsin K and DC-STAMP expression may be dependent on the pharmacological effects of Glycyrrhizin and Naringin. In contrast, Glycyrrhizin inhibited the expression of NFATc1, which is a member of the NFAT transcription factor family and a key regulator of RANKL-induced osteoclast differentiation [8]. Baicalein also suppressed NFATc1 expression and inhibited RANKL-induced bone resorption activity in RAW264.7 cells [40]. However, in the present study, GLY extracts did not influence RANKL-induced NFATc1 expression (Figure 3). We did not know the reason for this difference in NFATc1 expression because GLY extracts contain some components. NFATc1 is located in a cytoplasm of osteoclasts as an inactive form that does not have transcriptional activity [8]. When pre-osteoclastic cells are stimulated by RANKL/RANK interaction, NFATc1 in cells translocates from a cytoplasm into a nucleus and exhibits transcriptional activity as an activated form, and then regulates the expression of TRAP, DC-STAMP and cathepsin K, etc. [8,41]. An activation of NFATc1 induced a differentiation of osteoclast precursor to mature osteoclasts, and inhibition of NFATc1 activity resulted in suppressing osteoclast differentiation [36]. Wogonin, a component in GLY formula, inhibited osteoclast differentiation through the inhibition of NFATc1 translocation from the cytoplasm to the nucleus and the downregulation of genes associated with osteoclast differentiation [42]. Since GLY extracts did not suppress RANKL-induced NFATc1 production, but inhibited cathepsin K and DC-STAMP expression and osteoclast differentiation in RAW264.7 cells, GLY extracts might suppress NFATc1 activity by inhibiting the translocation of NFATc1 to a nucleus, but not NFATc1 production. Both our research and Mao et al. [43] found that ephrin B2 was expressed during RANKL-induced osteoclast differentiation in RAW264.7 cells. Ephrin B2 is identified as a target of NFATc1, and its expression is dependent on RANKL-induced NFATc1 transcription during osteoclast differentiation [44,45]. We speculate that a translocation of NFATc1 to a nucleus, but not NFATc1 production, may be suppressed by GLY extracts, since ephrin B2 expression was also down-regulated by GLY extracts in RANKL-induced RAW264.7 cells.

In the present study, GLY extracts showed an inhibitory effect on the ligature-induced alveolar bone resorption by an oral administration for 20 days (Figure 4), suggesting that GLY extracts have a pharmacological effect on a pathological stimulation to periodontal tissues. It was reported that oral administration of Bu-Shen-Gu-Chi-Wan, a traditional Chinese medicine, for four weeks is detected significant changes in alveolar bone volume and density in experimental periodontitis by micro-CT analysis in spite of no improvement of alveolar bone height by stereomicroscopy [46]. It was also reported that the intra-gastric administration of curcumin, an active ingredient of turmeric, for 30 days decrease the alveolar bone resorption in experimental periodontitis [47]. The inhibitory effect of GLY extracts on alveolar bone resorption showed even after a shorter administration period than that of Bu-Shen-Gu-Chi-Wan and curcumin, suggesting that GLY extracts an effective candidate to prevent alveolar bone resorption in periodontitis.

GLY extracts show not only inhibitory effect of bone resorption, but also anti-inflammatory action, and has been prescribed for treatments of stomatitis and glossodynia, as well as periodontitis. GLY extracts contain some components with anti-inflammatory action on some cells and tissues. Baicalin suppressed *P. gingivalis* LPS-induced IL-6 and IL-8 expressions in human oral keratinocytes [48], and baicalin, baicalein, and wogonin inhibited high glucose-induced inflammatory responses, including vascular permeability, expression of cell adhesion molecules, and production of reactive oxygen species (ROS) in HUVECs [49]. Liquiritigenin suppressed the productions of IL-1β, IL-6, and TNF-α when RAW264.7 cells were stimulated by LPS and carrageenan-induced paw edema in rats [50]. Naringin down-regulated the expressions of IL-1β, IL-6, and TNF-α and up-regulated the productions of anti-oxidants, such as glutathione, superoxide dismutase, and catalase in retinal tissues of diabetic rats [51]. Furthermore, GLY extracts suppressed TNF-α level in oral carcinoma cells [20]. The ligature stimulation caused infiltration of inflammatory cells in periodontal tissues of rat experimental periodontitis, and calcitonin, a calcium regulatory hormone, decreased the number of inflammatory cells and the ligature-induced alveolar bone resorption [11]. The relationships between inflammatory responses and bone resorption, osteoclast formation are generally known in periodontitis [52]. Although we did not evaluate the effect of GLY extracts on inflammatory responses in rat periodontal tissues in this study, we suggest that GLY extracts suppressed osteoclast differentiation and alveolar bone resorption by inhibiting inflammatory responses in periodontal tissues.

NTx-1 levels in sera from the rats administered GLY extracts was significantly lower than that of the control (non-administration), whereas osteocalcin level did not change between the administration groups and control (Figure 6A,B). NTx-1 and osteocalcin are biomarkers of bone resorption by osteoclasts and bone formation by osteoblasts, respectively [37]. These results suggest the possibility that GLY extracts systemically influenced the osteoclast activity, but not the osteoblast activity. However, GLY extracts did not show effects on the body weight, and bone volume fraction (BV/TV), bone mineral contents (BMC), and bone mineral density (BMD) in rat femurs used in this study (Figure 7). Thus, the side-effects of GLY extracts on systemic bone metabolism did not occur during the experimental period in this study. We do not know the exact reason why bone quality in the femur does not change in spite of a decrease in serum NTx-1 level. GLY extracts appear to inhibit the osteoclast activity at inflamed periodontal sites, but not systemically influence. But a longer period of administration and observation is needed to investigate systemic effects definitely.

This study suggests the possibility that herbal medicine that internally applies, but not locally administration, prevents alveolar bone resorption in periodontitis. Periodontitis is caused by infection of periodontopathic bacteria and decline of immunological functions in the human host, and results in periodontal tissue degradation. Plaque control, scaling of dental calculus, root planing, curettage of infectious materials, and inflammatory tissues are usually performed for treatments, and prevention of periodontal diseases. Although a few drugs, such as antibiotics and anti-inflammatory agents, are locally and systemically used to suppress inflammation of periodontal tissues, there are few medicines that targeted the prevention of alveolar bone resorption in periodontitis. Gokhale et al. [53] proposed that "systemic host modulatory agents", such as anti-cytokine agents, inhibitors of mitogen-activated protein kinase (MAPK), and NF-κB, and nitric oxide synthase inhibitors may be useful for periodontal therapies in combination with conventional periodontal treatment. By doing a human study of the combination of conventional periodontal treatment and the administration of GLY extracts, we hope that GLY extracts may become one of the host modulatory agents for the prevention and therapy of periodontitis to encourage better clinical outcomes. To our knowledge, we are the first group to demonstrate a predominant inhibitory function of GLY extracts on bone resorption in periodontitis model rats. We also hope that GLY extracts become an efficacious therapeutic agent to prevent bone destruction in various bone diseases and think that further basic and clinical studies will be necessary to evidence the usefulness of GLY extracts as a medical agent for periodontitis and other bone diseases.

5. Conclusions

The present study showed that GLY extracts significantly inhibited sRANKL-induced osteoclast differentiation and decreased the expressions of osteoclast differentiation markers in vitro, and that the administration of GLY extracts for rat experimental periodontitis inhibited the ligature-induced alveolar bone resorption and significantly decreased the number of TRAP-positive osteoclasts around a ligature in the alveolar bone in vivo. We conclude that GLY extracts suppress the alveolar bone resorption by inhibiting the osteoclast differentiation in experimental periodontitis, suggesting that GLY extracts are potentially useful for oral care to suppress the alveolar bone resorption in periodontitis. Further studies with the involvement of a long-term administration of GLY extracts would be beneficial in evaluating the effect of GLY extracts on bone metabolism in periodontitis.

Supplementary Materials: The following are available online at https://www.mdpi.com/2077-0383/10/3/386/s1, Figure S1: Effects of high concentration of GLY extracts on the RANKL-induced osteoclast differentiation. Figure S2: Effect of GLY extracts on osteoclast differentiation in palatal side of ligature-induced periodontitis.

Author Contributions: Y.I., J.K., Y.N., R.K., E.S. and M.B. performed the experiments, analyzed the data, contributed reagents/materials/analytical tools, prepared the figures, and reviewed drafts of the article. K.N. and T.N. contributed to the discussion, analyzed the data, and reviewed the drafts of the article. Y.I. and H.Y. conceived, designed the experiments, authored, or reviewed drafts of the article. All authors have read and agreed to the published version of the manuscript.

Funding: This study was supported in part by grants of the Grant-in-Aid for Scientific Research 16K11833 from the Japan Society for the Promotion of Science, and in part by research funds of Sunstar (Osaka, Japan).

Institutional Review Board Statement: Experiments using animal was approved by the Animal Research Control Committee of the Tokushima University Graduate School (T28-90). The study complied with the Animal Research: Reporting in vivo Experiments guidelines developed by the National Centre for the Replacement, Refinement, and Reduction of Animals in Research (NC3Rs).

Acknowledgments: We thank Kohei Nonaka and Ryosuke Takagi for their skillful technical assistance of the induction of experimental periodontitis and the administration of GLY extracts for rats.

Conflicts of Interest: The authors declare no conflict of interest.

References

1. Isola, G.; Polizzi, A.; Alibrandi, A.; Williams, R.C.; Leonardi, R. Independent impact of periodontitis and cardiovascular disease on elevated soluble urokinase-type plasminogen activator receptor (suPAR) levels. *J. Periodontol.* **2020**. [CrossRef]
2. Isola, G.; Polizzi, A.; Patini, R.; Ferlito, S.; Alibrandi, A.; Palazzo, G. Association among serum and salivary A. actinomycetemcomitans specific immunoglobulin antibodies and periodontitis. *BMC Oral Health* **2020**, *20*, 283. [CrossRef] [PubMed]
3. Bartold, P.M.; Cantley, M.D.; Haynes, D.R. Mechanisms and control of pathologic bone loss in periodontitis. *Periodontol. 2000* **2010**, *53*, 55–69. [CrossRef] [PubMed]
4. Zhao, Q.; Jia, Y.; Xiao, Y. Cathepsin K: A therapeutic target for bone diseases. *Biochem. Biophys. Res. Commun.* **2009**, *380*, 721–723. [CrossRef] [PubMed]
5. Miyamoto, T. Regulators of osteoclast differentiation and cell-cell fusion. *Keio J. Med.* **2011**, *60*, 101–105. [CrossRef]
6. Zhang, C.; Dou, C.E.; Xu, J.; Dong, S. DC-STAMP, the key fusion-mediating molecule in osteoclastogenesis. *J. Cell Physiol.* **2014**, *229*, 1330–1335. [CrossRef]
7. Yagi, M.; Miyamoto, T.; Toyama, Y.; Suda, T. Role of DC-STAMP in cellular fusion of osteoclasts and macrophage giant cells. *J. Bone Miner. Metab.* **2006**, *24*, 355–358. [CrossRef]
8. Zhao, Q.; Wang, X.; Liu, Y.; He, A.; Jia, R. NFATc1: Functions in osteoclasts. *Int. J. Biochem. Cell Biol.* **2010**, *42*, 576–579. [CrossRef]
9. Seto, H.; Ohba, H.; Tokunaga, K.; Hama, H.; Horibe, M.; Nagata, T. Topical administration of simvastatin recovers alveolar bone loss in rats. *J. Periodontal. Res.* **2008**, *43*, 261–267. [CrossRef]
10. Tokunaga, K.; Seto, H.; Ohba, H.; Mihara, C.; Hama, H.; Horibe, M.; Yoneda, S.; Nagata, T. Topical and intermittent application of parathyroid hormone recovers alveolar bone loss in rat experimental periodontitis. *J. Periodontal. Res.* **2011**, *46*, 655–662. [CrossRef]
11. Wada-Mihara, C.; Seto, H.; Ohba, H.; Tokunaga, K.; Kido, J.I.; Nagata, T.; Naruishi, K. Local administration of calcitonin inhibits alveolar bone loss in an experimental periodontitis in rats. *Biomed. Pharmacother.* **2018**, *97*, 765–770. [CrossRef] [PubMed]
12. An, J.; Hao, D.; Zhang, Q.; Chen, B.; Zhang, R.; Wang, Y.; Yang, H. Natural products for treatment of bone erosive diseases: The effects and mechanisms on inhibiting osteoclastogenesis and bone resorption. *Int. Immunopharmacol.* **2016**, *36*, 118–131. [CrossRef] [PubMed]
13. Martínez, C.C.; Gómez, M.D.; Oh, M.S. Use of traditional herbal medicine as an alternative in dental treatment in Mexican dentistry: A review. *Pharmaceutic. Biol.* **2017**, *55*, 1992–1998. [CrossRef] [PubMed]
14. Safiaghdam, H.; Oveissi, V.; Bahramsoltani, R.; Farzaei, M.H.; Rahimi, R. Medicinal plants for gingivitis: A review of clinical trials. *Iran J. Basic Med. Sci.* **2018**, *21*, 978–991. [PubMed]
15. Yonezawa, T.; Hasegawa, S.; Asai, M.; Ninomiya, T.; Sasaki, T.; Cha, B.Y.; Teruya, T.; Ozawa, H.; Yagasaki, K.; Nagai, K.; et al. Harmine, a β-carboline alkaloid, inhibits osteoclast differentiation and bone resorption *in vitro* and *in vivo*. *Eur. J. Pharmacol.* **2011**, *650*, 511–518. [CrossRef]
16. Lee, J.W.; Kobayashi, Y.; Nakamichi, Y.; Udagawa, N.; Takahashi, N.; Im, N.K.; Seo, H.J.; Jeon, W.B.; Yonezawa, T.; Cha, B.Y.; et al. Alisol-B, a novel phyto-steroid, suppresses the RANKL-induced osteoclast formation and prevents bone loss in mice. *Biochem. Pharmacol.* **2010**, *80*, 352–361. [CrossRef]
17. Kim, H.J.; Hong, J.; Jung, J.W.; Kim, T.H.; Kim, J.A.; Kim, Y.H.; Kim, S.Y. Gymnasterkoreayne F inhibits osteoclast formation by suppressing NFATc1 and DC-STAMP expression. *Int. Immunopharmacol.* **2010**, *10*, 1440–1447. [CrossRef]
18. Surathu, N.; Kurumathur, A.V. Traditional therapies in the management of periodontal disease in India and China. *Periodontol. 2000* **2011**, *56*, 14–24. [CrossRef]
19. Veilleux, M.P.; Moriyama, S.; Yoshioka, M.; Hinode, D.; Grenier, D. A review of evidence for a therapeutic application of traditional Japanese Kampo medicine for oral diseases/disorders. *Medicines* **2018**, *5*, 35. [CrossRef]
20. Takeda, O.; Toyama, T.; Watanabe, K.; Sato, T.; Sasaguri, K.; Akimoto, S.; Sato, S.; Kawata, T.; Hamada, N. Ameliorating effects of Juzentaihoto on restraint stress and *P. gingivalis*-induced alveolar bone loss. *Arch. Oral Biol.* **2014**, *59*, 1130–1138. [CrossRef]
21. Fournier-Larente, J.; Azelmat, J.; Yoshioka, M.; Hinode, D.; Grenier, D. The daiokanzoto (TJ-84) kampo formulation reduces virulence factor gene expression in *Porphyromonas gingivalis* and possesses anti-inflammatory and anti-protease activities. *PLoS ONE* **2016**, *11*, e0148860. [CrossRef] [PubMed]
22. Yang, J.S.; Wu, C.C.; Lee, H.Z.; Hsieh, W.T.; Tang, F.Y.; Bau, D.T.; Lai, K.C.; Lien, J.C.; Chung, J.G. Suppression of the TNF-alpha level is mediated by Gan-Lu-Yin (traditional Chinese medicine) in human oral cancer cells through the NF-kappa B, AKT, and ERK-dependent pathways. *Environ. Toxicol.* **2016**, *31*, 1196–1205. [CrossRef] [PubMed]
23. Liu, F.C.; Pan, C.H.; Lai, M.T.; Chang, S.J.; Chung, J.G.; Wu, C.H. Gan-Lu-Yin Inhibits Proliferation and Migration of Murine WEHI-3 Leukemia Cells and Tumor Growth in BALB/C Allograft Tumor Model. *Evid. Based Compl. Alter. Med.* **2013**, *2013*, 684071. [CrossRef] [PubMed]
24. Chien, Y.C.; Sheu, M.J.; Wu, C.H.; Lin, W.H.; Chen, Y.Y.; Cheng, P.L.; Cheng, H.C. A Chinese herbal formula "Gan-Lu-Yin" suppresses vascular smooth muscle cell migration by inhibiting matrix metalloproteinase-2/9 through the PI3K/AKT and ERK signaling pathways. *BMC Compl. Altern. Med.* **2012**, *12*, 137. [CrossRef]
25. Pan, C.H.; Hsieh, I.C.; Liu, F.C.; Hsieh, W.T.; Sheu, M.J.; Koizumi, A.; Wu, C.H. Effects of a Chinese herbal health formula, "Gan-Lu-Yin", on angiogenesis. *J. Agric. Food Chem.* **2010**, *58*, 7685–7692. [CrossRef] [PubMed]
26. Zhang, Y.; Wang, X.C.; Bao, X.F.; Hu, M.; Yu, W.X. Effects of *Porphyromonas gingivalis* lipopolysaccharide on osteoblast-osteoclast bidirectional EphB4-EphrinB2 signaling. *Exp. Ther. Med.* **2014**, *7*, 80–84. [CrossRef] [PubMed]

27. Messora, M.R.; Pereira, L.J.; Foureaux, R.; Oliveira, L.F.; Sordi, C.G.; Alves, A.J.; Napimoga, M.H.; Nagata, M.J.; Ervolino, E.; Furlaneto, F.A. Favourable effects of *Bacillus subtilis* and *Bacillus licheniformis* on experimental periodontitis in rats. *Arch. Oral Biol.* **2016**, *66*, 108–119. [CrossRef]

28. Seto, H.; Toba, Y.; Takada, Y.; Kawakami, H.; Ohba, H.; Hama, H.; Horibe, M.; Nagata, T. Milk basic protein increases alveolar bone formation in rat experimental periodontitis. *J. Periodontal. Res.* **2007**, *42*, 85–89. [CrossRef]

29. Hashimura, S.; Kido, J.; Matsuda, R.; Yokota, M.; Matsui, H.; Inoue-Fujiwara, M.; Inagaki, Y.; Hidaka, M.; Tanaka, T.; Tsutsumi, T.; et al. A low level of lysophosphatidic acid in human gingival crevicular fluid from patients with periodontitis due to high soluble lysophospholipase activity: Its potential protective role on alveolar bone loss by periodontitis. *Biochim. Biophys. Acta Mol. Cell Biol. Lipids* **2020**, *1865*, 158698. [CrossRef]

30. Vargas-Sanchez, P.K.; Moro, M.G.; Santos, F.A.D.; Anbinder, A.L.; Kreich, E.; Moraes, R.M.; Padilha, L.; Kusiak, C.; Scomparin, D.X.; Franco, G.C.N. Agreement, correlation, and kinetics of the alveolar bone-loss measurement methodologies in a ligature-induced periodontitis animal model. *J. Appl. Oral Sci.* **2017**, *25*, 490–497. [CrossRef]

31. De Molon, R.S.; Park, C.H.; Jin, Q.; Sugai, J.; Cirelli, J.A. Characterization of ligature-induced experimental periodontitis. *Microsc. Res. Tech.* **2018**, *81*, 1412–1421. [CrossRef] [PubMed]

32. Lin, I.H.; Lee, M.C.; Chuang, W.C. Application of LC/MS and ICP/MS for establishing the fingerprint spectrum of the traditional Chinese medicinal preparation Gan-Lu-Yin. *J. Sep. Sci.* **2006**, *29*, 172–179. [CrossRef] [PubMed]

33. Cai, X.; Li, C.; Du, G.; Cao, Z. Protective effects of baicalin on ligature-induced periodontitis in rats. *J. Periodontal. Res.* **2008**, *43*, 14–21. [CrossRef] [PubMed]

34. Yin, Z.; Zhu, W.; Wu, Q.; Zhang, Q.; Guo, S.; Liu, T.; Li, S.; Chen, X.; Peng, D.; Ouyang, Z. Glycyrrhizic acid suppresses osteoclast differentiation and postmenopausal osteoporosis by modulating the NF-κB, ERK, and JNK signaling pathways. *Eur. J. Pharmacol.* **2019**, *859*, 172550. [CrossRef] [PubMed]

35. Ang, E.S.; Yang, X.; Chen, H.; Liu, Q.; Zheng, M.H.; Xu, J. Naringin abrogates osteoclastogenesis and bone resorption via the inhibition of RANKL-induced NF-κB and ERK activation. *FEBS Lett.* **2011**, *585*, 2755–2762. [CrossRef] [PubMed]

36. Jang, S.; Bak, E.J.; Kim, M.; Kim, J.M.; Chung, W.Y.; Cha, J.H.; Yoo, Y.J. Wogonin inhibit osteoclast formation induced by lipopolysaccharide. *Phytother. Res.* **2010**, *24*, 964–968. [CrossRef] [PubMed]

37. Eyre, D.R. Bone biomarkers as tools in osteoporosis management. *Spine (Phila Pa 1976)* **1997**, *22*, 17S–24S. [CrossRef]

38. Wen, X.; Yi, L.Z.; Liu, F.; Wei, J.H.; Xue, Y. The role of cathepsin K in oral and maxillofacial disorders. *Oral Dis.* **2016**, *22*, 109–115. [CrossRef]

39. Li, Z.; Chen, C.; Zhu, X.; Li, Y.; Yu, R.; Xu, W. Glycyrrhizin suppresses RANKL-induced osteoclastogenesis and oxidative stress through inhibiting NF-kB and MAPK and activating AMPK/Nrf2. *Calcif. Tissue Int.* **2018**, *103*, 324–337. [CrossRef]

40. Kim, M.H.; Ryu, S.Y.; Bae, M.A.; Choi, J.S.; Min, Y.K.; Kim, S.H. Baicalein inhibits osteoclast differentiation and induces mature osteoclast apoptosis. *Food Chem. Toxicol.* **2008**, *46*, 3375–3382. [CrossRef]

41. Negishi-Koga, T.; Takayanagi, H. Ca2+-NFATc1 signaling is an essential axis of osteoclast differentiation. *Immunol. Rev.* **2009**, *231*, 241–256. [CrossRef] [PubMed]

42. Geng, X.; Yang, L.; Zhang, C.; Qin, H.; Liang, Q. Wogonin inhibits osteoclast differentiation by inhibiting NFATc1 translocation into the nucleus. *Exp. Ther. Med.* **2015**, *10*, 1066–1070. [CrossRef] [PubMed]

43. Mao, Y.; Huang, X.; Zhao, J.; Gu, Z. Preliminary identification of potential PDZ-domain proteins downstream of ephrin B2 during osteoclast differentiation of RAW264.7 cells. *Int. J. Mol. Med.* **2011**, *27*, 669–677. [PubMed]

44. Ikeda, K.; Takeshita, S. Factors and mechanisms involved in the coupling from bone resorption to formation: How osteoclasts talk to osteoblasts. *J. Bone Metab.* **2014**, *21*, 163–167. [CrossRef] [PubMed]

45. Zhao, C.; Irie, N.; Takada, Y.; Shimoda, K.; Miyamoto, T.; Nishiwaki, T.; Suda, T.; Matsuo, K. Bidirectional ephrinB2-EphB4 signaling controls bone homeostasis. *Cell Metab.* **2006**, *4*, 111–121. [CrossRef]

46. Yang, H.; Wen, Q.; Xue, J.; Ding, Y. Alveolar bone regeneration potential of a traditional Chinese medicine, Bu-Shen-Gu-Chi-Wan, in experimental periodontitis. *J. Periodontal. Res.* **2014**, *49*, 382–389. [CrossRef]

47. Zhou, T.; Chen, D.; Li, Q.; Sun, X.; Song, Y.; Wang, C. Curcumin inhibits inflammatory response and bone loss during experimental periodontitis in rats. *Acta Odontol. Scand.* **2013**, *71*, 349–356. [CrossRef]

48. Luo, W.; Wang, C.Y.; Jin, L. Baicalin downregulates *Porphyromonas gingivalis* lipopolysaccharide-upregulated IL-6 and IL-8 expression in human oral keratinocytes by negative regulation of TLR signaling. *PLoS ONE* **2012**, *7*, e51008. [CrossRef]

49. Ku, S.K.; Bae, J.S. Baicalin, baicalein and wogonin inhibits high glucose-induced vascular inflammation in vitro and in vivo. *BMB Rep.* **2015**, *48*, 519–524. [CrossRef]

50. Kim, Y.W.; Zhao, R.J.; Park, S.J.; Lee, J.R.; Cho, I.J.; Yang, C.H.; Kim, S.G.; Kim, S.C. Anti-inflammatory effects of liquiritigenin as a consequence of the inhibition of NF-kB-dependent iNOS and proinflammatory cytokines production. *Br. J. Pharmacol.* **2008**, *154*, 165–173. [CrossRef]

51. Liu, L.; Zuo, Z.; Lu, S.; Liu, A.; Liu, X. Naringin attenuates diabetic retinopathy by inhibiting inflammation, oxidative stress and NF-kB activation *in vivo* and *in vitro*. *Iran J. Basic Med. Sci.* **2017**, *20*, 813–821. [PubMed]

52. Hienz, S.A.; Paliwal, S.; Ivanovski, S. Mechanisms of Bone Resorption in Periodontitis. *J. Immunol. Res.* **2015**, *2015*, 615486. [CrossRef] [PubMed]

53. Gokhale, S.R.; Padhye, A.M. Future prospects of systemic host modulatory agents in periodontal therapy. *Br. Dent. J.* **2013**, *214*, 467–471. [CrossRef] [PubMed]

Journal of
Clinical Medicine

Article

Optimal Examination Sites for Periodontal Disease Evaluation: Applying the Item Response Theory Graded Response Model

Yoshiaki Nomura [1], Toshiya Morozumi [2,*], Mitsuo Fukuda [3], Nobuhiro Hanada [1], Erika Kakuta [4], Hiroaki Kobayashi [5], Masato Minabe [2], Toshiaki Nakamura [6], Yohei Nakayama [7], Fusanori Nishimura [8], Kazuyuki Noguchi [6], Yukihiro Numabe [9], Yorimasa Ogata [7], Atsushi Saito [10], Soh Sato [11], Satoshi Sekino [9], Naoyuki Sugano [12], Tsutomu Sugaya [13], Fumihiko Suzuki [14], Keiso Takahashi [15], Hideki Takai [7], Shogo Takashiba [16], Makoto Umeda [17], Hiromasa Yoshie [18], Atsutoshi Yoshimura [19], Nobuo Yoshinari [20] and Taneaki Nakagawa [21]

1   Department of Translational Research, Tsurumi University School of Dental Medicine, Yokohama 230-8501, Japan; nomura-y@tsurumi-u.ac.jp (Y.N.); hanada-n@tsurumi-u.ac.jp (N.H.)
2   Division of Periodontology, Department of Oral Interdisciplinary Medicine, Graduate School of Dentistry, Kanagawa Dental University, Yokosuka 238-8580, Japan; minabe@kdu.ac.jp
3   Department of Periodontology, School of Dentistry, Aichi Gakuin University, Nagoya 464-8650, Japan; fukuda-m@dpc.agu.ac.jp
4   Department of Oral Microbiology, Tsurumi University School of Dental Medicine, Yokohama 230-8501, Japan; kakuta-erika@tsurumi-u.ac.jp
5   Department of Periodontology, Graduate School of Medical and Dental Sciences, Tokyo Medical and Dental University, Tokyo 113-8510, Japan; h-kobayashi.peri@tmd.ac.jp
6   Department of Periodontology, Kagoshima University Graduate School of Medical and Dental Sciences, Kagoshima 890-8544, Japan; toshi-n@dent.kagoshima-u.ac.jp (T.N.); kazuperi@dent.kagoshima-u.ac.jp (K.N.)
7   Department of Periodontology, Nihon University School of Dentistry at Matsudo, Matsudo 271-8587, Japan; nakayama.youhei@nihon-u.ac.jp (Y.N.); ogata.yorimasa@nihon-u.ac.jp (Y.O.); takai.hideki@nihon-u.ac.jp (H.T.)
8   Section of Periodontology, Division of Oral Rehabilitation, Faculty of Dental Science, Kyushu University, Fukuoka 819-0395, Japan; fusanori@dent.kyushu-u.ac.jp
9   Department of Periodontology, School of Life Dentistry at Tokyo, The Nippon Dental University, Tokyo 102-8159, Japan; numabe-y@tky.ndu.ac.jp (Y.N.); sekino-s@tky.ndu.ac.jp (S.S.)
10   Department of Periodontology, Tokyo Dental College, Tokyo 101-0061, Japan; atsaito@tdc.ac.jp
11   Department of Periodontology, School of life Dentistry at Niigata, The Nippon Dental University, Niigata 951-8580, Japan; s-sato@ngt.ndu.ac.jp
12   Department of Periodontology, Nihon University School of Dentistry, Tokyo 101-8310, Japan; sugano.naoyuki@nihon-u.ac.jp
13   Division of Periodontology and Endodontology, Department of Oral Health Science, Hokkaido University Graduate School of Dental Medicine, Sapporo 060-8586, Japan; sugaya@den.hokudai.ac.jp
14   Division of Dental Anesthesiology, Department of Oral Surgery, Ohu University School of Dentistry, Koriyama 963-8611, Japan; f-suzuki@den.ohu-u.ac.jp
15   Division of Periodontics, Department of Conservative Dentistry, Ohu University School of Dentistry, Koriyama 963-8611, Japan; ke-takahashi@den.ohu-u.ac.jp
16   Department of Pathophysiology-Periodontal Science, Okayama University Graduate School of Medicine, Dentistry and Pharmaceutical Sciences, Okayama 700-8525, Japan; stakashi@okayama-u.ac.jp
17   Department of Periodontology, Osaka Dental University, Hirakata 573-1121, Japan; umeda-m@cc.osaka-dent.ac.jp
18   Division of Periodontology, Department of Oral Biological Science, Niigata University Graduate School of Medical and Dental Sciences, Niigata 951-8514, Japan; yoshie119@galaxy.ocn.ne.jp
19   Department of Periodontology and Endodontology, Nagasaki University Graduate School of Biomedical Sciences, Nagasaki 852-8588, Japan; ayoshi@nagasaki-u.ac.jp
20   Department of Periodontology, School of Dentistry, Matsumoto Dental University, Shiojiri 399-0781, Japan; nobuo.yoshinari@mdu.ac.jp

J. Clin. Med. **2020**, *9*, 3754

21   Department of Dentistry and Oral Surgery, School of Medicine, Keio University, Tokyo 60-8582, Japan;
      tane@z6.keio.jp
*   Correspondence: morozumi@kdu.ac.jp; Tel.: +81-46-822-8855

Received: 9 October 2020; Accepted: 19 November 2020; Published: 21 November 2020

Abstract: Periodontal examination data have a complex structure. For epidemiological studies, mass screenings, and public health use, a simple index that represents the periodontal condition is necessary. Periodontal indices for partial examination of selected teeth have been developed. However, the selected teeth vary between indices, and a justification for the selection of examination teeth has not been presented. We applied a graded response model based on the item response theory to select optimal examination teeth and sites that represent periodontal conditions. Data were obtained from 254 patients who participated in a multicenter follow-up study. Baseline data were obtained from initial follow-up. Optimal examination sites were selected using item information calculated by graded response modeling. Twelve sites—maxillary 2nd premolar (palatal-medial), 1st premolar (palatal-distal), canine (palatal-medial), lateral incisor (palatal-central), central incisor (palatal-distal) and mandibular 1st premolar (lingual, medial)—were selected. Mean values for clinical attachment level, probing pocket depth, and bleeding on probing by full mouth examinations were used for objective variables. Measuring the clinical parameters of these sites can predict the results of full mouth examination. For calculating the periodontal index by partial oral examination, a justification for the selection of examination sites is essential. This study presents an evidence-based partial examination methodology and its modeling.

Keywords: periodontitis; epidemiological index; item response theory; oral examination; diagnosis; bleeding on probing

---

1. Introduction

Periodontal examination should be carried out precisely for the evaluation of periodontal diseases, especially their clinical parameters. Improvement in or progression of periodontitis should be monitored at the site-level along with periodontal treatment [1]. One of the important characteristics of periodontal disease is the localization of infectious processes at specific sites, which eventually leads to tissue destruction [2]. Therefore, the accumulated data obtained during periodontal examination is copious and structurally complex [1]. Each of the 28 teeth has six examination sites for measuring parameters such as bleeding on probing (BOP), periodontal pocket probing depth (PD), and clinical attachment level (CAL). These measurements are evaluated several times throughout the course of periodontal disease treatment. Representative summary statistics include the mean value of CAL and PD, maximum value of CAL and PD, and percentage of sites with BOP (BOP%). Conventionally, summary statistics have been used in several clinical trials for the patient-level evaluation of clinical parameters [3]. However, aggregating summary statistics like the mean value or maximum value can lead to the loss of information [4]. Full-mouth protocols are proven to be the most effective [5]. When time and labor are limited, a simplified index may be necessary to represent the periodontal condition.

Several periodontal indices have been developed for epidemiological surveys and periodontal disease screening: Periodontal disease index (PDI) [6,7], Periodontal Index (PI) [8], Community periodontal index (CPI or CPITN) [8,9], Gingival bone count [10], PMA index [11], Gingival index [12]. They are based on a partial-examination method in which the target teeth for examination vary between the indices. A justification for the selection of these target teeth is not always clear. However, the developers of epidemiological indices may notice that there are several teeth or examination sites within the oral cavity that represent periodontal conditions at an individual level.

In many educational and psychological studies, a latent variable is often used as the outcome variable. These variables cannot be measured directly and are estimated by their response to the observational items. Item response theory (IRT) modeling is an important methodology commonly used for the development of tests to measure the ability by total score of test consisted of items with weighted scores [13]. Through the application of IRT, we can examine each item's reliability and whether it contributes to an overall construct [14,15]. Total score correspond to the sum of the values of periodontal examinations. Items correspond to the values of periodontal examinations. Therefore, IRT is applicable for the indexes used in dental search.

IRT can be applied to the clinical parameters of periodontal disease. The progress of periodontal disease at an individual level corresponds to ability. Susceptibility to the progress of PD, CAL, and BOP corresponds to item difficulty. Discrimination parameter corresponds to the predictability of each site to represent all examination sites or teeth. Previous studies have shown that IRT can efficiently characterize dental caries susceptibility [16,17]. IRT graded response modeling has also been applied in the evaluation of existing measures in several clinical areas, such as those related to swallowing and communication disorders [18,19]. Furthermore, by using IRT models with clinical diagnoses from electronic health records, a constellation of high-risk patients could be identified [20]. Therefore, by applying the IRT model to periodontal data, evidence-based target sites or teeth can be selected to represent and reflect the periodontal conditions of all sites or teeth in the oral cavity.

This study aimed to identify the most reliable subset of teeth able to represent a full-mouth periodontal diagnosis.

2. Materials and Methods

2.1. Study Design

2.1.1. Setting

This study was part of a clinical research project by the Japanese Society of Periodontology, in cooperation with 17 facilities (one clinic and 16 university hospitals) in Japan for the diagnosis of periodontitis [1,21,22]. Two-hundred-fifty-four patients with chronic periodontitis were chosen between February 2009 and February 2012 for this study, who had completed their active treatment regulated by the Japanese health insurance system. All 254 patients who registered the study were analyzed.

2.1.2. Diagnosis

Each patient was diagnosed according to the guideline at the time (Guidelines of the American Academy of Periodontology) [23]. One examiner from each institute (T.M., M.F., H.K., M.M., T.N., Y.N., K.N., S.S., N.S., S.S., T.S., F.S., H.T., H.Y., A.Y., N.Y. and T.N.) was chosen to carry out the oral examinations. Each examiner was a periodontist licensed by the Japanese Society of Periodontology.

Intra- and inter-examiner calibration session were conducted at the beginning and middle of the study period. Diagnosis of periodontitis was based on the proposed criteria by the Center for Disease Control and Prevention (CDC) in partnership with the American Academy of Periodontology (AAP) [24].

2.1.3. Patients

Each patient was ≥ 30 years of age, possessed at least 20 teeth, was systemically healthy, and had not been administered immunosuppressive or anti-inflammatory drugs or systemic antibiotics within 3 months before the initiation of the investigation.

2.2. Research Data

In this study, we analyzed CAL, PD, BOP, plaque index (PII), and tooth mobility. CAL was measured at six sites for all of the remaining teeth (mesiobuccal, buccal, distobuccal, mesiolingual, lingual, and distolingual). The data of CAL were categorized as < 4, 4–5, and > 5 mm.

2.3. Statistical Analysis

2.3.1. IRT Modeling

Based on the IRT model for ordinal polytomous data, we applied a Graded Response Model [25–29]. Item difficulty, item discrimination, item information for the examined sites, and ability of the subjects were calculated [30–34]. The R software with the ltm package was used to perform the IRT analysis [27]. To reduce the total number of examination sites, sites with small item information were removed from the IRT model. This procedure were based on a step-by-step analysis. Using the data for CAL, a model was constructed for all examination sites (Model 1). Next, out of all 168 examination sites, 28 sites representing the highest information for each tooth (sum of left and right side) were selected. An IRT model was constructed using these 28 sites (Model 2). Out of these 28 sites, 12 sites in six teeth were selected for depicting a higher information (Model 3). Finally, the data from the right and left side were categorized as follows: at least one site with >5 mm CAL; at least one site with 4–5 mm CAL; or both sites with <4 mm CAL. Even though there may be optimal examination sites for each clinical parameter, the examination of numerous sites for each clinical parameter may be a laborious procedure for an epidemiological examiner or clinician. IRT models for BOP and PD were constructed in the same manner as for CAL.

2.3.2. Model Evaluation

For the scatterplot, regression analysis was carried out. Generalized linear models were applied. For optimal link functions, models were evaluated using Akaike's information criteria [35]. Receiver operating characteristic (ROC) curve was used to analyze sensitivities and specificities. The cutoff points were determined as the minimum difference between specificity and sensitivity [36,37]. The mean CAL of all examination sites and community periodontal index (CPI) were used as reference. Diagnostic criteria by the CDC-AAP [24] was used. Statistical Package for the Social Sciences version 24.0 (IBM, Tokyo, Japan) was used to perform the analyses.

To compare the model to other studies, Sensitivity, relative bias (Severity) relative bias (Extent) were calculated [38–40].

2.4. Ethical Approval

The study was conducted in compliance with the principles outlined in the Helsinki Declaration. Informed written consent was obtained from each subject, and the protocol was approved by the Institutional Review Board of each participating institution. The ethics committee members' names and reference numbers are listed in Appendix A.

3. Results

3.1. Descriptive Statistics of the Subjects Participated in this Study

Descriptive statistics of periodontal clinical parameters were the 3.1 mm for mean of CAL, 2.5 mm for mean of PD, 15.0% for BOP%, and 0.3 for PII.

3.2. Optimal Site Selection by IRT Modeling

The final model for CAL (Model 4) is shown in Table 1, accompanied with models for the remaining clinical parameters. Item information and item response curves of Model 4 are shown in Figure S1. Using these steps, 168 examination sites were narrowed down to six variables located at 12 sites (same sites on the right and left side). The results of each step from Model 1 to 4 are shown in Table S1. A quick reference for the calculation of ability by Model 4 is presented in Appendix B.

Table 1. Final model (Model 4) for the clinical attachment level.

		Maxilla					Mandibular		
		2nd Premolar	1st Premolar	Canine	Lateral Incisor	Central Incisor	1st Premolar	AIC	BIC
		Palatal	Palatal	Palatal	Palatal	Palatal	Lingual		
		Medial	Distal	Medial	Central	Distal	Medial		
CAL	Extrmt1	0.28	0.45	0.65	1.13	0.75	0.53	1796.06	1859.73
	Extrmt2	1.28	1.24	1.31	1.59	1.38	1.57		
	Discrimination	3.07	3.42	4.51	4.83	3.16	2.03		
	Item information	5.47	5.92	7.97	7.93	5.02	3.27		
PD	Extrmt1	0.72	0.88	0.96	1.35	1.01	1.03	1198.73	1262.41
	Extrmt2	1.46	1.48	1.39	1.79	1.45	1.63		
	Discrimination	3.89	3.38	4.89	4.63	3.90	3.39		
	Item information	6.82	5.37	7.87	7.41	5.94	5.40		
BOP	Extrmt1	1.30	1.22	1.56	1.45	1.63	1.95	686.60	729.04
	Discrimination	3.70	7.71	2.93	16.96	2.48	2.09		
	Item information	3.70	7.62	2.93	16.65	2.48	2.09		
PlI	Extrmt1	0.38	0.47	0.35	0.27	0.47	0.82	1965.74	2029.41
	Extrmt2	1.85	1.70	1.65	1.81	1.96	2.03		
	Discrimination	2.78	3.85	3.86	2.66	2.14	2.19		
	Item information	5.25	7.44	7.50	5.03	3.85	3.76		
Tooth Mobility	Extrmt1	0.87	0.77	1.22	0.69	0.72	1.19	1445.90	1509.57
	Extrmt2	1.90	1.85	2.21	1.79	1.99	2.56		
	Discrimination	2.41	2.53	2.98	3.81	3.23	2.07		
	Item information	4.06	4.39	5.28	7.24	6.11	3.62		

Extrmt: extremity parameters; CAL: clinical attachment level; PD: probing depth; BOP: bleeding on probing; PlI: plaque index. For CAL, it shows the cutoff to discriminate CAL < 4 mm, CAL 4–5 mm, and CAL > 5 mm. Extrmt1 discriminates CAL < 4 mm and (CAL 4–5 mm and CAL > 5 mm), and Extrmt 2 discriminates (CAL < 4 mm and CAL 4–5 mm) and CAL > 5 mm. Discrimination: This parameter shows the height of item characteristic curves. For the item response theory (IRT) analysis, CAL and PD were categorized as at least one site with >6 mm, at least one site with 4–6 mm, or both sites with <4 mm on the left or right side. IRT analysis was carried out using a graded response model. AIC: Akaike's information criterion; BIC: Bayesian information criterion; both are fitness indices, in which small values are more suitable for a model fit.

3.3. Model Evaluation

3.3.1. Evaluation of Selected Sites

In clinical practice, the mean values of all examination sites are often used as summary statistics. The selected 12 sites were evaluated using a scatter plot by plotting the mean values of each clinical parameter against the mean values of the selected 12 sites. The results are shown in Figure 1. For each clinical parameter, adequate co-relations were obtained.

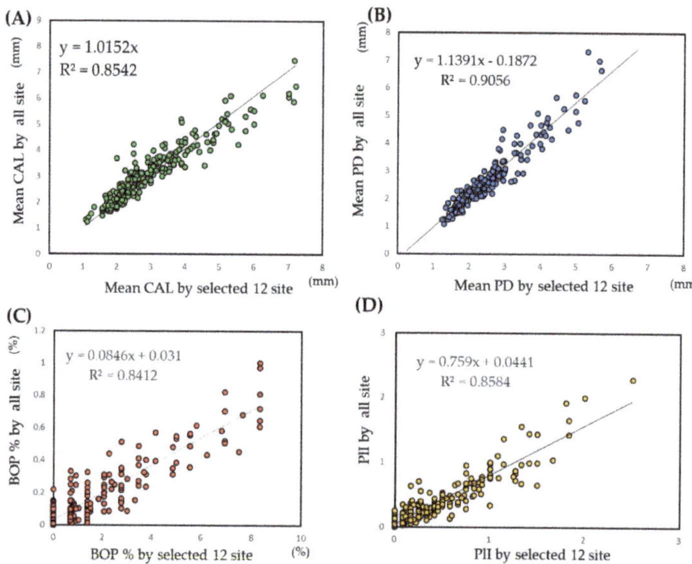

Figure 1. Scatter plot of the mean values of clinical parameters against the mean values of the selected 12 sites. (**A**) CAL: clinical attachment level. (**B**) PD: probing depth. (**C**) BOP: bleeding on probing. (**D**) PlI: plaque index. The selected 12 sites were the same sites that are listed in the Figure 1 legend.

3.3.2. Model Evaluation

The models were evaluated using two methods: correlation between predictive values and observed values and ROC curve analysis. Ability calculated by IRT analysis indicates the predictive value of the sample. The scatter plot of the ability calculated using Model 4 against the mean values of CAL is illustrated in Figure 2. A scatter plot of all 168 examination sites (Model 1) is also presented as a reference. As the plot appears to be a curve, the generalized linear model was applied. The coefficient and intercept were statistically significant. The scatter plot of the result of the generalized linear model against the models for other clinical parameters is shown in Table S2.

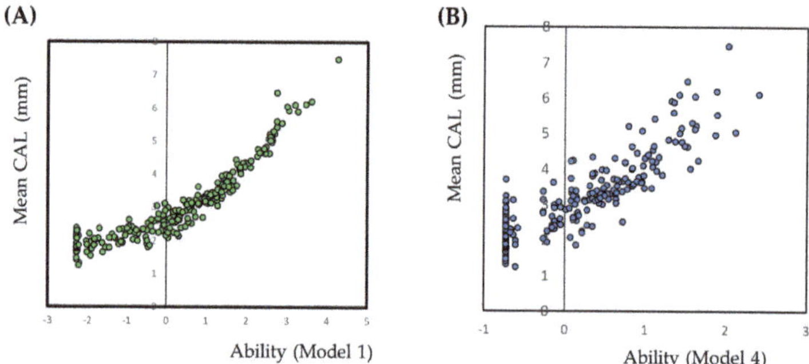

Figure 2. Scatter plot of the mean value of the clinical attachment level against the ability calculated by item response theory. Abilities are calculated using a graded response model under the item response theory approach. Abilities are calculated for all the 168 examined sites (**A**) and the selected six variables at the 12 sites (**B**). Ability means sum of the weighted scores of each items. In this case, ability indicate the sum of the weighted scores of positive for periodontal examination.

The data from the same location site are combined with at least one site with >6 mm, at least one site with 4–6 mm, or both sites <4 mm.

The plot appears to be a curve. Therefore, the generalized linear model is applied for the relationship.

The selected sites were maxillary 2nd premolar (palatal-distal), maxillary 1st premolar (palatal-medial), maxillary Canine (palatal-distal), maxillary lateral incisor (palatal-central), maxillary central incisor (palatal-medial), and mandibular 1st premolar (lingual-medial).

Based on the ROC analysis, sensitivity, specificity, likelihood, and area under ROC curve (AUR) are presented in Table 2. The results of the models for other clinical parameters are also shown in Table 2. For the mean value of CAL >3 mm, sensitivity and specificity were 0.832 and 0.852, respectively, and for >5 mm, they were 0.895 and 0.911, respectively. The ROC curve of each clinical parameter and various cutoff points are presented in Figure S2.

Table 2. Sensitivity, specificity, and area under the receiver operating characteristic curve for the six variables.

		Cutoff Point	**Sensitivity**	**Specificity**	**AUR**
	>3 mm	−0.190	0.832	0.852	0.922
	>3.5 mm	0.193	0.865	0.878	0.939
Mean CAL	>4 mm	0.546	0.891	0.885	0.960
	>4.5 mm	1.033	0.929	0.929	0.978
	>5 mm	1.082	0.895	0.911	0.977
	>3 mm	0.703	0.897	0.912	0.974
	>3.5 mm	0.891	0.926	0.930	0.983
Mean PD	>4 mm	1.202	0.947	0.949	0.985
	>4.5 mm	1.391	1.000	0.971	0.989
	>5 mm	1.548	1.000	0.984	0.992
	>2.5%	−0.097	0.453	0.913	0.671
	>5%	−0.097	0.388	0.939	0.698
BOP	>10%	−0.097	0.571	0.886	0.755
	>20%	−0.097	0.687	0.834	0.804
	>25%	−0.097	0.765	0.813	0.847
	>10	−0.141	0.849	0.829	0.890
	>20	0.313	0.838	0.839	0.915
PlI	>30	0.540	0.824	0.822	0.915
	>40	0.695	0.831	0.776	0.924
	>50	0.710	0.825	0.838	0.912

CAL: clinical attachment level; PD: probing depth; BOP: bleeding on probing; PlI: plaque index; AUR: area under receiver operating characteristic curve. Cutoff points are set in abilities calculated using graded response theory. Graded response theory is one of the models of item response theory.

3.4. Application of the CAL Model for Diagnosis of Periodontal Disease

The cutoff point, sensitivity, specificity, and AUR for the diagnosis of periodontal disease by the CDC-AAP are presented in Table 3. ROC curves are shown in Figure S3. For moderate periodontitis, the simple mean CAL of all examination sites is most useful, followed by the mean CAL of optimal examination sites. For severe periodontitis, the CPI is most useful. Simple mean CAL were not obtained similar AUR for CPI.

Table 3. Receiver operating characteristic analysis for the selected optimal examination sites for the diagnosis of periodontal disease.

Diagnosis	Moderate Periodontitis				Severe Periodontitis			
	Cutoff	Sensitivity	Specificity	AUR	Cutoff	Sensitivity	Specificity	AUR
CAL mean (12 sites)	2.20	0.82	0.88	0.91	2.65	0.75	0.72	0.85
CAL mean (All sites)	2.21	0.91	0.92	0.96	3.04	0.78	0.78	0.88
CAL ability (6 value)	−0.72	0.71	0.88	0.82	−0.47	0.86	0.62	0.83
CAL ability (12 sites)	−0.79	0.79	0.79	0.87	0.05	0.81	0.81	0.87
CPI	2	0.84	0.92	0.88	4	0.81	0.94	0.91

CPI: community periodontal index; CAL: clinical attachment level.

3.5. Prediction of Conventional Periodontal Indices by the CAL Model

The model was applied to conventionally use summary statistics, i.e., mean value of CAL, PD, and BOP%. Clinically useful cutoff points were set for each index. The results are presented in Table 4. The best predictors were the mean values of the clinical parameters by the mean values of the 12 selected sites (e.g., mean CAL by the mean CAL of the 12 sites). For obtaining the mean values of PD and BOP%, we were able to obtain higher values of the AUR compared to the CPI by calculating the simple mean CAL of the 12 selected sites. Ability of CAL, weighted CAL, could obtain higher AUR. All ROC curves are shown in Figure S4. By measuring the CAL, all other clinical parameters can be predicted.

3.6. Model Evaluation by Prevalence, Severity, and Extent

The prevalence, relative bias for severity and extent of model 4 were 62%, −0.0035, and −0.017, respectively.

Table 4. Receiver operating characteristic curve analysis of the selected optimal examination sites for the mean values of pocket probing depth, clinical attachment level, and bleeding on probing.

	Cutoff	Sensitivity	Specificity	AUR	Cutoff	Sensitivity	Specificity	AUR	Cutoff	Sensitivity	Specificity	AUR
Mean CAL	Mean CAL > 3.5 mm				Mean CAL > 4 mm				Mean CAL > 4.5 mm			
CAL mean (12 sites)	3.04	0.92	0.90	0.96	3.52	0.91	0.91	0.98	3.95	0.93	0.94	0.99
CAL ability (6 values)	0.16	0.89	0.88	0.95	0.55	0.89	0.88	0.96	1.03	0.93	0.93	0.98
CAL ability (12 site)	0.35	0.89	0.90	0.96	0.69	0.93	0.93	0.97	0.90	0.93	0.93	0.99
CPI	4	0.72	0.77	0.77	2	0.96	0.36	0.81	4	0.86	0.69	0.78
Mean PD	Mean PD > 3.5 mm				Mean PD > 4 mm				Mean PD > 5 mm			
CAL mean (12 sites)	3.90	0.08	0.11	0.96	4.13	0.95	0.92	0.97	5.04	1.00	0.96	0.99
CAL ability (6 values)	4.03	0.89	0.91	0.97	0.99	0.89	0.89	0.96	1.44	1.00	0.96	0.99
CAL ability (12 sites)	0.70	0.85	0.87	0.95	0.89	0.89	0.89	0.96	1.20	1.00	0.93	0.98
CPI	4	1.00	0.71	0.85	4	1.00	0.68	0.84	4	1.00	0.65	0.82
PD mean (12 sites)	0.68	0.89	0.86	0.95	3.78	1.00	0.95	0.99	4.46	1.00	0.97	0.99
BOP	BOP > 10%				BOP > 20%				BOP > 30%			
CAL mean (12 sites)	2.65	0.64	0.65	0.71	2.96	0.76	0.75	0.83	3.26	0.82	0.80	0.89
CAL ability (6 values)	−0.26	0.65	0.62	0.69	0.12	0.74	0.77	0.80	0.46	0.77	0.82	0.86
CAL ability (12 sites)	0.03	0.63	0.66	0.67	0.31	0.74	0.75	0.80	0.60	0.75	0.85	0.85
CPI	4	0.59	0.79	0.72	4	0.75	0.76	0.80	4	0.84	0.72	0.81
BOP% mean (12 sites)	0.73	0.84	0.87	0.91	1.97	0.81	0.94	0.95	2.18	0.91	0.92	0.97

CAL: clinical attachment level; PD: probing depth; BOP: bleeding on probing; AUR: area under receiver operating characteristic curve; CPI: community periodontal index.

4. Discussion

Public health applications of periodontal examination such as epidemiological surveys, mass screenings, and community diagnosis, simplified indices that represent an indivisible disease status are indispensable. For this purpose, several indices have been developed [6–12]. Developed by the World Health Organization, the CPI has been utilized in epidemiological surveys not just for community diagnosis but also for the screening of periodontal disease. Originally, the index was calculated through the examination of eight teeth [41]. It has currently been revised to include the examination of all teeth [42]; however, the original method of examining just eight teeth is also still applied. According to Japan's Survey of Dental Diseases, national oral health surveys, which are conducted every six years, still use the original CPI examination method. However, we were unable to find any justification for the tooth-selection methodology or periodontal indices used in these surveys [43].

Several methods that do not require oral examinations for the screening of periodontal disease have been proposed [36,44–48]. These include questionnaires [49,50] and biochemical analysis of the saliva [47] or gingival crevicular fluid [51]. However, the sensitivity and specificity of questionnaires used for periodontitis screening is not high enough, and using biochemical analyses requires a special measuring device. Therefore, oral examinations are still widely used for periodontal disease screenings.

In comparison to the CPI, our model was superior in predicting clinical parameters and almost equivalent in diagnosing moderate periodontitis. Further, partial examination using the CPI requires the examination of 60 sites; our model requires the examination of only 6–12 sites. Furthermore, the examination sites presented by our model are more representative of the periodontal conditions in an oral cavity. As shown in Table 4, the partial examination of these sites represents the mean value of each examined index by the mean value of all examination sites in the oral cavity. By simply measuring the CAL, all other clinical parameters can be predicted. The model presented in this study was derived using the IRT approach. The IRT model is very useful for the selection of items that have high information. However, IRT models can only process dichotomous variables or ordinal scale; they are unable to process contentious variables. At this stage, a loss of information can occur. Therefore, the simple mean of 12 selected sites is more suitable for calculating some of the predictions presented in this study.

When the 12 selected sites are compared with other partial examination protocol, sampling sites are predominantly small. Sampling sites of other protocols were 84 [38,39,52,53], 60 [54], and 56 [38,39,52]. Partial examination protocol by high number of examination site can detect small number of deep CAL or PD. The sensitivity indicate to detect subject with at least one site of CAL > 4 mm. The sensitivity by 84 examination site was 92% [52], by 56 site was 66% [52] and by 28 site was 57% [52]. The sensitivity by the 12 site in this study was 62%. In addition, relative bias of severity of the 12 sites, which estimate the difference of mean value of CAL between full mouth examination, was −0.0035. This value protocols by 84 site were 0.009 [52], −0.046 [38] and −0.01 [53]. The 12 sites based on statistical model may equal, in other partial examination protocol, more than 5 times higher numbers of examination sites.

In this study, the teeth selected for examination included premolars and anterior teeth; the molars were excluded. The molar is a double-rooted tooth with complex anatomical root morphology, including root length, furcation area, and divergence of root and root trunk. Cervical enamel projections and enamel pearls also occur commonly in molars and are considered to be risk factors for periodontal disease; however, their occurrence varies among different individuals [55]. Additionally, molars have to withstand high occlusal forces, which can contribute to periodontal tissue destruction. Therefore, we excluded molars from representing oral examination sites in our analysis.

The periodontal disease index is used for assessing the periodontal status in epidemiological surveys; six target teeth (maxillary right 1stmolar, maxillary left central incisor, maxillary left 1st premolar, mandibular left 1st molar, mandibular left 1st molar, and mandibular right 1st premolar) are scored for the assessment of the disease. However, a sufficient justification for the selection of these teeth has not been provided [6]. For the index, evidence is indispensable.

In this study, optimal sites were selected based in the item information through the models presented in Table S1, and Table 1. Twelve sites: maxillary 2nd premolar (palatal-medial), 1st premolar (palatal-distal), 3 canine (palatal-medial), lateral incisor (palatal-central), central incisor (palatal-distal) and mandibular 1st premolar (lingual-medial) selected in this study were based on statistical modeling and represented the periodontal conditions. a full mouth examination is a best method; however, as time and labor are limited, partial examination may be applicable. Partial examination of these sites may be useful tool for epidemiological studies, mass screenings, and public health use.

There are several limitations in this study. The study population was consisted of the patients who experienced active periodontal treatment. The wider population is necessary to confirm the robustness of the model presented in this study. However, several partial-mouth assessments were not based on the statistical modeling. The strength of the model presented in this study was based on the IRT model, and weights for the site were calculated to improve the predictive values.

5. Conclusions

For calculating the periodontal index by partial oral examination, a justification for the selection of examination sites is necessary. This study presents an evidence-based partial examination method and its modeling. The 12 sites presented in this study almost equal to other partial examination protocol, which have more than 5 times the number of sampling sites.

Supplementary Materials: The attached supplementary materials are available online at http://www.mdpi.com/2077-0383/9/11/3754/s1, Figure S1: Item response curve and item information curves by the selected six values of the clinical attachment level, Figure S2: ROC curves by the selected six values, Figure S3: ROC curves for the diagnosis of periodontal disease by the selected site, Figure S4: ROC curves for clinical parameters by the selected site, Table S1: Constructed models, Table S2: Generalized linear model to predict the mean value of the clinical attachment level by the selected site

Author Contributions: Conceptualization, Y.N. (Yoshiaki Nomura) and T.M.; methodology, Y.N. (Yoshiaki Nomura); software, Y.N. (Yoshiaki Nomura); validation, E.K. and N.H.; formal analysis, Y.N. (Yoshiaki Nomura); investigation, T.M., M.F., H.K., M.M., T.N. (Taneaki Nakagawa), Y.N. (Yukihiro Numabe), F.N., K.N., Y.N. (Yohei Nakayama), Y.O., A.S., S.S. (Soh Sato), N.S., S.S. (Satoshi Sekino), T.S., F.S., K.T., H.T., M.U., H.Y., A.Y., N.Y. and T.N. (Toshiaki Nakamura); data curation, S.T.; writing—original draft preparation, Y.N. (Yoshiaki Nomura); writing—review and editing, Y.N. (Yoshiaki Nomura) and T.M.; visualization, Y.N. (Yoshiaki Nomura); supervision, T.M. and H.Y.; project administration, T.M., T.N. (Taneaki Nakagawa) and H.Y.; funding acquisition, T.N. (Taneaki Nakagawa) and H.Y. All authors have read and agreed to the published version of the manuscript.

Funding: This research was funded by a clinical research project grant from the Japanese Society of Periodontology for the diagnosis of periodontitis.

Acknowledgments: The authors thank Toshihide Noguchi, Masamitsu Kawanami, Koichi Ito, Yuichi Izumi, Yoshitaka Hara, Osamu Fujise, Yuzo Abe, Tomoo Kono, Asako Makino-Oi, and Chie Fukaya, for their advice and useful comments.

Conflicts of Interest: The authors declare no conflict of interest.

Appendix A

List of ethical committees and their approval numbers.

Institution	Name of the Ethics Committee	Reference Number
Niigata University	The regional ethical committee of the Faculty of Dentistry, Niigata University	20-R17-08-06
Keio University	Keio University School of Medicine, Ethics Committee	20080096
Hokkaido University	Institutional Review Board for Clinical Research of Hokkaido University Hospital	008-0113
Ohu University	Ohu University Research Ethics Committee	52

Institution	Name of the Ethics Committee	Reference Number
School of life Dentistry at Niigata, The Nippon Dental University	The Ethical Review Committee of The Nippon Dental University School of Life Dentistry at Niigata	151
Tokyo Dental College	Ethics Committee of Tokyo Dental College	208
Bunkyo-Dori Dental Clinic	The regional ethical committee of the Faculty of Dentistry, Niigata University	20-R17-08-06
Nihon University School of Dentistry at Matsudo	Ethics Committee in Nihon University School of Dentistry at Matsudo	EC 08-014
Tokyo Medical and Dental University	Dental Research Ethics Committee of Tokyo Medical and Dental University	660
Nihon University School of Dentistry	Ethical Committee of Nihon University School of Dentistry	EP08D016
School of Life Dentistry at Tokyo, The Nippon Dental University	The Institutional Review Board of Nippon Dental University.	2-1-22
Matsumoto Dental University	The Ethics Committee of Matsumoto Dental University	0090
Aichi Gakuin University	Ethics Committee: Aichi Gakuin University, School of Dentistry	158
Osaka Dental University	the Ethics Committee of Osaka Dental University	80712
Kyushu University	Ethical Committee of Kyushu University Faculty of Dental Science	20-11
Nagasaki University	The Ethics Committee, Nagasaki University Graduate School of Biomedical Sciences	0846-2
Kagoshima University	Ethical Committee of Kagoshima University Medical and Dental Hospital.	20-58

Appendix B

Quick reference for the calculation of ability by Model 4.

Maxilla					Mandibular	
5	4	3	2	1	4	
Palatal	Palatal	Palatal	Palatal	Palatal	Lingual	Ability
Medial	Distal	Medial	Central	Distal	Medial	
1	1	1	1	1	1	−0.932
1	1	1	1	1	2	−0.442
1	1	1	1	2	1	−0.223
1	1	1	1	2	2	0.052
1	1	1	1	3	2	0.159
1	1	1	2	1	1	0.16
1	1	2	1	1	1	−0.013
1	1	2	1	1	2	0.21
1	1	2	1	2	1	0.321

	Maxilla			Mandibular		
5	**4**	**3**	**2**	**1**	**4**	
Palatal	Palatal	Palatal	Palatal	Palatal	Lingual	Ability
Medial	Distal	Medial	Central	Distal	Medial	
1	1	3	1	1	1	0.108
1	1	3	3	2	2	1.315
1	2	1	1	1	1	−0.506
1	2	1	1	1	2	−0.156
1	2	1	1	1	3	−0.066
1	2	1	1	2	1	0.009
1	2	2	1	1	3	0.434
1	2	2	1	2	2	0.575
1	2	2	1	3	1	0.566
1	2	2	2	2	1	0.828
1	3	1	1	2	2	0.293
1	3	2	2	2	1	0.887
2	1	1	1	1	1	−0.306
2	1	1	1	1	2	−0.008
2	1	1	1	2	1	0.135
2	1	1	2	2	1	0.65
2	1	2	1	1	1	0.28
2	1	2	1	1	3	0.522
2	1	2	2	2	3	1.038
2	1	2	3	3	2	1.334
2	1	3	1	1	1	0.468
2	1	3	1	3	1	0.887
2	2	1	1	1	1	−0.053
2	2	1	1	1	2	0.176
2	2	1	1	1	3	0.262
2	2	1	1	2	2	0.449
2	2	1	1	2	3	0.53
2	2	1	2	1	1	0.542
2	2	2	1	1	1	0.411
2	2	2	1	1	2	0.549
2	2	2	1	2	2	0.727
2	2	2	1	3	2	0.847
2	2	2	2	2	2	1.027
2	2	2	2	3	3	1.228
2	2	2	3	2	1	1.121
2	2	3	1	1	1	0.613

Maxilla					Mandibular	Ability
5	**4**	**3**	**2**	**1**	**4**	
Palatal	Palatal	Palatal	Palatal	Palatal	Lingual	
Medial	Distal	Medial	Central	Distal	Medial	
2	2	3	2	2	2	1.229
2	2	3	2	2	3	1.31
2	2	3	2	3	2	1.373
2	3	1	1	1	1	0.023
2	3	1	1	1	3	0.339
2	3	2	1	2	1	0.68
2	3	2	1	2	2	0.785
2	3	2	1	3	3	0.991
2	3	2	2	2	2	1.083
2	3	2	2	3	2	1.213
2	3	3	1	2	3	1.081
2	3	3	1	3	1	1.048
2	3	3	2	1	1	1.097
2	3	3	2	2	3	1.38
2	3	3	3	2	3	1.701
3	1	1	1	2	3	0.61
3	1	1	1	3	2	0.663
3	1	3	3	3	3	1.979
3	2	1	1	1	2	0.393
3	2	2	1	2	1	0.773
3	2	2	1	2	2	0.868
3	2	2	1	3	2	0.995
3	2	2	2	3	1	1.209
3	2	2	3	3	3	1.684
3	2	3	2	1	1	1.181
3	2	3	2	3	2	1.521
3	2	3	3	3	1	1.801
3	2	3	3	3	3	2.042
3	3	1	1	1	2	0.468
3	3	2	1	2	2	0.927
3	3	2	2	1	3	1.185
3	3	2	2	2	1	1.137
3	3	2	2	3	3	1.442
3	3	2	3	2	2	1.475
3	3	2	3	3	2	1.664
3	3	3	1	2	3	1.243
3	3	3	1	3	1	1.222

Maxilla					Mandibular	
5	**4**	**3**	**2**	**1**	**4**	
Palatal	Palatal	Palatal	Palatal	Palatal	Lingual	**Ability**
Medial	Distal	Medial	Central	Distal	Medial	
3	3	3	2	1	3	1.439
3	3	3	2	2	1	1.362
3	3	3	2	2	3	1.53
3	3	3	3	2	3	1.9
3	3	3	3	3	2	2.012
3	3	3	3	3	3	2.228

1: CAL <4mm; 2: CAL = 4–6mm; 3: CAL >6mm.

References

1. Nomura, Y.; Morozumi, T.; Nakagawa, T.; Sugaya, T.; Kawanami, M.; Suzuki, F.; Takahashi, K.; Abe, Y.; Sato, S.; Makino-Oi, A.; et al. Site-level progression of periodontal disease during a follow-up period. *PLoS ONE* **2017**, *12*, e0188670. [CrossRef] [PubMed]
2. Socransky, S.S.; Haffajee, A.D. The nature of periodontal diseases. *Ann. Periodontol.* **1997**, *2*, 3–10. [CrossRef] [PubMed]
3. Sharma, P.; Cockwell, P.; Dietrich, T.; Ferro, C.; Ives, N.; Chapple, I.L.C. Influence of successful periodontal intervention in renal disease (INSPIRED): Study protocol for a randomized controlled pilot clinical trial. *Trials* **2017**, *18*, 535. [CrossRef] [PubMed]
4. Cho, Y.; Kim, H.Y. Analysis of periodontal data using mixed effects models. *J. Periodontal. Implant. Sci.* **2015**, *45*, 2–7. [CrossRef] [PubMed]
5. Botelho, J.; Machado, V.; Proença, L.; Mendes, J.J. The 2018 periodontitis case definition improves accuracy performance of full-mouth partial diagnostic protocols. *Sci. Rep.* **2020**, *10*, 7093. [CrossRef] [PubMed]
6. Ramfjord, S.P. The Periodontal Disease Index (PDI). *J. Periodontol.* **1967**, *38*, S602–S610. [CrossRef]
7. Wang, W.J.; Liu, C.Y.; Liu, D.Z.; Lee, C.J. Survey of periodontal disease among workers in Tianjin using Ramfjord's Periodontal Disease Index (PDI). *Community Dent. Oral Epidemiol.* **1987**, *15*, 98–99. [CrossRef] [PubMed]
8. Cutress, T.W.; Hunter, P.B.; Hoskins, D.I. Comparison of the Periodontal Index (PI) and Community Periodontal Index of Treatment Needs (CPITN). *Community Dent. Oral Epidemiol.* **1986**, *14*, 39–42. [CrossRef]
9. Ainamo, J.; Barmes, D.; Beagrie, G.; Cutress, T.; Martin, J.; Infirri, S.J. Development of the World Health Organization (WHO) community periodontal index of treatment needs (CPITN). *Int. Dent. J.* **1982**, *32*, 281–291.
10. Dunning, J.M.; Leach, L.B. Gingival-bone count: A method for epidemiological study of periodontal disease. *J. Dent. Res.* **1960**, *39*, 506–513. [CrossRef]
11. Massler, M. The P-M-A index for the assessment of gingivitis. *J. Periodontol.* **1967**, *38*, S592–S601. [CrossRef] [PubMed]
12. Löe, H. The Gingival Index, the Plaque Index and the Retention Index Systems. *J. Periodontol.* **1967**, *38*, S610–S616. [CrossRef]
13. Edelen, M.O.; Reeve, B.B. Applying item response theory (IRT) modeling to questionnaire development; evaluation; and refinement. *Qual. Life Res.* **2007**, *16*, 5–18. [CrossRef] [PubMed]
14. Thomas, M.L. Advances in applications of item response theory to clinical assessment. *Psychol. Assess.* **2019**, *12*, 1442–1455. [CrossRef]
15. Reise, S.P.; Waller, N.G. Item response theory and clinical measurement. *Annu. Rev. Clin. Psychol.* **2009**, *5*, 27–48. [CrossRef]
16. Nomura, Y.; Otsuka, R.; Wint, W.Y.; Okada, A.; Hasegawa, R.; Hanada, N. Tooth-Level Analysis of Dental Caries in Primary Dentition in Myanmar Children. *Int. J. Environ. Res. Public Health* **2020**, *17*, 7613. [CrossRef]

17. Nomura, Y.; Maung, K.; Khine, E.M.K.; Sint, K.M.; Lin, M.P.; Myint, M.K.W.; Aung, T.; Sogabe, K.; Otsuka, R.; Okada, A.; et al. Prevalence of dental caries in 5- and 6-year-old Myanmar children. *Int. J. Dent.* **2019**, *2019*, 5948379. [CrossRef]

18. Donovan, N.J.; Rosenbek, J.C.; Ketterson, T.U.; Velozo, C.A. Adding meaning to measurement: Initial rasch analysis of the ASHA FACS social communication subtest. *Aphasiology* **2006**, *20*, 362–373. [CrossRef]

19. Cordier, R.; Joosten, A.; Clavé, P.; Schindler, A.; Bülow, M.; Demir, N.; Arslan, S.S.; Speyer, R. Evaluating the psychometric properties of the eating assessment tool (EAT-10) using rasch analysis. *Dysphagia* **2017**, *32*, 250–260. [CrossRef]

20. Prenovost, K.M.; Fihn, S.D.; Maciejewski, M.L.; Nelson, K.; Vijan, S.; Rosland, A.M. Using item response theory with health system data to identify latent groups of patients with multiple health conditions. *PLoS ONE* **2018**, *13*, e0206915. [CrossRef]

21. Morozumi, T.; Nakagawa, T.; Nomura, Y.; Sugaya, T.; Kawanami, M.; Suzuki, F.; Takahashi, K.; Abe, Y.; Sato, S.; Makino-Oi, A.; et al. Salivary pathogen and serum antibody to assess the progression of chronic periodontitis: A 24-mo prospective multicenter cohort study. *J. Periodontal. Res.* **2016**, *51*, 768–778. [CrossRef] [PubMed]

22. Kakuta, E.; Nomura, Y.; Morozumi, T.; Nakagawa, T.; Nakamura, T.; Noguchi, K.; Yoshimura, A.; Hara, Y.; Fujise, O.; Nishimura, F.; et al. Assessing the progression of chronic periodontitis using subgingival pathogen levels: A 24-month prospective multicenter cohort study. *BMC Oral Health* **2017**, *17*, 46. [CrossRef] [PubMed]

23. Krebs, K.A.; Clem, D.S., 3rd. American Academy of Periodontology, Guidelines for the management of patients with periodontal diseases. *J. Periodontol.* **2006**, *77*, 1607–1611. [CrossRef] [PubMed]

24. Eke, P.I.; Page, R.C.; Wei, L.; Evans, G.T.; Genco, R.J. Update of the case definitions for population-based surveillance of periodontitis. *J. Periodontol.* **2012**, *83*, 1449–1454. [CrossRef]

25. Yang, R.; Donaldson, G.W.; Edelman, L.S.; Cloyes, K.G.; Sanders, N.A.; Pepper, G.A. Fear of older adult falling questionnaire for caregivers (FOAFQ-CG): Evidence from content validity and item-response theory graded-response modelling. *J. Adv. Nurs.* **2020**, *76*, 2768–2780. [CrossRef]

26. Khazaal, Y.; Breivik, K.; Billieux, J.; Zullino, D.; Thorens, G.; Achab, S.; Gmel, G.; Chatton, A. Game addiction scale assessment through a nationally representative sample of young adult men: Item response heory graded-response modeling. *J. Med. Internet Res.* **2018**, *20*, e10058. [CrossRef]

27. Rizopoulos, D. An R package for latent variable modelling and item response theory analyses. *J. Stat. Softw.* **2006**, *17*, 1–25. [CrossRef]

28. Molenaar, D.; Dolan, C.V.; Boeck, P.D. The Heteroscedastic graded response model with a skewed latent trait: Testing statistical and substantive hypotheses related to skewed item category functions. *Psychometrika* **2012**, *77*, 455–478. [CrossRef]

29. Kang, H.A.; Su, Y.H.; Chang, H.H. A note on monotonicity of item response functions for ordered polytomous item response theory models. *Br. J. Math. Stat. Psychol.* **2018**, *71*, 523–535. [CrossRef]

30. Nomura, Y.; Okada, A.; Kakuta, E.; Otsuka, R.; Saito, H.; Maekawa, H.; Daikoku, H.; Hanada, N.; Sato, T. Workforce and Contents of Home Dental Care in Japanese Insurance System. *Int. J. Dent.* **2020**, *2020*, 7316796. [CrossRef]

31. Nomura, Y.; Kakuta, E.; Okada, A.; Yamamoto, Y.; Tomonari, H.; Hosoya, N.; Hanada, N.; Yoshida, N.; Takei, N. Prioritization of the Skills to Be Mastered for the Daily Jobs of Japanese Dental Hygienists. *Int. J. Dent.* **2020**, *2020*, 4297646. [CrossRef] [PubMed]

32. Nomura, Y.; Kakuta, E.; Okada, A.; Otsuka, R.; Shimada, M.; Tomizawa, Y.; Taguchi, C.; Arikawa, K.; Daikoku, H.; Sato, T.; et al. Effects of self-assessed chewing ability, tooth loss and serum albumin on mortality in 80-year-old individuals: A 20-year follow-up study. *BMC Oral Health.* **2020**, *20*, 122. [CrossRef] [PubMed]

33. Nomura, Y.; Tsutsumi, I.; Nagasaki, M.; Tsuda, H.; Koga, F.; Kashima, N.; Uraguchi, M.; Okada, A.; Kakuta, E.; Hanada, N. Supplied Food Consistency and Oral Functions of Institutionalized Elderly. *Int. J. Dent.* **2020**, *2020*, 3463056. [CrossRef] [PubMed]

34. Nomura, Y.; Matsuyama, T.; Fukai, K.; Okada, A.; Ida, M.; Yamauchi, N.; Hanamura, H.; Yabuki, Y.; Watanabe, K.; Sugawara, M.; et al. PRECEDE-PROCEED model based questionnaire and saliva tests for oral health checkup in adult. *J. Oral Sci.* **2019**, *61*, 544–548. [CrossRef] [PubMed]

35. Zheng, B.; Agresti, A. Summarizing the predictive power of a generalized linear model. *Stat. Med.* **2000**, *19*, 1771–1781. [CrossRef]

36. Nomura, Y.; Okada, A.; Kakuta, E.; Gunji, T.; Kajiura, S.; Hanada, N. A new screening method for periodontitis: An alternative to the community periodontal index. *BMC Oral Health* **2016**, *16*, 64. [CrossRef]
37. Kudo, C.; Naruishi, K.; Maeda, H.; Abiko, Y.; Hino, T.; Iwata, M.; Mitsuhashi, C.; Murakami, S.; Nagasawa, T.; Nagata, T.; et al. Assessment of use of plasma/serum IgG test to screen for periodontitis. *J. Dent. Res.* **2012**, *91*, 1190–1195. [CrossRef]
38. Kingman, A.; Susin, C.; Albandar, J.M. Effect of partial recording protocols on severity estimates of periodontal disease. *J. Clin. Periodontol.* **2008**, *35*, 659–667. [CrossRef]
39. Susin, C.; Kingman, A.; Albandar, J.M. Effect of partial recording protocols on estimates of prevalence of periodontal disease. *J. Periodontol.* **2005**, *76*, 262–267. [CrossRef]
40. Tran, D.T.; Gay, I.; Du, X.L.; Fu, Y.; Bebermeyer, R.D.; Neumann, A.S.; Streckfus, C.; Chan, W.; Walji, M.F. Assessing periodontitis in populations: A systematic review of the validity of partial-mouth examination protocols. *J. Clin. Periodontol.* **2013**, *40*, 1064–1071. [CrossRef]
41. World Health Organization. *Oral Health Surveys Basic Methods*, 4th ed.; World Health Organization: Geneva, Switzerland, 1997; p. 39.
42. World Health Organization. *Oral Health Surveys Basic Methods*, 5th ed.; World Health Organization: Geneva, Switzerland, 2013; pp. 47–49.
43. Ministry of Health, Labour and Welfare Japan. Survey of Dental Diseases. Available online: https://www.mhlw.go.jp/toukei/list/62-28.html (accessed on 1 September 2020).
44. Okada, A.; Nomura, Y.; Sogabe, K.; Oku, H.; Gillbreath, S.A.; Hino, F.; Hayashi, H.; Yoshino, H.; Utsunomiya, H.; Suzuki, K.; et al. Comparison of salivary hemoglobin measurements for periodontitis screening. *J. Oral Sci.* **2017**, *59*, 63–69. [CrossRef] [PubMed]
45. Kakuta, E.; Nomura, Y.; Naono, Y.; Koresawa, K.; Shimizu, K.; Hanada, N. Correlation between health-care costs and salivary tests. *Int. Dent. J.* **2013**, *63*, 249–253. [CrossRef] [PubMed]
46. Nomura, Y.; Shimada, Y.; Hanada, N.; Numabe, Y.; Kamoi, K.; Sato, T.; Gomi, K.; Arai, T.; Inagaki, K.; Fukuda, M.; et al. Salivary biomarkers for predicting the progression of chronic periodontitis. *Arch. Oral Biol.* **2012**, *57*, 413–420. [CrossRef] [PubMed]
47. Nomura, Y.; Okada, A.; Tamaki, Y.; Miura, H. Salivary Levels of Hemoglobin for Screening Periodontal Disease: A Systematic Review. *Int. J. Dent.* **2018**, *2018*, 2541204. [CrossRef]
48. Nomura, Y.; Tamaki, Y.; Tanaka, T.; Arakawa, H.; Tsurumoto, A.; Kirimura, K.; Sato, T.; Hanada, N.; Kamoi, K. Screening of periodontitis with salivary enzyme tests. *J. Oral Sci.* **2006**, *48*, 177–183. [CrossRef]
49. Chatzopoulos, G.S.; Tsalikis, L.; Konstantinidis, A.; Kotsakis, G.A. A two-domain self-report measure of periodontal disease has good accuracy for periodontitis screening in dental school outpatients. *J. Periodontol.* **2016**, *87*, 1165–1173. [CrossRef]
50. Wright, N.M.; Cheng, B.; Tafreshi, S.N.; Lamster, I.B. A simple self-report health assessment questionnaire to identify oral diseases. *Int. Dent. J.* **2018**, *68*, 428–432. [CrossRef]
51. Preianò, M.; Savino, R.; Villella, C.; Pelaia, C.; Terracciano, R. Gingival crevicular fluid peptidome profiling in healthy and in periodontal diseases. *Int. J. Mol. Sci.* **2020**, *21*, 5270. [CrossRef]
52. Kingman, A.; Albandar, J.M. Methodological aspects of epidemiological studies of periodontal diseases. *Periodontology* **2002**, *29*, 11–30. [CrossRef]
53. Vettore, M.V.; de Lamarca, G.A.; Leao, A.T.; Sheiham, A.; do Leal, M.C. Partial recording protocols for periodontal disease assessment in epidemiological surveys. *Cademos Saude Publica* **2007**, *23*, 33–42. [CrossRef]
54. Benigeri, M.; Brodeur, J.M.; Payette, M.; Charbonneau, A.; Ismail, A.I. Community periodontal index of treatment needs and prevalence of periodontal conditions. *J. Clin. Periodontol.* **2000**, *27*, 308–312. [CrossRef] [PubMed]
55. Newman, M.G.; Takei, H.H.; Klokkevod, P.R.; Carranza, F.A. Clinical risk assessment. In *Carranza's Clinical Periodontology*; Elsevier Saunders: Philadelphia, PA, USA, 2012; pp. 370–372.

Publisher's Note: MDPI stays neutral with regard to jurisdictional claims in published maps and institutional affiliations.

Article

Estimation of the Periodontal Inflamed Surface Area by Simple Oral Examination

Yoshiaki Nomura [1], Toshiya Morozumi [2,*], Yukihiro Numabe [3], Yorimasa Ogata [4], Yohei Nakayama [4], Tsutomu Sugaya [5], Toshiaki Nakamura [6], Soh Sato [7], Shogo Takashiba [8], Satoshi Sekino [3], Nobuo Yoshinari [9], Nobuhiro Hanada [1], Naoyuki Sugano [10], Mitsuo Fukuda [11], Masato Minabe [2], Makoto Umeda [12], Koichi Tabeta [13], Keiso Takahashi [14], Kazuyuki Noguchi [6], Hiroaki Kobayashi [15], Hideki Takai [4], Fusanori Nishimura [16], Fumihiko Suzuki [17], Erika Kakuta [18], Atsutoshi Yoshimura [19], Atsushi Saito [20] and Taneaki Nakagawa [21]

1 Department of Translational Research, Tsurumi University School of Dental Medicine, Yokohama 230-8501, Japan; nomura-y@tsurumi-u.ac.jp (Y.N.); hanada-n@tsurumi-u.ac.jp (N.H.)
2 Division of Periodontology, Department of Oral Interdisciplinary Medicine, Graduate School of Dentistry, Kanagawa Dental University, Yokosuka 238-8580, Japan; minabe@kdu.ac.jp
3 Department of Periodontology, School of Life Dentistry at Tokyo, The Nippon Dental University, Tokyo 102-8159, Japan; numabe-y@tky.ndu.ac.jp (Y.N.); sekino-s@tky.ndu.ac.jp (S.S.)
4 Department of Periodontology, Nihon University School of Dentistry at Matsudo, Matsudo 271-8587, Japan; nakayama.youhei@nihon-u.ac.jp (Y.N.); ogata.yorimasa@nihon-u.ac.jp (Y.O.); takai.hideki@nihon-u.ac.jp (H.T.)
5 Division of Periodontology and Endodontology, Department of Oral Health Science, Hokkaido University Graduate School of Dental Medicine, Sapporo 060-8586, Japan; sugaya@den.hokudai.ac.jp
6 Department of Periodontology, Kagoshima University Graduate School of Medical and Dental Sciences, Kagoshima 890-8544, Japan; toshi-n@dent.kagoshima-u.ac.jp (T.N.); kazuperi@dent.kagoshima-u.ac.jp (K.N.)
7 Department of Periodontology, School of Life Dentistry at Niigata, The Nippon Dental University, Niigata 951-8580, Japan; s-sato@ngt.ndu.ac.jp
8 Department of Pathophysiology-Periodontal Science, Okayama University Graduate School of Medicine, Dentistry and Pharmaceutical Sciences, Okayama 700-8525, Japan; stakashi@okayama-u.ac.jp
9 Department of Periodontology, School of Dentistry, Matsumoto Dental University, Shiojiri 399-0781, Japan; nobuo.yoshinari@mdu.ac.jp
10 Department of Periodontology, Nihon University School of Dentistry, Tokyo 101-8310, Japan; sugano.naoyuki@nihon-u.ac.jp
11 Department of Periodontology, School of Dentistry, Aichi Gakuin University, Nagoya 464-8650, Japan; fukuda-m@dpc.agu.ac.jp
12 Department of Periodontology, Osaka Dental University, Hirakata 573-1121, Japan; umeda-m@cc.osaka-dent.ac.jp
13 Division of Periodontology, Department of Oral Biological Science, Niigata University Graduate School of Medical and Dental Sciences, Niigata 951-8514, Japan; koichi@dent.niigata-u.ac.jp
14 Division of Periodontics, Department of Conservative Dentistry, Ohu University School of Dentistry, Koriyama 963-8611, Japan; ke-takahashi@den.ohu-u.ac.jp
15 Department of Periodontology, Graduate School of Medical and Dental Sciences, Tokyo Medical and Dental University, Tokyo 113-8510, Japan; h-kobayashi.peri@tmd.ac.jp
16 Section of Periodontology, Division of Oral Rehabilitation, Faculty of Dental Science, Kyushu University, Fukuoka 812-8582, Japan; fusanori@dent.kyushu-u.ac.jp
17 Division of Dental Anesthesiology, Department of Oral Surgery, Ohu University School of Dentistry, Koriyama 963-8611, Japan; f-suzuki@den.ohu-u.ac.jp
18 Department of Oral Microbiology, Tsurumi University School of Dental Medicine, Yokohama 230-8501, Japan; kakuta-erika@tsurumi-u.ac.jp
19 Department of Periodontology and Endodontology, Nagasaki University Graduate School of Biomedical Sciences, Nagasaki 852-8588, Japan; ayoshi@nagasaki-u.ac.jp
20 Department of Periodontology, Tokyo Dental College, Tokyo 101-0061, Japan; atsaito@tdc.ac.jp
21 Department of Dentistry and Oral Surgery, School of Medicine, Keio University, Tokyo 160-8582, Japan; tane@z6.keio.jp
* Correspondence: morozumi@kdu.ac.jp; Tel.: +81-46-822-8855

check for **updates**

Citation: Nomura, Y.; Morozumi, T.; Numabe, Y.; Ogata, Y.; Nakayama, Y.; Sugaya, T.; Nakamura, T.; Sato, S.; Takashiba, S.; Sekino, S.; et al. Estimation of the Periodontal Inflamed Surface Area by Simple Oral Examination. *J. Clin. Med.* **2021**, *10*, 723. https://doi.org/10.3390/jcm10040723

Academic Editor: Camile S. Farah
Received: 5 January 2021
Accepted: 8 February 2021
Published: 12 February 2021

Publisher's Note: MDPI stays neutral with regard to jurisdictional claims in published maps and institutional affiliations.

Abstract: The periodontal inflamed surface area (PISA) is a useful index for clinical and epidemiological assessments, since it can represent the inflammation status of patients in one contentious variable. However, calculation of the PISA is difficult, requiring six point probing depth measurements with or without bleeding on probing on 28 teeth, followed by data input in a calculation program. More

117

simple methods are essential for screening periodontal disease or in epidemiological studies. In this study, we tried to establish a convenient partial examination method to estimate PISA. Cross-sectional data of 254 subjects who completed active periodontal therapy were analyzed. Teeth that represent the PISA value were selected by an item response theory approach. The maxillary second molar, first premolar, and lateral incisor and the mandibular second molar and lateral incisor were selected. The sum of the PISAs of these teeth was significantly correlated with the patient's PISA (R^2 = 0.938). More simply, the sum of the maximum values of probing pocket depth with bleeding for these teeth were also significantly correlated with the patient's PISA (R^2 = 0.6457). The simple model presented in this study may be useful to estimate PISA.

Keywords: periodontal diseases; periodontal pocket; health status indicators; periodontal inflamed surface area (PISA); item response theory

1. Introduction

Several periodontal indexes have been proposed and used for clinical or epidemiological assessments. The indexes used in epidemiological studies include the commonality periodontal index (CPI) [1–3], clinical attachment level (CAL) [4,5], and bleeding on probing (BOP) [6], while other studies used combinations of indexes, such as CPI, CAL, BOP, and probing pocket depth (PD) [7], CAL, PD, gingival index (GI), and plaque index (PlI) [8], CAL, PD, and GI [9], and CAL and BOP [10]. For clinical assessments, PD, CAL, and BOP have been commonly used for evaluation of diagnoses or therapeutic effects. Although summary statistics of these parameters, like the mean or maximum value, have been conventionally used, these summary statistics involve loss of information. The periodontal inflamed surface area (PISA) was developed to address these issues [11]. PISA is a very convenient index that reflects the surface area of bleeding pocket epithelium in square millimeters and is calculated using conventional clinical parameters of periodontal health, namely BOP combined with either PD, or CAL and gingival recession [11]. Thus, PISA can represent the inflammation status of subjects in one contentious variable, for which calculation of mean or maximum value is not necessary. Recent study have reported that PISA is effectively associated with systemic markers of low-grade inflammation, such as C-reactive protein [12]. Periodontal disease is a risk factor for non-communicable diseases through the localized inflammation and periodontal pathogens [13–20]. Therefore, PISA may be an optimal index to investigate the correlation between periodontal disease and non-communicable diseases.

However, calculation of PISA is an extremely difficult task, requiring six probing depth measurements of all the teeth and BOP and PD data for 168 sites. A simpler method is indispensable for screening or epidemiological studies. Therefore, in this study, we tried to establish a convenient partial examination method to estimate PISA. Our simple, convenient, and evidence-based partial oral examination method to estimate PISA may be a useful tool like other periodontal indexes for screening or epidemiological studies.

2. Materials and Methods

2.1. Study Design

2.1.1. Setting

This study was a part of a clinical research project conducted by the Japanese Society of Periodontology in cooperation with 17 facilities (one clinic and 16 university hospitals) in Japan for the diagnosis of periodontitis. In our previous reports, we had analyzed 124 participants who successfully completed the study protocol [21–23]. For this study, we selected 254 patients with chronic periodontitis who had completed their active treatment under the regulations of the Japanese health insurance system. After registering for a screening examination before their follow-up, all 254 patients were analyzed in the study.

The inclusion criteria were age greater than 30 years, number of remaining teeth more than 20, and systemically healthy status.

2.1.2. Diagnosis

Each patient was diagnosed according to the guidelines used at the time (Guidelines of the American Academy of Periodontology) [24]. Oral examinations were carried out by one examiner at each institute (T.M., Y.N., T.S., T.N., S.S., S.S., N.Y., N.S., M.F., M.M., K.N., H.K., H.T., F.S., A.Y., and T.N.). Each examiner was a periodontist licensed by the Japanese Society of Periodontology. Intra- and interexaminer calibration sessions were conducted using periodontal disease models (P15FE-500HPRO-S2A1-GSF, NISSIN, Kyoto, Japan) at the beginning and middle of the study period. In brief, full-mouth PD and recessions were measured twice, and repeatability for CAL was assessed. The examiner was judged to have made reproducible measurements after reaching a percentage of agreement within ± 1 mm between repeated measurements of at least 95% of measurements.

2.2. Research Data

In this study, we analyzed PD and BOP data for calculation of PISA. A freely downloadable spreadsheet is available to calculate the PISA [11], and CAL and PlI were used as associated factors for PISA. Details of the data are described in our previous reports [21–23].

2.3. Statistical Analysis

The number of bleeding sites and PDs were measured at six sites for all of the remaining teeth with a periodontal probe (CP-12 Color-Coded Probe; Hu-Friedy, Chicago, IL, USA). To find out the optimal value for IRT analysis, 6 kinds of cut-off points were set for number of bleeding site per teeth: (0, 1–6), (0 and 1, 2–6), (0–2, 3–6), (0–3, 4–6), (0–4, 5 and 6), and (0–5, 6). For each cut-off point, models were constructed. For the PD, the maximum value (mm) for one tooth in the six sites were used.

By using the dichotomized data, a three-parameter logistic model based on the item response theory (IRT) was applied. For the IRT analysis, R software with the ltm and irtoys package was used.

2.4. Ethical Approval

The study was conducted in compliance with the principles outlined in the Helsinki Declaration. Written informed consent was obtained from each participant, and the protocol was approved by the Institutional Review Board of each participating institution. The ethics committee members' names and reference numbers are as follows: The regional ethical committee of the Faculty of Dentistry, Niigata University (20-R17-08-06); Keio University School of Medicine, Ethics Committee (20080096); Ethical Committee of Kagoshima University Medical and Dental Hospital (20-58); The Ethics Committee, Nagasaki University Graduate School of Biomedical Sciences (0846-2); Ethical Committee of Kyushu University Faculty of Dental Science (20-11); The Ethics Committee of Osaka Dental University (80712); Ethics Committee of Aichi Gakuin University, School of Dentistry (158); The Ethics Committee of Matsumoto Dental University (0090); The Institutional Review Board of Nippon Dental University (2-1-22); Ethical Committee of Nihon University School of Dentistry (EP08D016); Dental Research Ethics Committee of Tokyo Medical and Dental University (660); Ethics Committee in Nihon University School of Dentistry at Matsudo (EC 08-014); Ethics Committee of Tokyo Dental College (208); The Ethical Review Committee of The Nippon Dental University School of Life Dentistry at Niigata (151); Ohu University Research Ethics Committee (52); Institutional Review Board for Clinical Research of Hokkaido University Hospital (008-0113).

3. Results

3.1. Participant Characteristics

The study population consisted of 114 men and 140 women. Their mean age was 55.6 +/− 10.3 years, and their mean number of remaining teeth was 25.4 +/− 2.61.

3.2. Prediction of PISA by IRT Analysis Based on the Number of BOP Sites

Since PISA is calculated on the basis of BOP and PD measurements, to reduce the number of BOP measurements, IRT analysis was performed. Cut-off points were set as at least 1, 2, 3, and 4 sites for BOP. The ability, which indicates the weighted sum of the total number of bleeding sites, was calculated for each participant. Scatter plots were obtained for the PISA against the ability calculated by IRT analysis. After $\log_{10}$ transformation of PISA, linear relationships were observed. The results are shown in Figure 1, and the models are shown in Table S1. Item response curves and item information curves are shown in Figure S1.

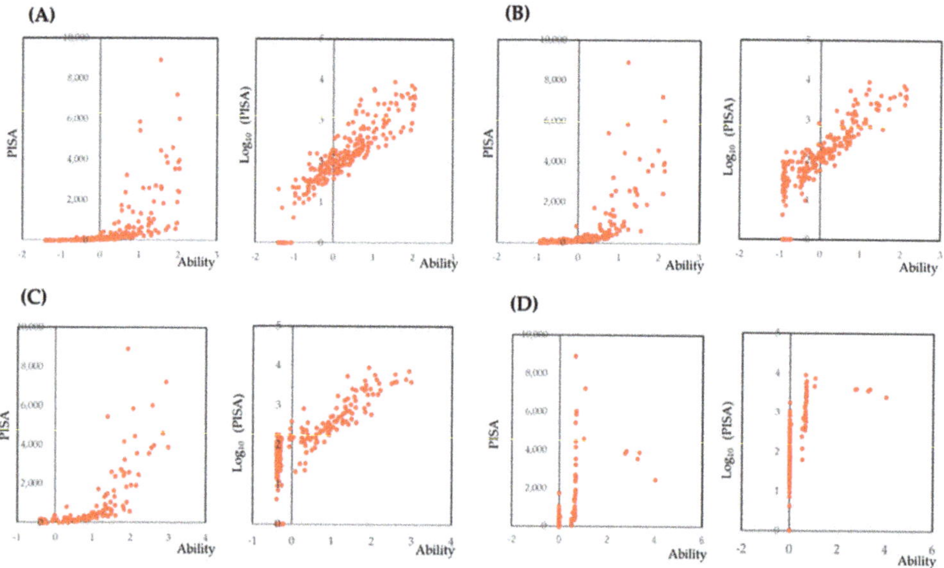

Figure 1. Scatter plots of PISA against the number of bleeding sites within each tooth. (**A–D**) Scheme 1. 2, 3, and 4 bleeding sites within each tooth. Ability, which indicates the weighted sum of the number of sites with bleeding on probing, was calculated by the three-parameter logistic model based on item response theory analysis. When PISA was $\log_{10}$ transformed, linear relationships were observed.

The results indicated that at least one site with bleeding for each tooth may be enough for evaluation of bleeding on probing.

3.3. Prediction of PISA by the Maximum Value of the PD at Each Tooth by IRT Analysis

In clinical settings, even the calculation of mean values of PD may be laborious. Thus, assessments based on the maximum value are simpler and more suitable for rapid evaluations. Therefore, the maximum PD value for each tooth was used as the variable. To investigate which site represents the PISA, a three-parameter logistic model based on IRT was used. To transform the maximum values for dichotomous variable, cut-off values were set as >3 mm, >4 mm, 5 mm, and 6 mm. The results are shown in Figure 2, the models are indicated in Table S2, and the item response curves and item information curves are shown in Figure S2.

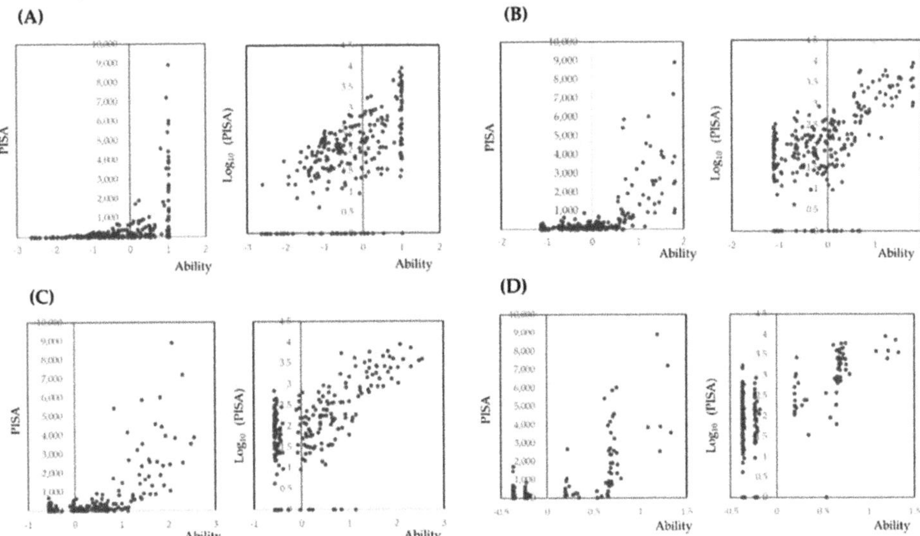

Figure 2. Scatter plots of PISA against the maximum value of PD within each teeth. (**A–D**) shows the findings with cut-off points of 3, 4, 5, and 6 mm of PD within one tooth. When PISA was $\log_{10}$ transformed, linear relationships were observed.

3.4. Prediction of PISA Based on the Selected Sites

The item information curves in Figure S1 and S2 provided valuable information for tooth selection. The area under the information curves for each tooth were calculated, and the results were presented by the mean values for left and right teeth (Table 1). Teeth with relatively higher information were the maxillary second molar, maxillary first premolar, maxillary lateral incisor, mandibular second molar, and mandibular lateral incisors. To confirm that the selected teeth were optimal, scatterplots of the overall PISA against the PISAs of these teeth were illustrated. The results are presented in Figure S3. R^2 values of these teeth were relatively higher.

Table 1. Item information for tooth type for the number of bleeding sites and the maximum value of probing depth.

		Bleeding on Probing			Maximum Value of Probing Depth		
		1 Site	2 Sites	3 Sites	3 Mm	4 mm	5 mm
Maxillary	2nd Molar	27.0	35.9	128.7	22.4	29.9	32.1
	1st Molar	25.4	23.3	35.9	24.0	35.9	43.8
	2nd Premolar	30.7	32.3	47.0	35.1	41.0	47.8
	1st Premolar	33.7	39.9	53.8	30.5	40.1	43.7
	Canine	28.5	35.1	48.0	32.2	43.5	429.5
	Lateral incisor	27.8	48.4	71.7	30.8	44.8	49.8
	Central incisor	30.7	32.1	42.4	28.7	38.5	41.4
Mandibular	2nd Molar	28.0	27.3	36.6	24.6	34.5	30.5
	1st Molar	27.8	27.6	39.1	15.0	37.7	47.6
	2nd Premolar	26.7	27.2	32.8	32.3	34.6	31.7
	1st Premolar	32.6	27.9	33.6	31.1	43.9	39.0
	Canine	24.8	30.3	35.1	35.3	47.7	51.6
	Lateral incisor	32.0	32.6	32.6	43.1	48.3	41.3
	Central incisor	36.8	32.7	34.8	42.8	46.2	35.5

3.5. Prediction of PISA Based on the Selected Teeth

3.5.1. Correlation of PISA of Selected Teeth with PISA

On the basis of the data presented in Table 1 and Figure S2, five teeth were selected: the maxillary second molar, maxillary first molar, maxillary lateral incisor, mandibular second molar, and mandibular lateral incisor. The scatter plot for the PISA against the sum of PISAs of these 10 teeth are provided in Figure 3. Since linear regression is affected by outliers, PISA less than 1000 is illustrated again in Figure 3B. The R^2 was 0.938 for all the data and 0.817 for PISA less than 1000.

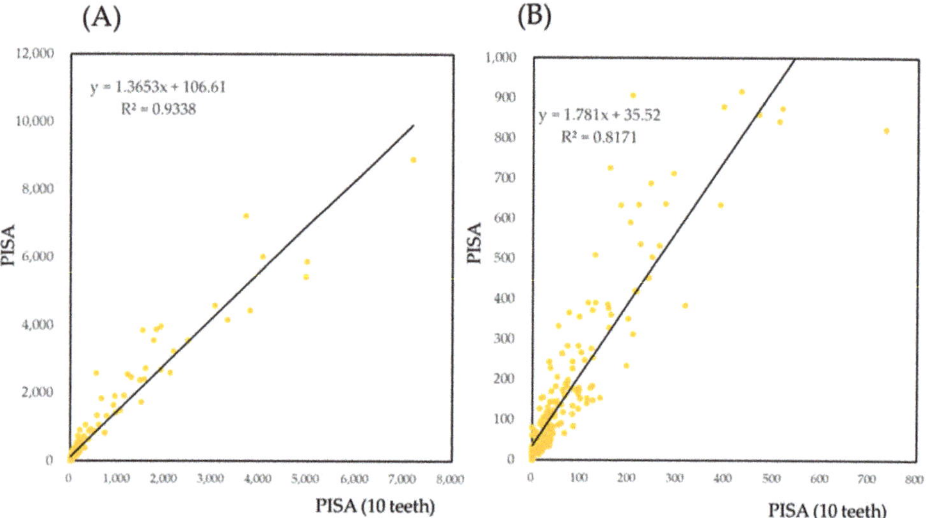

Figure 3. Scatter plot of PISA against the maximum value of PD in each tooth. (**A,B**) are the same figure. (**B**) is magnified PISA for less than 1000. Regression line (**B**) was calculated by PISA less than 1000.

3.5.2. Prediction of PISA by the Maximum Value of PD

To create a simple model for clinical convenience, the sum of the maximum PD values with BOP were used to generate a summary score to predict PISA. The scatter plot with the net values showed an exponential curve (Figure 4A). When the values were $\log_{10}$ transformed, a linear relationship was observed. The PISA against a value of 0 on the X-axis was less than 2. This indicated that when the 10 teeth did not show any bleeding, PISA is less than 100.

The effect of missing teeth is uncertain. In addition, a selected site of 0 may not guarantee a PISA value of 0. Figure 4 was categorized by the groups; Group 0: no missing teeth and no bleeding in 10 teeth, Group 1: one missing teeth and no bleeding in nine teeth, Group 2: two missing teeth and no bleeding in eight teeth, Group 3: three missing teeth and no bleeding in seven teeth, Group 10: one missing teeth in 10 teeth, Group 11: one missing teeth in 10 teeth, Group 12: two missing teeth in 10 teeth, Group 13: three missing teeth in ten teeth, Group 14: four or five missing teeth in ten teeth (Figure S4).

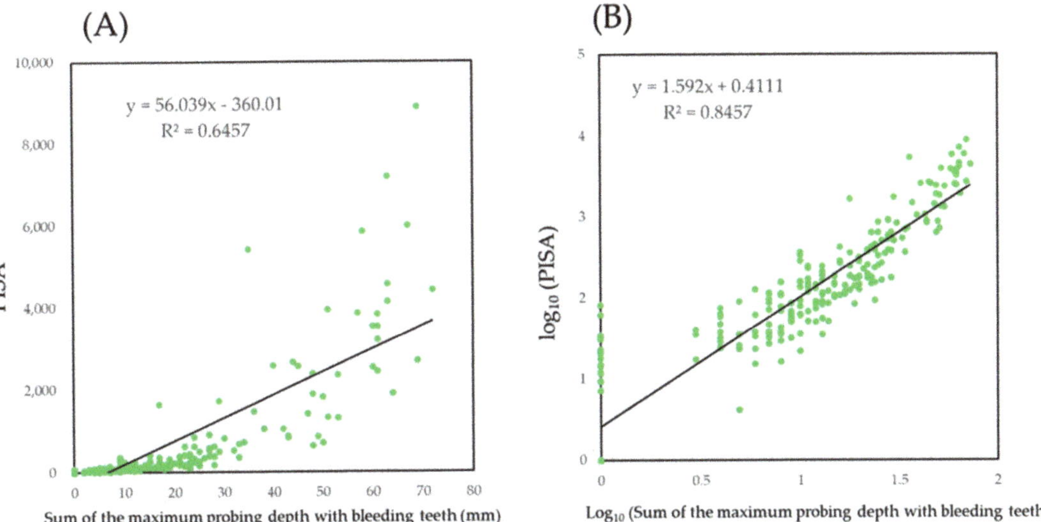

Figure 4. Scatter plot of PISA against sum of the maximum values of PD for each of the bleeding teeth. (**A**) shows the scatter plot by net values. (**B**) shows the values were log_{10} transformed.

If there was no bleeding in ten teeth, PISA was estimated a less than 100. Thus, there may be little effect of missing teeth on PISA (Figure 4B).

The final model is presented in the following formula:

$$Log_{10} (PISA) = 1.592 \times log_{10}\Sigma(\text{maximum value of probing depth with bleeding of selected teeth}) + 0.4111 \tag{1}$$

Selected teeth are maxillary second molar, maxillary first premolar, maxillary lateral incisor, mandibular second molar, and mandibular lateral incisor.

When separately predicted PISA by bleeding of approximal surface and flat surface, effects of bleeding in approximal surface was larger than flat surface. The results are shown in Figure S5.

Additionally, for clinical convenience, quick reference for the predictive values of PISA calculated by the formula described above is presented in Table S3.

4. Discussion

In this study, we tried to establish a partial examination method using the selected teeth to estimate PISA. The final simple model that uses the sum of the maximum PD values of the 10 selected teeth with BOP may be useful to estimate PISA.

Many studies have used PISA for evaluation of periodontal conditions. Since PISA reflects the inflammation status of periodontal tissue, these studies focused on the correlation of PISA and systemic diseases [25–43], or the correlation of PISA with novel disease markers [41–48]. In the present study, IRT analysis was used to set a cut-off point for the number of bleeding sites and the maximum values of PD in a single tooth. IRT is based on the relationship between the performance of the subjects on a test item and the overall measured ability. It can calculate the weight of each item and the ability of individuals. These values correspond to the weight of clinical symptoms of each teeth and PISA. Therefore, selection of the teeth as items and calculation of the ability as PISA with IRT is a reasonable approach.

As shown in Figure 1, the cut-off point of one site of bleeding in one tooth may be suitable. When the cut-off point was set to more than two sites, the number of participants with an estimated PISA of 0 increased. The number of participants who missed with less than 100 PISA was also increased. If the PD is transformed to a dichotomous variable by an optimal cut-off point, several statistical methods for the screening such as ROC analysis and evaluations of sensitivity, specificity, positive predictive value, and negative predictive value become available. However, the optimal cut-off points were not presented by previous studies. In addition, the scatterplots shown in Figure 2 were not as clear as the scatter plot shown in Figure 1A. Therefore, in the model shown in Figure 4, a simple sum of the maximum value of PD in the selected teeth with BOP was applied as an independent value. However, this simple model was sufficient to estimate PISA. Missing teeth had little effect on PISA in the simple model shown in Figure 4. Since PISA is a sum of the PISAs of each tooth, a value of 0 as a result of missing teeth would not include the PISA. However, the simple model could not detect participants with PISA less than 100.

The PISA value for the diagnosis of periodontal disease has been defined previously [48]. However, calculation of PISA is more laborious than examination of periodontal tissue for the diagnosis. The value of PISA was correlated with the clinical parameters of periodontal tissue and periodontal pathogens [49] and dental plaque metabolic byproducts [50]. Only a few studies have investigated the characteristics of PISA. Thus, additional studies are necessary to understand the characteristics of PISA.

Representative six teeth by Ramfjord was used in epidemiological studies. Although it reflect the entire periodontal disease, those sites were selected from studies by numerous investigators using (a) P.M.A. index, (b) formation of pocket depth and bone loss, and (c) teeth extraction record [51]. While PISA focus on degree of periodontal inflammation, as an advantage of the study, we calculated using our raw data for selection.

There is a limitation of this study. Study population of this study was consisted of the patients who finished periodontal treatment. There may be bias in periodontal conditions. Further study is necessary to confirm the availability the model presented in this study by epidemiological study which included the periodontal healthy subjects and subjects with severe periodontal conditions. In addition, further study is necessary to confirm application of the model presented in this study for the patients with systemic diseases.

In this study, by using patients' data, we tried to establish a convenient partial examination method to estimate PISA. Simple, convenient, and evidence-based partial examination methods to estimate PISA may be useful tools for the screening or epidemiological studies, such as other periodontal indexes.

Supplementary Materials: The following are available online at https://www.mdpi.com/2077-0383/10/4/723/s1, Figure S1: Item response curves and item information curves of the number of bleeding sites, Figure S2: Item response curves and item information curves of the maximum value of probing depth, Figure S3. Scatter plot of PISA against the PISA of each tooth type, Figure S4: Scatter plot of PISA against the PISA of each tooth type. Figure S5: Scatter plot of PISA against the PISA calculated by bleeding with approximal surface and flat surface. Table S1: Three-parameter logistic model for the number of bleeding sites, Table S2: Three-parameter logistic model for the maximum value of probing depth. Table S3: Quick reference of the predictive PISA values.

Author Contributions: Conceptualization, Y.N. (Yoshiaki Nomura) and T.M.; methodology, Y.N. (Yoshiaki Nomura); software, Y.N. (Yoshiaki Nomura); validation, E.K. and N.H.; formal analysis, Y.N. (Yoshiaki Nomura); investigation, T.M., Y.N. (Yukihiro Numabe), Y.O., Y.N. (Yohei Nakayama), T.S., T.N. (Toshiaki Nakamura), S.S. (Soh Sato), S.S. (Satoshi Sekino), N.Y., N.S., M.F., M.M., M.U., K.T., K.N., H.K., H.T., F.N., F.S., A.Y., A.S., and T.N. (Taneaki Nakagawa); data curation, S.T.; writing—original draft preparation, Y.N. (Yoshiaki Nomura); writing—review and editing, Y.N. (Yoshiaki Nomura), T.M. and K.T.; visualization, Y.N. (Yoshiaki Nomura); supervision, T.M.; project administration, T.M. and T.N.; funding acquisition, T.N. All authors have read and agreed to the published version of the manuscript.

Funding: This research was funded by a clinical research project grant from the Japanese Society of Periodontology for the diagnosis of periodontitis.

Informed Consent Statement: Informed consent was obtained from all subjects involved in the study.

Data Availability Statement: The data presented in this study are available on request from the corresponding author. The data are not publicly available due to privacy.

Acknowledgments: The authors thank Hiromasa Yoshie, Toshihide Noguchi, Masamitsu Kawanami, Koichi Ito, Yuichi Izumi, Yoshitaka Hara, Osamu Fujise, Yuzo Abe, Tomoo Kono, Asako Makino-Oi, and Chie Fukaya, for their advice and useful comments.

Conflicts of Interest: The authors declare no conflict of interest.

References

1. Thomas, J.T.; Thomas, T.; Ahmed, M.; Kannan, S.K.; Abdullah, Z.; Alghamdi, S.A.; Joseph, B. Prevalence of Periodontal Disease among Obese Young Adult Population in Saudi Arabia-A Cross-Sectional Study. *Medicina* **2020**, *56*, 197. [CrossRef]
2. Lee, K.; Kim, J. Dairy Food Consumption is Inversely Associated with the Prevalence of Periodontal Disease in Korean Adults. *Nutrients* **2019**, *11*, 1035. [CrossRef]
3. Cengiz, M.İ.; Zengin, B.; İçen, M.; Köktürk, F. Prevalence of periodontal disease among mine workers of Zonguldak, Kozlu District, Turkey: A cross-sectional study. *BMC Public Health* **2018**, *18*, 361. [CrossRef] [PubMed]
4. Sun, H.Y.; Jiang, H.; Du, M.Q.; Wang, X.; Feng, X.P.; Hu, Y.; Lin, H.C.; Wang, B.; Si, Y.; Wang, C.X.; et al. The Prevalence and Associated Factors of Periodontal Disease among 35 to 44-year-old Chinese Adults in the 4th National Oral Health Survey. *Chin. J. Dent. Res.* **2018**, *21*, 241–247. [CrossRef] [PubMed]
5. Figueiredo, A.; Soares, S.; Lopes, H.; Santos, J.N.; Ramalho, L.M.; Cangussu, M.C.; Cury, P.R. Destructive periodontal disease in adult Indians from Northeast Brazil: Cross-sectional study of prevalence and risk indicators. *J. Clin. Periodontol.* **2013**, *40*, 1001–1006. [CrossRef] [PubMed]
6. Salih, Y.; Nasr, A.M.; Ahmed, A.B.A.; Sharif, M.E.; Adam, I. Prevalence of and risk factors for periodontal disease among pregnant women in an antenatal care clinic in Khartoum, Sudan. *BMC Res. Notes* **2020**, *13*, 147. [CrossRef]
7. García, B.M.; Plana, C.J.M.; González, A.P.I. Prevalence and severity of periodontal disease among Spanish military personnel. *BMJ Mil. Health* **2020**. [CrossRef]
8. Bostancı, V.; Toker, H.; Senel, S.; Sahin, S. Prevalence of periodontal disease in patients with Familial Mediterranean Fever: A cohort study from central Turkey. *Quintessence Int.* **2014**, *45*, 743–748. [CrossRef] [PubMed]
9. Singh, M.; Bains, V.K.; Jhingran, R.; Srivastava, R.; Madan, R.; Maurya, S.C.; Rizvi, I. Prevalence of Periodontal Disease in Type 2 Diabetes Mellitus Patients: A Cross-sectional Study. *Contemp Clin. Dent.* **2019**, *10*, 349–357. [CrossRef] [PubMed]
10. Chikte, U.; Pontes, C.C.; Karangwa, I.; Dhansay, K.F.; Erasmus, R.T.; Kengne, A.P.; Matsha, T.E. Periodontal Disease Status among Adults from South Africa-Prevalence and Effect of Smoking. *Int. J. Environ. Res. Public Health* **2019**, *16*, 3662. [CrossRef]
11. Nesse, W.; Abbas, F.; Ploeg, I.; Spijkervet, F.K.L.; Dijkstra, P.U.; Vissink, A. Periodontal inflamed surface area: Quantifying inflammatory burden. *J. Clin. Periodontol.* **2008**, *35*, 668–673. [CrossRef]
12. Susanto, H.; Nesse, W.; Dijkstra, P.U.; Hoedemaker, E.; van Reenen, Y.H.; Agustina, D.; Vissink, A.; Abbas, F. Periodontal inflamed surface area and C-reactive protein as predictors of HbA1c: A study in Indonesia. *Clin. Oral Investig.* **2012**, *16*, 1237–1242. [CrossRef]
13. Orlandi, M.; Graziani, F.; D'Aiuto, F. Periodontal therapy and cardiovascular risk. *Periodontol. 2000* **2020**, *83*, 107–124. [CrossRef] [PubMed]
14. Jepsen, S.; Suvan, J.; Deschner, J. The association of periodontal diseases with metabolic syndrome and obesity. *Periodontol. 2000* **2020**, *83*, 125–153. [CrossRef]
15. Genco, R.J.; Sanz, M. Clinical and public health implications of periodontal and systemic diseases: An overview. *Periodontol. 2000* **2020**, *83*, 7–13. [CrossRef]
16. Sanz, M.; Castillo, M.D.A.; Jepsen, S.; Juanatey, G.J.R.; D'Aiuto, F.; Bouchard, P.; Chapple, I.; Dietrich, T.; Gotsman, I.; Graziani, F.; et al. Periodontitis and cardiovascular diseases: Consensus report. *J. Clin. Periodontol.* **2020**, *47*, 268–288. [CrossRef]
17. Bourgeois, D.; Inquimbert, C.; Ottolenghi, L.; Carrouel, F. Periodontal Pathogens as Risk Factors of Cardiovascular Diseases, Diabetes, Rheumatoid Arthritis, Cancer, and Chronic Obstructive Pulmonary Disease-Is There Cause for Consideration? *Microorganisms* **2019**, *7*, 424. [CrossRef]
18. Blaschke, K.; Seitz, M.W.; Schubert, I.; Listl., S. Methodological approaches for investigating links between dental and chronic diseases with claims data: A scoping study. *J. Public Health Dent.* **2019**, *79*, 334–342. [CrossRef] [PubMed]
19. Sanz, M.; Ceriello, A.; Buysschaert, M.; Chapple, I.; Demmer, R.T.; Graziani, F.; Herrera, D.; Jepsen, S.; Lione, L.; Madianos, P.; et al. Scientific evidence on the links between periodontal diseases and diabetes: Consensus report and guidelines of the joint workshop on periodontal diseases and diabetes by the International Diabetes Federation and the European Federation of Periodontology. *J. Clin. Periodontol.* **2018**, *45*, 138–149. [CrossRef] [PubMed]
20. Jin, L.J.; Lamster, I.B.; Greenspan, J.S.; Pitts, N.B.; Scully, C.; Warnakulasuriya, S. Global burden of oral diseases: Emerging concepts, management and interplay with systemic health. *Oral Dis.* **2016**, *22*, 609–619. [CrossRef]
21. Nomura, Y.; Morozumi, T.; Nakagawa, T.; Sugaya, T.; Kawanami, M.; Suzuki, F.; Takahashi, K.; Abe, Y.; Sato, S.; Makino-Oi, A.; et al. Site-level progression of periodontal disease during a follow-up period. *PLoS ONE* **2017**, *12*, e0188670. [CrossRef] [PubMed]

22. Morozumi, T.; Nakagawa, T.; Nomura, Y.; Sugaya, T.; Kawanami, M.; Suzuki, F.; Takahashi, K.; Abe, Y.; Sato, S.; Makino-Oi, A.; et al. Salivary pathogen and serum antibody to assess the progression of chronic periodontitis: A 24-mo prospective multicenter cohort study. *J. Periodontal Res.* **2016**, *51*, 768–778. [CrossRef]
23. Kakuta, E.; Nomura, Y.; Morozumi, T.; Nakagawa, T.; Nakamura, T.; Noguchi, K.; Yoshimura, A.; Hara, Y.; Fujise, O.; Nishimura, F.; et al. Assessing the progression of chronic periodontitis using subgingival pathogen levels: A 24-month prospective multicenter cohort study. *BMC Oral Health* **2017**, *17*, 46. [CrossRef]
24. Krebs, K.A.; Clem, D.S., 3rd. American Academy of Periodontology, Guidelines for the management of patients with periodontal diseases. *J. Periodontol.* **2006**, *77*, 1607–1611. [CrossRef]
25. Schöffer, C.; Oliveira, L.M.; Santi, S.S.; Antoniazzi, R.P.; Zanatta, F.B. C-reactive protein levels are associated with periodontitis and periodontal inflamed surface area in adults with end-stage renal disease. *J. Periodontol.* **2020**. [CrossRef]
26. Smojver, B.K.; Altabas, K.; Knotek, M.; Jukić, N.B.; Aurer, A. Periodontal inflamed surface area in patients on haemodialysis and peritoneal dialysis: A Croatian cross-sectional study. *BMC Oral Health* **2020**, *20*, 95. [CrossRef]
27. Yamashita, M.; Kobayashi, T.; Ito, S.; Kaneko, C.; Murasawa, A.; Ishikawa, H.; Tabeta, K. The periodontal inflamed surface area is associated with the clinical response to biological disease-modifying antirheumatic drugs in rheumatoid arthritis: A retrospective study. *Mod. Rheumatol.* **2020**, *30*, 990–996. [CrossRef] [PubMed]
28. Teke, E.; Kırzıoğlu, E.Y.; Korkmaz, H.; Calapoğlu, M.; Orhan, H. Does metabolic control affect salivary adipokines in type 2 diabetes mellitus? *Dent. Med. Probl.* **2019**, *56*, 11–20. [CrossRef] [PubMed]
29. Matsuda, S.; Okanobu, A.; Hatano, S.; Kajiya, M.; Sasaki, S.; Hamamoto, Y.; Iwata, T.; Ouhara, K.; Takeda, K.; Mizuno, N.; et al. Relationship between periodontal inflammation and calcium channel blockers induced gingival overgrowth-a cross-sectional study in a Japanese population. *Clin. Oral Investig.* **2019**, *23*, 4099–4105. [CrossRef] [PubMed]
30. Leira, Y.; Yáñez, M.R.; Arias, S.; Dequidt, I.L.; Campos, F.; Sobrino, T.; D'Aiuto, F.; Castillo, J.; Blanco, J. Periodontitis is associated with systemic inflammation and vascular endothelial dysfunction in patients with lacunar infarct. *J. Periodontol.* **2019**, *90*, 465–474. [CrossRef] [PubMed]
31. Leira, Y.; Yáñez, M.R.; Arias, S.; Dequidt, I.L.; Campos, F.; Sobrino, T.; D'Aiuto, F.; Castillo, J.; Blanco, J. Periodontitis as a risk indicator and predictor of poor outcome for lacunar infarct. *J. Clin. Periodontol.* **2019**, *46*, 20–30. [CrossRef]
32. Iwasaki, M.; Kimura, Y.; Ogawa, H.; Yamaga, T.; Ansai, T.; Wada, W.; Sakamoto, R.; Ishimoto, Y.; Fujisawa, M.; Okumiya, K.; et al. Periodontitis, periodontal inflammation, and mild cognitive impairment: A 5-year cohort study. *J. Periodontal Res.* **2019**, *54*, 233–240. [CrossRef] [PubMed]
33. Temelli, B.; Ay, Z.Y.; Aksoy, F.; Büyükbayram, H.İ.; Doğuç, D.K.; Uskun, E.; Varol, E. Platelet indices (mean platelet volume and platelet distribution width) have correlations with periodontal inflamed surface area in coronary artery disease patients: A pilot study. *J. Periodontol.* **2018**, *89*, 1203–1212. [CrossRef] [PubMed]
34. Epping, L.; Miesbach, W.; Nickles, K.; Eickholz, P. Is gingival bleeding a symptom of type 2 and 3 von Willebrand disease? *PLoS ONE* **2018**, *13*, e0191291. [CrossRef] [PubMed]
35. Punj, A.; Santhosh, B.; Shenoy, A.B.; Subramanyam, K. Comparison of endothelial function in healthy patients and patients with chronic periodontitis and myocardial infarction. *J. Periodontol.* **2017**, *88*, 1234–1243. [CrossRef]
36. Yoshihara, A.; Sugita, N.; Iwasaki, M.; Wang, Y.; Miyazaki, H.; Yoshie, H.; Nakamura, K. Relationship between renal function and periodontal disease in community-dwelling elderly women with different genotypes. *J. Clin. Periodontol.* **2017**, *44*, 484–489. [CrossRef] [PubMed]
37. Yoshihara, A.; Iwasaki, M.; Miyazaki, H.; Nakamura, K. Bidirectional relationship between renal function and periodontal disease in older Japanese women. *J. Clin. Periodontol.* **2016**, *43*, 720–726. [CrossRef] [PubMed]
38. Khanuja, P.K.; Narula, S.C.; Rajput, R.; Sharma, R.K.; Tewar, I.S. Association of periodontal disease with glycemic control in patients with type 2 diabetes in Indian population. *Front. Med.* **2017**, *11*, 110–119. [CrossRef]
39. Sato, M.; Iwasaki, M.; Yoshihara, A.; Miyazaki, M. Association between periodontitis and medical expenditure in older adults: A 33-month follow-up study. *Geriatr. Gerontol. Int.* **2016**, *16*, 856–864. [CrossRef] [PubMed]
40. Jaedicke, K.M.; Bissett, S.M.; Finch, T.; Thornton, J.; Preshaw, P.M. Exploring changes in oral hygiene behaviour in patients with diabetes and periodontal disease: A feasibility study. *Int. J. Dent. Hyg.* **2019**, *17*, 55–63. [CrossRef]
41. Weickert, L.; Miesbach, W.; Alesci, S.R.; Eickholz, P.; Nickles, K. Is gingival bleeding a symptom of patients with type 1 von Willebrand disease? A case-control study. *J. Clin. Periodontol.* **2014**, *41*, 766–771. [CrossRef]
42. Iwasaki, M.; Taylor, G.W.; Nesse, W.; Vissink, A.; Yoshihara, A.; Miyazaki, H. Periodontal disease and decreased kidney function in Japanese elderly. *Am. J. Kidney Dis.* **2012**, *59*, 202–209. [CrossRef] [PubMed]
43. Nesse, W.; Linde, A.; Abbas, F.; Spijkervet, F.K.L.; Dijkstra, P.U.; Brabander, E.C.; Gerstenbluth, I.; Vissink, A. Dose-response relationship between periodontal inflamed surface area and HbA1c in type 2 diabetics. *J. Clin. Periodontol.* **2009**, *36*, 295–300. [CrossRef]
44. Tan, A.; Gürbüz, N.; Özbalci, F.İ.; Koşkan, Ö.; Yetkin, A.Z. Increase in serum and salivary neutrophil gelatinase-associated lipocalin levels with increased periodontal inflammation. *J. Appl. Oral Sci.* **2020**, *28*, e20200276. [CrossRef]
45. Kasai, S.; Onizuka, S.; Katagiri, S.; Nakamura, T.; Hanatani, T.; Kudo, T.; Sugata, Y.; Ishimatsu, M.; Usui, M.; Nakashima, K. Associations of cytokine levels in gingival crevicular fluid of mobile teeth with clinical improvement after initial periodontal treatment. *J. Oral Sci* **2020**, *62*, 189–196. [CrossRef]

46. Leira, Y.; Ameijeira, P.; Domínguez, C.; Arias, E.L.; Gómez, Á.P.; Mato, M.P.; Sobrino, T.; Campos, F.; D'Aiuto, F.; Leira, R.; et al. Periodontal inflammation is related to increased serum calcitonin gene-related peptide levels in patients with chronic migraine. *J. Periodontol.* **2019**, *90*, 1088–1095. [CrossRef]
47. Chaparro, A.; Zúñiga, E.; Godoy, M.V.; Albers, D.; Ramírez, V.; Hernández, M.; Kusanovic, J.P.; Gallardo, S.A.; Rice, G.; Illanes, S.E. Periodontitis and placental growth factor in oral fluids are early pregnancy predictors of gestational diabetes mellitus. *J. Periodontol.* **2018**, *89*, 1052–1060. [CrossRef] [PubMed]
48. Leira, Y.; Lancharro, P.M.; Blanco, J. Periodontal inflamed surface area and periodontal case definition classification. *Acta Odontol. Scand* **2018**, *76*, 195–198. [CrossRef]
49. Park, S.Y.; Ahn, S.; Lee, J.T.; Yun, P.Y.; Lee, Y.J.; Lee, J.Y.; Song, Y.W.; Chang, Y.S.; Lee, H.J. Periodontal inflamed surface area as a novel numerical variable describing periodontal conditions. *J. Periodontal Implant. Sci.* **2017**, *47*, 328–338. [CrossRef] [PubMed]
50. Sakanaka, A.; Kuboniwa, M.; Hashino, E.; Bamba, T.; Fukusaki, E.; Amano, A. Distinct signatures of dental plaque metabolic by products dictated by periodontal inflammatory status. *Sci. Rep.* **2017**, *7*, 42818. [CrossRef]
51. Ramfjord, S.P. Indices for prevalence and incidence of periodontal disease. *J. Periodontol.* **1959**, *30*, 51–59. [CrossRef]

Journal of
Clinical Medicine

Article

Prospective Longitudinal Changes in the Periodontal Inflamed Surface Area Following Active Periodontal Treatment for Chronic Periodontitis

Yoshiaki Nomura [1], Toshiya Morozumi [2,*], Atsushi Saito [3], Atsutoshi Yoshimura [4], Erika Kakuta [5], Fumihiko Suzuki [6], Fusanori Nishimura [7], Hideki Takai [8], Hiroaki Kobayashi [9], Kazuyuki Noguchi [10], Keiso Takahashi [11], Koichi Tabeta [12], Makoto Umeda [13], Masato Minabe [2], Mitsuo Fukuda [14], Naoyuki Sugano [15], Nobuhiro Hanada [1], Nobuo Yoshinari [16], Satoshi Sekino [17], Shogo Takashiba [18], Soh Sato [19], Toshiaki Nakamura [10], Tsutomu Sugaya [20], Yohei Nakayama [8], Yorimasa Ogata [8], Yukihiro Numabe [17] and Taneaki Nakagawa [21]

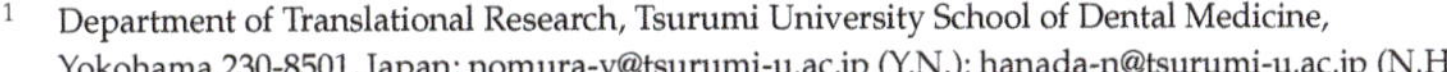

1   Department of Translational Research, Tsurumi University School of Dental Medicine, Yokohama 230-8501, Japan; nomura-y@tsurumi-u.ac.jp (Y.N.); hanada-n@tsurumi-u.ac.jp (N.H.)
2   Division of Periodontology, Department of Oral Interdisciplinary Medicine, Graduate School of Dentistry, Kanagawa Dental University, Yokosuka 238-8580, Japan; minabe@kdu.ac.jp
3   Department of Periodontology, Tokyo Dental College, Tokyo 101-0061, Japan; atsaito@tdc.ac.jp
4   Department of Periodontology and Endodontology, Nagasaki University Graduate School of Biomedical Sciences, Nagasaki 852-8588, Japan; ayoshi@nagasaki-u.ac.jp
5   Department of Oral Microbiology, Tsurumi University School of Dental Medicine, Yokohama 230-8501, Japan; kakuta-erika@tsurumi-u.ac.jp
6   Division of Dental Anesthesiology, Department of Oral Surgery, Ohu University School of Dentistry, Koriyama 963-8611, Japan; f-suzuki@den.ohu-u.ac.jp
7   Section of Periodontology, Division of Oral Rehabilitation, Faculty of Dental Science, Kyushu University, Fukuoka 812-8582, Japan; fusanori@dent.kyushu-u.ac.jp
8   Department of Periodontology, Nihon University School of Dentistry at Matsudo, Matsudo 271-8587, Japan; takai.hideki@nihon-u.ac.jp (H.T.); nakayama.youhei@nihon-u.ac.jp (Y.N.); ogata.yorimasa@nihon-u.ac.jp (Y.O.)
9   Department of Periodontology, Graduate School of Medical and Dental Sciences, Tokyo Medical and Dental University, Tokyo 113-8510, Japan; h-kobayashi.peri@tmd.ac.jp
10 Department of Periodontology, Kagoshima University Graduate School of Medical and Dental Sciences, Kagoshima 890-8544, Japan; kazuperi@dent.kagoshima-u.ac.jp (K.N.); toshi-n@dent.kagoshima-u.ac.jp (T.N.)
11 Division of Periodontics, Department of Conservative Dentistry, Ohu University School of Dentistry, Koriyama 963-8611, Japan; ke-takahashi@den.ohu-u.ac.jp
12 Division of Periodontology, Department of Oral Biological Science, Niigata University Graduate School of Medical and Dental Sciences, Niigata 951-8514, Japan; koichi@dent.niigata-u.ac.jp
13 Department of Periodontology, Osaka Dental University, Hirakata 573-1121, Japan; umeda-m@cc.osaka-dent.ac.jp
14 Department of Periodontology, School of Dentistry, Aichi Gakuin University, Nagoya 464-8650, Japan; fukuda-m@dpc.agu.ac.jp
15 Department of Periodontology, Nihon University School of Dentistry, Tokyo 101-8310, Japan; sugano.naoyuki@nihon-u.ac.jp
16 Department of Periodontology, School of Dentistry, Matsumoto Dental University, Shiojiri 399-0781, Japan; nobuo.yoshinari@mdu.ac.jp
17 Department of Periodontology, School of Life Dentistry at Tokyo, The Nippon Dental University, Tokyo 102-8159, Japan; sekino-s@tky.ndu.ac.jp (S.S.); numabe-y@tky.ndu.ac.jp (Y.N.)
18 Department of Pathophysiology-Periodontal Science, Okayama University Graduate School of Medicine, Dentistry and Pharmaceutical Sciences, Okayama 700-8525, Japan; stakashi@okayama-u.ac.jp
19 Department of Periodontology, School of life Dentistry at Niigata, The Nippon Dental University, Niigata 951-8580, Japan; s-sato@ngt.ndu.ac.jp
20 Division of Periodontology and Endodontology, Department of Oral Health Science, Hokkaido University Graduate School of Dental Medicine, Sapporo 060-8586, Japan; sugaya@den.hokudai.ac.jp
21 Department of Dentistry and Oral Surgery, School of Medicine, Keio University, Tokyo 160-8582, Japan; tane@z6.keio.jp
*   Correspondence: morozumi@kdu.ac.jp; Tel.: +81-46-822-8855

check for **updates**

Citation: Nomura, Y.; Morozumi, T.; Saito, A.; Yoshimura, A.; Kakuta, E.; Suzuki, F.; Nishimura, F.; Takai, H.; Kobayashi, H.; Noguchi, K.; et al. Prospective Longitudinal Changes in the Periodontal Inflamed Surface Area Following Active Periodontal Treatment for Chronic Periodontitis. *J. Clin. Med.* **2021**, *10*, 1165. https://doi.org/10.3390/jcm10061165

Academic Editor: Denis Bourgeois

Received: 22 February 2021
Accepted: 8 March 2021
Published: 10 March 2021

Publisher's Note: MDPI stays neutral with regard to jurisdictional claims in published maps and institutional affiliations.

Abstract: Periodontal disease is a chronic inflammatory disease of the periodontal tissue. The periodontal inflamed surface area (PISA) is a proposed index for quantifying the inflammatory burden resulting

129

from periodontitis lesions. This study aimed to investigate longitudinal changes in the periodontal status as evaluated by the PISA following the active periodontal treatment. To elucidate the prognostic factors of PISA, mixed-effect modeling was performed for clinical parameters, tooth-type, and levels of periodontal pathogens as independent variables. One-hundred-twenty-five patients with chronic periodontitis who completed the active periodontal treatment were followed-up for 24 months, with evaluations conducted at 6-month intervals. Five-times repeated measures of mean PISA values were $130+/-173$, $161+/-276$, $184+/-320$, $175+/-417$, and $209+/-469$ mm^2. Changes in clinical parameters and salivary and subgingival periodontal pathogens were analyzed by mixed-effect modeling. Plaque index, clinical attachment level, and salivary levels of *Porphyromonas gingivalis* were associated with changes in PISA at the patient- and tooth-level. Subgingival levels of *P. gingivalis* and *Prevotella intermedia* were associated with changes in PISA at the sample site. For most patients, changes in PISA were within 10% of baseline during the 24-month follow-up. However, an increase in the number of bleeding sites in a tooth with a deep periodontal pocket increased the PISA value exponentially.

Keywords: periodontal inflamed surface area; periodontal pathogen; mixed effect modeling; follow-up study

1. Introduction

Periodontal disease is a chronic inflammatory disease of the periodontal tissue. Its progression has been associated with a regression of the mean clinical parameters of most of the oral sites, including teeth. Most patients are generally stable or exhibit a linear type of progression [1–4]. However, a small fraction exhibit burst-type progressions [1,5,6].

The active periodontal treatment aims to reduce inflammatory response by removing pathogenic bacterial deposits. Following the active treatment, supportive periodontal therapy (SPT) is employed to reduce the probability of periodontal disease progression. The long-term successful SPT prevents tooth loss [7]. The reported rate of tooth loss is 10% of teeth [8–15] in 20% of patients [16] over a period of 10 years. Further, tooth loss has been shown to occur in a small fraction of patients during SPT [8,17]. Therefore, monitoring of periodontal conditions during SPT is important for a successful treatment [18]. Several indices have been routinely used for this purpose. The probing pocket depth (PD) and clinical attachment level (CAL) of the periodontal pocket are morphological outcomes of periodontal disease. However, these outcomes do not quantify the proportion of inflamed periodontal tissue. Additionally, summary statistics of these outcomes using the mean or maximum values results in information loss [1]. The mean values of these outcomes obscure the small number of sites that exhibit a substantial progress, and the maximum value represents only one of 168 probing sites. Therefore, site-level evaluation is important in the clinical setting. During the follow-up period, the periodontal tissue should be evaluated in terms of the morphological and inflammatory response status.

The periodontal inflamed surface area (PISA) is a convenient index that quantifies the surface area of the bleeding pocket epithelium in square millimeters and is calculated using conventional clinical parameters of periodontal health, namely, bleeding on probing (BOP) combined with either PD or CAL, and gingival recession [19]. Thus, PISA may represent the inflammation status of the patients as one continuous variable, in which the calculation of mean or maximum value is not necessary. Several studies have utilized the PISA for evaluating the periodontal status. However, these studies have focused on the cross-sectional correlation of the PISA and non-communicable diseases [20–28]. There have only been a few longitudinal studies regarding the changes in PISA [29].

This study aimed to investigate the changes in periodontal status, as evaluated by the PISA during the longitudinal follow-up care after an active treatment, and to identify the factors that affect the PISA.

2. Materials and Methods

2.1. Study Design and Ethics Approval

This study is part of a clinical research project performed by The Japanese Society of Periodontology. A 24-month follow-up study was performed with 163 patients who had completed an active periodontal treatment (such as initial therapy or periodontal surgery) at 17 facilities in Japan. The end point of those active periodontal treatments was periodontal examination. If the lesion was considered to be dormant, such as a probing pocket depth of 4 mm or more without BOP, the disease was considered stable [30]. In addition, those patients were the candidates for this study. The follow-up patients were seen trimonthly, and the supragingival plaque and calculus were removed if detected. The details of changes in the clinical parameters during the follow-ups have been described in our pervious report [1,31,32]. The study was conducted in compliance with the principles outlined in the Helsinki Declaration. A written informed consent was obtained from each study participant, and the protocol was approved by the Institutional Review Board of each participating institution.

2.2. Inclusion Criteria

The inclusion criteria were as follows: The severity of periodontitis on the initial visit was generalized moderate–to-severe chronic periodontitis (generalized periodontitis, stage III or IV, grade B) [33,34], patients who were systemically healthy; aged >30 years; >20 remaining teeth; and not taking antibiotics, anti-inflammatory drugs or immunosuppressive drugs for 3 months before the start of follow-up.

2.3. Diagnosis and Evaluation

Each patient was diagnosed according to the Guidelines of the American Academy of Periodontology [33]. Oral examinations were carried out by one examiner from each institute (T.M., A.Y., F.S., H.T., H.K., K.N., M.M., M.F., N.S., N.Y., S.S., S.S., T.N., T.S., Y.O., Y.N. and T.N.). Each examiner was a periodontist licensed by the Japanese Society of Periodontology. Intra- and inter-examiner calibration sessions were conducted using periodontal disease models (P15FE-500HPRO-S2A1-GSF) at the beginning and middle of the study period. In brief, full-mouth PD and gingival recession were measured twice, and repeatability for the CAL was assessed. The examiner was judged to have made reproducible measurements after reaching a percentage of agreement within ±1 mm between repeated measurements of at least 95% of measurements.

The PD, BOP, and CAL were measured at six sites per tooth (mesiobuccal, buccal, distobuccal, mesiolingual, lingual, and distolingual) with a periodontal probe (CP-12 Color-Coded Probe; Hu-Friedy, Chicago, IL, USA). The PISA was calculated using the Excel sheet program [19]. By inputting the BOP and PD data from six sites per tooth, the PISA values of each tooth were calculated. The sum of the PISA values is automatically calculated. This value is the summary statistics of PISA for each individual.

The plaque index (PlI) was recorded at four sites per tooth (mesial, buccal, distal, and lingual). The degree of tooth mobility was scored on a four-point scale (0–3) at the tooth level. All clinical parameters were recorded at baseline and after 6, 12, 18, and 24 months, the third molars were excluded.

2.4. Evaluation of Periodontal Pathogens

Three periodontal pathogens in the saliva and subgingival plaque were measured: *Aggregatibacter actinomycetemcomitans*, *Porphyromonas gingivalis*, and *Prevotella intermedia*. These samples were collected from each patient at every visit. Briefly, whole saliva was collected by asking the patient to chew on a gum base for 5 min [31], and the subgingival plaque from the deepest pockets (except for those at the third molars) was obtained by consecutive insertion of two sterile paper points into the periodontal pocket for 10 s per point [32]. These three periodontal pathogens and the total bacteria in the subgingival plaque and saliva were counted using a modification of the Invader PLUS assay [35,36]. Bacterial ratios (%) for each species were also determined.

Briefly, a quantitative analysis of the total bacterial count and periodontopathic bacterial counts, including *P. gingivalis*, *P. intermedia*, and *A. actinomycetemcomitans*, was performed using a modification of the Invader PLUS assay [31,32,35,36]. The bacterial DNA was extracted from the subgingival plaque samples from the deepest pockets by suspending each plaque sample in 1 mL of phosphate-buffered saline (pH 7.4) and processed using the MagNA Pure LC Total Nucleic Acid Isolation Kit (Roche, Basel, Switzerland) according to the manufacturer's instructions. Similarly, bacterial DNA was extracted from the 100-µL whole saliva samples using the MagNA Pure LC Total Nucleic Acid Isolation Kit. The individual sequences of each bacterial species were obtained from a public database (National Center for Biotechnology Information, Bethesda, MD, USA). Primers for each species were designed based on a region of the 16S rRNA gene. A pair of universal primers and a universal probe were used to determine the total number of bacteria. Primary probes and Invader oligos were designed using the Invader technology creator (HOLOGIC, Madison, WI, USA) and were based on sequences in the amplified regions [31,32,35–37].

The template DNA was added to a 15-µL reaction mixture containing primers for each species [50 µM deoxynucleoside triphosphate (dNTP), 700 nM primary probe, 70 nM Invader oligo, 2.5 U polymerase chain reaction (PCR) enzyme (EagleTaq DNA polymerase, Roche, Basel, Switzerland), and the Invader core reagent kit (Cleavase XI Invader core reagent kit, HOLOGIC, Madison, WI, USA) containing a fluorescence resonance energy transfer (FRET) mix and an enzyme/MgCl$_2$ solution]. The reaction mixture was preheated at 95 °C for 20 min, and a two-step polymerase chain reaction (95 °C for 1 s and 63 °C for 1 min) was performed for 35 cycles using an ABI PRISM 7900 thermocycler (Applied Biosystems, Foster City, CA, USA). The fluorescence values of carboxyfluorescein (FAM; wavelength/bandwidth: Excitation, 485/20 nm; emission, 530/25 nm) were measured at the end of the incubation/extension step at 63 °C for each cycle.

The limit of detection for this method was determined for each species with dilutions of bacterial DNA. Standard curves were constructed based on a crossing point determined by the fit-point method.

The proportions of the three pathogens within the total bacterial count were subsequently determined [38,39]. Bacterial ratios (%) and bacterial counts (log$_{10}$) for each species were also used in various comparison and diagnostic analyses.

2.5. Statistical Analysis

2.5.1. Mixed-Effect Modeling

Changes in the PISA were analyzed at the patient-level and tooth-level by mixed-effect modeling [1,40–44]. The data structure of the PISA was five-time repeated measures. The PISA values for each tooth had a hierarchical structure in the patients. Generalized mixed-effect modeling was applied using the following formula:

Model 1: Patient-level PISA

$$-L1 : PISA(Subject\ level)_i =$$

$$\pi_0 + \pi_1(Age)_i + \pi_2(Sex)_i + \pi_4(Salivary\ levels\ of\ A.\ a)_i + \pi_5(Salivary\ levels\ of\ P.\ g)_i \tag{1}$$

$$+\pi_6(Salivary\ levels\ of\ P.\ i)_i + \pi_7(mean\ CAL)_i + \pi_8(Pli)_i + \sum_{m=1}^{5} \pi_9^{(m)}(time)_j + \varepsilon_i$$

$$e_i \sim N\left(0, \delta_e^2\right)$$

Fixed effect;
Patient-level: Age, sex, time, salivary levels of *A. actinomycetemcomitans*, *P. gingivalis*, and *P. intermedia*, mean CAL, PlI, time
Random effect: Patients, random intercept
Covariance Type: AR1
Link functions: Gamma

Model 2: Tooth-level PISA

$$-L1 : PISA(Tooth\ level)_{ij} =$$
$$\pi_{0j} + \pi_{1ij}(Age)_{ij} + \pi_{2j}(Sex)_{ij} + \pi_{3j}(Salivary\ levels\ of\ A.\ a)_{ij} + \pi_{4j}(Salivary\ levels\ of\ P.\ g)_{ij} \tag{1}$$
$$+\pi_{5j}(Salivary\ levels\ of\ P.\ i)_{ij} + \varepsilon_{ij}$$

$$-L2 : \pi_{0j} = \beta_{0j} + \beta_{1j}(Mean\ CAL) + \beta_{2j}(Pli) + \sum_{m=1}^{3} \beta_{3j}^{(m)}(tooth\ mobility)_j + \sum_{m=1}^{6} \beta_{4j}^{(m)}(Tooth\ type)_j + \sum_{m=1}^{5} \beta_{05}^{(m)}(time)_j + r_j \tag{2}$$

$$e_{ij} \sim N\left(0, \delta_e^2\right), r_{oj} \sim N\left(0, \delta_r^2\right)$$

Fixed effect;
Patient-level: Age, sex, days, salivary levels of *A. actinomycetemcomitans*, *P. gingivalis*, and *P. intermedia*
Tooth-level: Mean CAL, PlI, tooth mobility, tooth type, time
Random effect: Patients, random intercept
Covariance Type: AR1
Link functions: Gamma

Model 3: PISA of the sampling teeth for subgingival periodontal pathogens

$$-L1 : PISA(Sampling\ teeth)_i =$$
$$\pi_0 + \pi_1(\%\ of\ A.a\ in\ subgingival\ taotal\ bacteria)_i$$
$$+\pi_2(\%\ of\ A.a\ in\ subgingival\ taotal\ bacteria)_i + \pi_3(\%\ of\ P.\ g\ in\ subgingival\ taotal\ bacteria)_i \tag{1}$$
$$+\pi_4(\%\ of\ P.\ i\ in\ subgingival\ taotal\ bacteria)_i + \sum_{m=1}^{5} \pi_5^{(m)}(time)_j + \varepsilon_i$$

$$e_i \sim N\left(0, \delta_e^2\right)$$

Fixed effect;
Patient-level: Proportion of *A. actinomycetemcomitans*, *P. gingivalis*, and *P. intermedia* in the total subgingival bacteria, time
Random effect: Patients
Covariance Type: AR1
Link functions: Gamma

2.5.2. Cluster Analysis

A cluster analysis was carried out using K-means. The data used for the cluster analysis were sequential data obtained from five-time repeated measures of PISA at the patient level. The parameter settings for the cluster analysis were: Distance measure: Log likelihood, and the criteria of clusters were Bayesian Information Criterion (BIC) [−2 ln (L) + k ln (L)]. The generated cluster was used for the analysis.

All analyses were performed using SPSS Statistics version 24.0 (IBM, Tokyo, Japan).

3. Results

3.1. Patient Characteristics

Among the 163 patients enrolled in the study, 38 patients were disqualified for lack of data due to a variety of causes such as earthquake, failure to comply with visit, use of antimicrobial agents for acute periodontal abscesses, and tooth extraction for root fracture. Thus, data from a total of 125 patients, 3107 teeth, and their five-time repeated measure PISA values were obtained. A total of 15,535 data were analyzed from the teeth. The study population comprised 50 men and 75 women and the mean age was 59.2 ± 8.7 years.

Clinical examinations and measurement of periodontal pathogens were carried out at every visit. Five-time repeated measures of these parameters were analyzed.

3.2. Analysis of the Change in PISA at the Patient-Level

Factors correlated with the PISA were identified using generalized mixed-effect modeling. The results are shown in Table 1. The salivary levels of *P. gingivalis*, mean CAL, and PlI were identified as significant factors affecting the PISA. Time was not statistically significant.

Table 1. Patient-level analysis of factors affecting the periodontal inflamed surface area (PISA).

	Coefficient (95% CI)	*p*-Value
Sex	−0.124 (−0.507–0.258)	0.524
Age (years)	0.012 (−0.010–0.033)	0.276
Salivary levels of *A. a* (%)	−3.275 (−39.368–32.818)	0.859
Salivary levels of *P. g* (%)	1.141 (0.148–2.136)	0.025
Salivary levels of *P. i* (%)	0.206 (−0.282–0.694)	0.408
Mean CAL (mm)	0.737 (0.540–0.934)	<0.001
PlI	1.069 (0.637–1.500)	<0.001
Time (6 months)	0.008 (−0.030–0.045)	0.693
Intercept	1.338 (0.009–2.677)	0.048

Fitness index, BIC: 1317; AICC, 1326; PlI: Plaque index; CAL: Clinical attachment level; *A. a*: *Aggregatibacter actinomycetemcomitans*; *P. g*: *Porphyromonas gingivalis*; *P. i*: *Prevotella intermedia*; CI: Confidence interval; $AICC = -2L + 2k(k-1)/(n-k-1)$; $BIC = -2L + k\ln(n)$; L: Likelihood; k: Number of parameters; n: Sample size.

3.3. Analysis of the Change in PISA at the Tooth-Level

The sum of the PISA of each tooth was calculated as a dependent variable. To determine the effect of tooth type, a multilevel mixed-effect model was used for the analysis. As shown in Table 2, the salivary levels of *P. gingivalis* were identified as a significant factor at the patient-level. The CAL, the PlI, and tooth mobility >2 were identified as significant factors at the tooth-level. Regarding the tooth type, the maxillary molar and premolar tended to increase the PISA during the follow-up period. Time was not statistically significant.

Table 2. Tooth-level analysis of factors affecting the PISA.

		Coefficient (95% CI)	*p*-Value
Patient-level variables			
Age		−0.002 (−0.010–0.007)	0.736
Sex		−0.076 (−0.225–0.074)	0.320
Salivary levels of *A. a* (%)		−3.624 (−17.577–10.329)	0.611
Salivary levels of *P. g* (%)		0.669 (0.323–1.015)	<0.001
Salivary levels of *P. i* (%)		−0.028 (−0.230–0.173)	0.782
Tooth-level variables			
Mean CAL/teeth (mm)		0.293 (0.266–0.320)	<0.001
PlI/teeth		0.234 (0.157–0.310)	<0.001
Tooth mobility	0	Reference	
	1	−0.044 (−0.136–0.048)	0.348
	2 and 3	0.697 (0.472–0.923)	<0.001
Tooth-type	Mandibular anterior	Reference	
	Mandibular premolar	−0.033 (−0.124–0.058)	0.477
	Mandibular molar	−0.016 (−0.105–0.074)	0.732
	Maxillary anterior	−0.116 (−0.197––0.035)	0.005
	Maxillary premolar	0.444 (0.359–0.530)	<0.001
	Maxillary molar	0.680 (0.590–0.770)	<0.001
Time		−0.004 (−0.022–0.014)	0.632
Intercept		1.516 (0.972–2.061)	<0.001

Fitness index, AICC: 11,732.53; BIC: 11,745.40; PlI: Plaque index; CAL: Clinical attachment level; *A. a*: *Aggregatibacter actinomycetemcomitans*; *P. g*: *Porphyromonas gingivalis*; *P. i*: *Prevotella intermedia*; $AICC = -2L + 2k(k-1)/(n-k-1)$; $BIC = -2L + k\ln(n)$; L: Likelihood; k: Number of parameters; n: Sample size.

3.4. Classification of the Changes in the PISA

The change in the PISA was dependent on the tooth-type (Figure S1). The PISA of maxillary molars increased throughout the follow-up period, while the PISA of other tooth-types were relatively stable.

A cluster analysis was carried out to characterize the changes in the PISA. The results of the cluster analysis and changes in the PISA in each cluster are shown in Figure 1. Most of the patients were classified as Cluster 1, and the PISA of this cluster was stable. A minority of patients were classified to other clusters, which exhibited drastic changes in the PISA during the follow-up period.

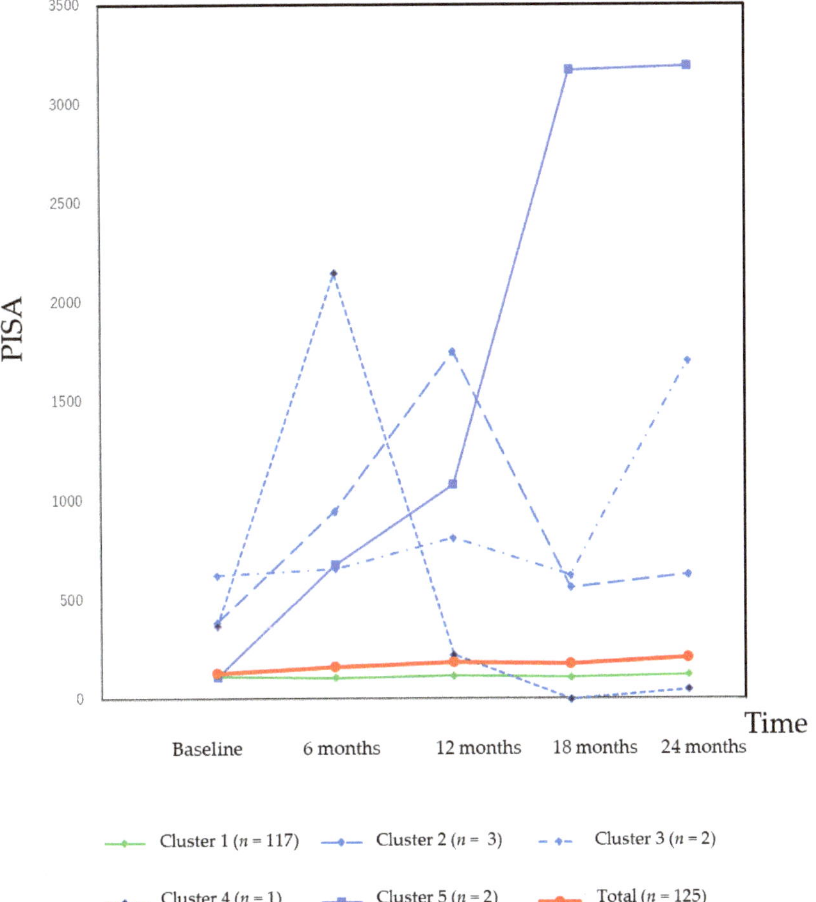

Figure 1. Changes in the PISA according to the clusters. For 117 (93.6%) patients, the PISA was stable during the 24-month follow-up period. However, a minority of patients exhibited a drastic increase or fluctuation in the PISA.

A minority of the teeth may have had an effect on the overall PISA. The teeth with PISA >500 were identified and the changes in the PISA of these teeth are illustrated in Figure 2. A total of seven teeth (maxillary 2nd molars or mandibular 2nd molars) had PISA >500 at least once during the follow-up period. The drastic increase in the overall PISA was primarily dependent on the increase in the number of BOP sites.

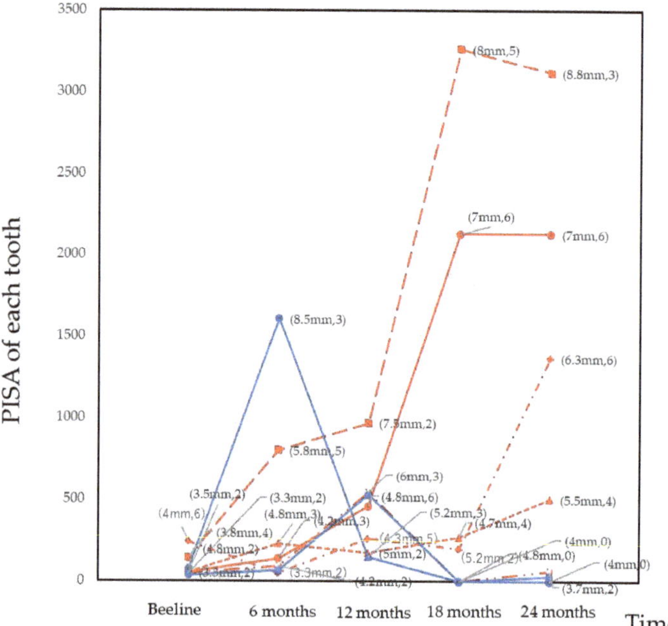

Figure 2. Changes in the PISA at the tooth-level in teeth with PISA > 500 during the 24-month follow-up period. Red lines indicate the maxillary 2nd molars and blue lines indicate mandibular 2nd molars. Seven teeth exceed 500 PISA. The numbers indicate the mean probing depth of each tooth and number of bleeding sites on probing. In these cases, drastic changes in the PISA were predominantly dependent on the increase in the number of bleeding sites on probing.

3.5. Association between Periodontal Pathogens and the Change in the PISA

Mixed-effect modeling was employed to evaluate the local association between periodontal pathogens and the tooth-level PISA. The results are shown in Table 3. The PISA of the sample teeth was included as a dependent variable. The periodontal pathogens, *P. gingivalis* and *P. intermedia*, were observed to be significantly associated with the PISA of the sample teeth. Time was not a significant factor for PISA. The list of the sample teeth is shown in Table S1.

Table 3. Association between subgingival periodontal pathogens and the changes in the PISA of the sample sites.

	PISA	
	Coefficient (95% CI)	***p*-Value**
A. a	−0.221 (−1.054–0.612)	0.602
P. g	0.033 (0.022–0.044)	<0.001
P. i	0.116 (0.024–0.208)	0.014
Time	0.033 (−0.010–0.077)	0.130
Intercept	2.847 (2.377–3.070)	<0.001

Fitness index: For PISA, AICC: 734, BIC: 741; CAL: Clinical attachment level, *A. a: Aggregatibacter actinomycetemcomitans, P. g: Porphyromonas gingivalis, P. i: Prevotella intermedia*, CI: Confidence interval. AICC = − 2 *L* + 2 *k* (*k* − 1)/ (n − *k* − 1), BIC = −2 *L* + *k* ln (n). *L*: Likelihood; *k*: Number of parameters; n: Sample size.

4. Discussion

This study aimed to investigate the changes in the periodontal status, as evaluated by the PISA during a longitudinal follow-up, and to identify the factors that associate with the PISA.

The mixed-effect models used in this study did not include BOP and PD as independent variables as the PISA was calculated using these variables. The PlI and CAL were observed to be significantly associated with the PISA at both the patient-level and tooth-level. Thus, the results indicated that the PISA was associated with the oral hygiene status and the morphology of periodontal tissue. A previous report has indicated that loss of CAL is associated with poor oral hygiene [45], change in the PD is associated with BOP [46], the PlI is associated with tooth loss [47], and changes in the PD are associated with tooth-type and tooth mobility [48]. These findings suggest that changes in the clinical parameters after an active periodontal treatment are interactive, and that deteriorations in these clinical parameters ultimately lead to tooth loss.

As shown in Tables 1–3, time was not significantly associated with the PISA. This finding suggests that the PISA was a stable clinical parameter during the follow-up period. In contrast, it has been previously reported that time is significantly associated with changes in the CAL, with a positive coefficient in the mixed-effect modeling [1]. During the follow-up period in our study, the CAL deteriorated gradually. In the site-level analysis, most of the CAL measurements were stable. A small fraction of sites with progressive loss of CAL affected the overall CAL. As shown in Figure 1, most of the patients were classified as Cluster 1 and the PISA of this cluster was stable. However, a drastic increase in the PISA in one tooth had a large effect on the PISA at the patient-level. In cases of deep periodontal pockets, the increase in the number of bleeding sites increased the PISA value exponentially. In this sense, the PISA is more sensitive than BOP%. As shown in Figure 2, the drastic increase in the PISA was derived from the increase in bleeding sites rather than deep periodontal pockets. Thus, PISA is a sensitive clinical parameter for identifying the progression of inflamed periodontal tissue.

The salivary levels of *P. gingivalis* were significantly associated with the PISA at the patient-level and tooth-level. Additionally, the subgingival levels of *P. gingivalis* and *P. intermedia* were significantly associated with the PISA of the periodontal pocket of the sample sites. Previous studies have shown that bacteria levels in the subgingival plaque and in whole saliva are also significantly associated with the PISA [49–54]. Further, levels of *P. gingivalis* in the subgingival plaque from the deepest pockets have shown to be associated with the progression of periodontitis [32], and levels of *P. gingivalis* and *P. intermedia* in the subgingival plaque have shown to be associated with the progression of periodontal inflammation [55]. *P. intermedia* is also correlated with BOP and deeper pockets [56,57]. One study reports that reinfection of the treated sites by *P. gingivalis* and/or *P. intermedia* diminishes the effect of therapy during the follow-up [58], and that an increase in periodontal pathogens increases the PISA. Several studies have also indicated that the adjunctive use of antimicrobial agents during the follow-up enhances the effect of mechanical debridement [59]. Therefore, when the PISA increases during the follow-up period, monitoring periodontal pathogen levels and the adjunctive use of antimicrobial agents may be useful for successful long-term treatment.

Our study has some limitations. First, there was no data regarding the initial mean PISA values of the patients before the periodontal treatment. Thus, we were unable to observe any change in the PISA over the entire treatment period of the patient. Second, all patients were systemically healthy. We enrolled patients with periodontitis and without systemic disease as the effect of systemic inflammation can serve as a bias for PISA. Therefore, the association between PISA and the etiology of periodontal disease due to the acquired risk factors was not elucidated. Third, intra- and inter-examiner calibration sessions were conducted using a periodontal disease model rather than in patients due to the large number of examiners required in a multicenter study [60].

5. Conclusions

For most of the patients, changes in the PISA during the follow-up period after an active periodontal treatment were within 10% of the baseline. However, an increase in the number of bleeding sites in a tooth with a deep periodontal pocket was associated with an exponential increase in the PISA.

Supplementary Materials: The following are available online at https://www.mdpi.com/2077-0383/10/6/1165/s1. Figure S1: Changes in PISA by tooth-type during the follow-up period; Table S1: List of sample teeth for the periodontal pathogen.

Author Contributions: Conceptualization, Y.N. (Yoshiaki Nomura) and T.M.; methodology, Y.N. (Yoshiaki Nomura); software, Y.N. (Yoshiaki Nomura); validation, E.K. and N.H.; formal analysis, Y.N. (Yoshiaki Nomura); investigation, T.M., A.S., A.Y., F.S., F.N., H.T., H.K., K.N., K.T., M.U., M.M., M.F., N.S., N.Y., S.S. (Satoshi Sekino), S.S. (Soh Sato), T.N. (Toshiaki Nakamura), T.S., Y.N. (Yohei Nakayama), Y.O., Y.N. (Yukihiro Numabe) and T.N. (Taneaki Nakagawa); data curation, S.T.; writing—original draft preparation, Y.N. (Yoshiaki Nomura); writing—review and editing, Y.N. (Yoshiaki Nomura), T.M. and K.T.; visualization, Y.N. (Yoshiaki Nomura); supervision, T.M.; project administration, T.M. and T.N. (Taneaki Nakagawa); funding acquisition, T.N. (Taneaki Nakagawa). All authors have read and agreed to the published version of the manuscript.

Funding: This research was funded by a clinical research project grant from the Japanese Society of Periodontology for the diagnosis of periodontitis.

Institutional Review Board Statement: The names of the ethics committee members and their reference numbers are as follows: The regional ethical committee of the Faculty of Dentistry, Niigata University (20-R17-08-06); Keio University School of Medicine, Ethics Committee (20080096); Ethical Committee of Kagoshima University Medical and Dental Hospital (20-58); The Ethics Committee, Nagasaki University Graduate School of Biomedical Sciences (0846-2); Ethical Committee of Kyushu University Faculty of Dental Science (20-11); The Ethics Committee of Osaka Dental University (80712); Ethics Committee of Aichi Gakuin University, School of Dentistry (158); The Ethics Committee of Matsumoto Dental University (0090); The Institutional Review Board of Nippon Dental University (2-1-22); Ethical Committee of Nihon University School of Dentistry (EP08D016); Dental Research Ethics Committee of Tokyo Medical and Dental University (660); Ethics Committee in Nihon University School of Dentistry at Matsudo (EC 08-014); Ethics Committee of Tokyo Dental College (208); The Ethical Review Committee of The Nippon Dental University School of Life Dentistry at Niigata (151); Ohu University Research Ethics Committee (52); Institutional Review Board for Clinical Research of Hokkaido University Hospital (008-0113).

Informed Consent Statement: Informed consent was obtained from all subjects involved in the study.

Data Availability Statement: The data presented in this study are available on request from the corresponding author. The data are not publicly available due to privacy.

Acknowledgments: The authors thank Hiromasa Yoshie, Toshihide Noguchi, Masamitsu Kawanami, Koichi Ito, Yuichi Izumi, Yoshitaka Hara, Osamu Fujise, Yuzo Abe, Tomoo Kono, Asako Makino-Oi, and Chie Fukaya, for their advice and helpful comments.

Conflicts of Interest: The authors declare no conflict of interest.

References

1. Nomura, Y.; Morozumi, T.; Nakagawa, T.; Sugaya, T.; Kawanami, M.; Suzuki, F.; Takahashi, K.; Abe, Y.; Sato, S.; Makino-Oi, A.; et al. Site-level progression of periodontal disease during a follow-up period. *PLoS ONE* **2017**, *12*, e0188670. [CrossRef] [PubMed]
2. Gilthorpe, M.S.; Zamzuri, A.T.; Griffiths, G.S.; Maddick, I.H.; Eaton, K.A.; Johnson, N.W. Unification of the "burst" and "linear" theories of periodontal disease progression: A multilevel manifestation of the same phenomenon. *J. Dent. Res.* **2003**, *82*, 200–205. [CrossRef]
3. Socransky, S.S.; Haffajee, A.D.; Goodson, J.M.; Lindhe, J. New concepts of destructive periodontal disease. *J. Clin. Periodontol.* **1984**, *11*, 21–32. [CrossRef] [PubMed]
4. Hujoel, P.P.; Leroux, B.G. Evaluating the Burst Hypothesis at a Site-Specific Level Using the Lack-of-Fit Test. *J. Periodontol.* **1998**, *69*, 357–362. [CrossRef] [PubMed]
5. Gilthorpe, M.S.; Griffiths, G.S.; Maddick, I.H.; Zamzuri, A.T. An application of multilevel modelling to longitudinal perio-dontal research data. *Community Dent. Health* **2001**, *18*, 79–86. [PubMed]
6. Park, S.Y.; Ahn, S.; Lee, J.T.; Yun, P.Y.; Lee, Y.J.; Lee, J.Y.; Song, Y.W.; Chang, Y.S.; Lee, H.J. Periodontal inflamed surface area as a novel numerical variable describing periodontal conditions. *J. Periodontal Implant. Sci.* **2017**, *47*, 328–338. [CrossRef] [PubMed]
7. Lamont, R.J.; Hajishengallis, G. Polymicrobial synergy and dysbiosis in inflammatory disease. *Trends Mol. Med.* **2015**, *21*, 172–183. [CrossRef]

8. Chambrone, L.A.; Chambrone, L. Tooth loss in well-maintained patients with chronic periodontitis during long-term supportive therapy in Brazil. *J. Clin. Periodontol.* **2006**, *33*, 759–764. [CrossRef]

9. Carnevale, G.; Cairo, F.; Tonetti, M.S. Long-term effects of supportive therapy in periodontal patients treated with fibre retention osseous resective surgery. II: Tooth extractions during active and supportive therapy. *J. Clin. Periodontol.* **2007**, *34*, 342–348. [CrossRef]

10. Fardal, Ø.; Johannessen, A.C.; Linden, G.J. Tooth loss during maintenance following periodontal treatment in a periodontal practice in Norway. *J. Clin. Periodontol.* **2004**, *31*, 550–555. [CrossRef] [PubMed]

11. McLeod, D.E.; Lainson, P.A.; Spivey, J.D. The predictability of periodontal treatment as measured by tooth loss: A retrospec-tive study. *Quintessence Int.* **1998**, *29*, 631–635.

12. McLeod, D.E.; Lainson, P.A.; Spivey, J.D. Tooth Loss Due to Periodontal Abscess: A Retrospective Study. *J. Periodontol.* **1997**, *68*, 963–966. [CrossRef]

13. Loesche, W.J.; Giordano, J.R.; Soehren, S.; Kaciroti, N. The nonsurgical treatment of patients with periodontal disease: Results after five years. *J. Am. Dent. Assoc.* **2002**, *133*, 311–320. [CrossRef] [PubMed]

14. Wilson, T.G.; Glover, M.E.; Malik, A.K.; Schoen, J.A.; Dorsett, D. Tooth Loss in Maintenance Patients in a Private Periodontal Practice. *J. Periodontol.* **1987**, *58*, 231–235. [CrossRef]

15. Wood, W.R.; Greco, G.W.; McFall, W.T. Tooth Loss in Patients with Moderate Periodontitis After Treatment and Long–Term Maintenance Care. *J. Periodontol.* **1989**, *60*, 516–520. [CrossRef]

16. Matthews, D.C.; Smith, C.G.; Hanscom, S.L. Tooth loss in periodontal patients. *J. Can. Dent. Assoc.* **2001**, *67*, 207–210. [PubMed]

17. Tonetti, M.S.; Steffen, P.; Muller-Campanile, V.; Suvan, J.; Lang, N.P. Initial extractions and tooth loss during supportive care in a periodontal population seeking comprehensive care. *J. Clin. Periodontol.* **2000**, *27*, 824–831. [CrossRef] [PubMed]

18. Manresa, C.; Sanz-Miralles, E.C.; Twigg, J.; Bravo, M. Supportive periodontal therapy (SPT) for maintaining the dentition in adults treated for periodontitis. *Cochrane Database Syst. Rev.* **2018**, *1*, CD009376. [CrossRef] [PubMed]

19. Nesse, W.; Abbas, F.; Van Der Ploeg, I.; Spijkervet, F.K.L.; Dijkstra, P.U.; Vissink, A. Periodontal inflamed surface area: Quantifying inflammatory burden. *J. Clin. Periodontol.* **2008**, *35*, 668–673. [CrossRef]

20. Teke, E.; Kırzıoğlu, F.Y.; Korkmaz, H.; Calapoğlu, M.; Orhan, H. Does metabolic control affect salivary adipokines in type 2 diabetes mellitus? *Dent. Med. Probl.* **2019**, *56*, 11–20. [CrossRef]

21. Matsuda, S.; Okanobu, A.; Hatano, S.; Kajiya, M.; Sasaki, S.; Hamamoto, Y.; Iwata, T.; Ouhara, K.; Takeda, K.; Mizuno, N.; et al. Relationship between periodontal inflammation and calcium channel blockers induced gingival overgrowth—A cross-sectional study in a Japanese population. *Clin. Oral Investig.* **2019**, *23*, 4099–4105. [CrossRef] [PubMed]

22. Leira, Y.; Rodríguez-Yáñez, M.; Arias, S.; López-Dequidt, I.; Campos, F.; Sobrino, T.; D'Aiuto, F.; Castillo, J.; Blanco, J. Periodontitis is associated with systemic inflammation and vascular endothelial dysfunction in patients with lacunar infarct. *J. Periodontol.* **2019**, *90*, 465–474. [CrossRef] [PubMed]

23. Temelli, B.; Ay, Z.Y.; Aksoy, F.; Büyükbayram, H.I.; Doguc, D.K.; Uskun, E.; Varol, E. Platelet indices (mean platelet volume and platelet distribution width) have correlations with periodontal inflamed surface area in coronary artery disease patients: A pilot study. *J. Periodontol.* **2018**, *89*, 1203–1212. [CrossRef] [PubMed]

24. Punj, A.; Shenoy, S.B.; Subramanyam, K. Comparison of Endothelial Function in Healthy Patients and Patients with Chronic Periodontitis and Myocardial Infarction. *J. Periodontol.* **2017**, *88*, 1234–1243. [CrossRef]

25. Khanuja, P.K.; Narula, S.C.; Rajput, R.; Sharma, R.K.; Tewari, S. Association of periodontal disease with glycemic control in patients with type 2 diabetes in Indian population. *Front. Med.* **2017**, *11*, 110–119. [CrossRef] [PubMed]

26. Nesse, W.; Linde, A.; Abbas, F.; Spijkervet, F.K.L.; Dijkstra, P.U.; De Brabander, E.C.; Gerstenbluth, I.; Vissink, A. Dose-response relationship between periodontal inflamed surface area and HbA1c in type 2 Diabetics. *J. Clin. Periodontol.* **2009**, *36*, 295–300. [CrossRef] [PubMed]

27. Iwasaki, M.; Kimura, Y.; Yamaga, T.; Yamamoto, N.; Ishikawa, M.; Wada, T.; Sakamoto, R.; Ishimoto, Y.; Fujisawa, M.; Okumiya, K.; et al. A population-based cross-sectional study of the association between periodontitis and arterial stiffness among the older Japanese population. *J. Periodontal Res.* **2020**. [CrossRef]

28. Aoyama, N.; Fujii, T.; Kida, S.; Nozawa, I.; Taniguchi, K.; Fujiwara, M.; Iwane, T.; Tamaki, K.; Minabe, M. Association of Periodontal Status, Number of Teeth, and Obesity: A Cross-Sectional Study in Japan. *J. Clin. Med.* **2021**, *10*, 208. [CrossRef]

29. Salhi, L.; Albert, A.; Seidel, L.; Lambert, F. Respective Effects of Oral Hygiene Instructions and Periodontal Nonsurgical Treatment (Debridement) on Clinical Parameters and Patient-Reported Outcome Measures with Respect to Smoking. *J. Clin. Med.* **2020**, *9*, 2491. [CrossRef]

30. Echeverría, J.J.; Manau, G.C.; Guerrero, A. Supportive care after active periodontal treatment: A review. *J. Clin. Periodontol.* **1996**, *23*. [CrossRef]

31. Morozumi, T.; Nakagawa, T.; Nomura, Y.; Sugaya, T.; Kawanami, M.; Suzuki, F.; Takahashi, K.; Abe, Y.; Sato, S.; Makino-Oi, A.; et al. Salivary pathogen and serum antibody to assess the progression of chronic periodontitis: A 24-mo prospective multicenter cohort study. *J. Periodontal Res.* **2016**, *51*, 768–778. [CrossRef] [PubMed]

32. Kakuta, E.; Nomura, Y.; Morozumi, T.; Nakagawa, T.; Nakamura, T.; Noguchi, K.; Yoshimura, A.; Hara, Y.; Fujise, O.; Nishimura, F.; et al. Assessing the progression of chronic periodontitis using subgingival pathogen levels: A 24-month prospective multicenter cohort study. *BMC Oral Health* **2017**, *17*, 46. [CrossRef] [PubMed]

33. Krebs, K.A.; Clem, D.S., 3rd. American Academy of Periodontology, Guidelines for the Management of Patients with Periodontal Diseases. *J. Periodontol.* **2006**, *77*, 1607–1611. [CrossRef] [PubMed]

34. Papapanou, P.N.; Sanz, M.; Buduneli, N.; Dietrich, T.; Feres, M.; Fine, D.H.; Flemmig, T.F.; Garcia, R.; Giannobile, W.V.; Graziani, F.; et al. Periodontitis: Consensus report of workgroup 2 of the 2017 World Workshop on the Classification of Periodontal and Peri-Implant Diseases and Conditions. *J. Periodontol.* **2018**, *89* (Suppl. 1), S173–S182. [CrossRef]

35. Tadokoro, K.; Yamaguchi, T.; Kawamura, K.; Shimizu, H.; Egashira, T.; Minabe, M.; Yoshino, T.; Oguchi, H. Rapid quantification of periodontitis-related bacteria using a novel modification of Invader PLUS technologies. *Microbiol. Res.* **2010**, *165*, 43–49. [CrossRef] [PubMed]

36. Tada, A.; Takeuchi, H.; Shimizu, H.; Tadokoro, K.; Tanaka, K.; Kawamura, K.; Yamaguchi, T.; Egashira, T.; Nomura, Y.; Hanada, N. Quantification of Periodontopathic Bacteria in Saliva Using the Invader Assay. *Jpn. J. Infect. Dis.* **2012**, *65*, 415–423. [CrossRef]

37. Lyons, S.R.; Griffen, A.L.; Leys, E.J. Quantitative real-time PCR for Porphyromonas gingivalis and total bacteria. *J. Clin. Microbiol.* **2000**, *38*, 2362–2365. [CrossRef]

38. Haffajee, A.D.; Socransky, S.S.; Smith, C.; Dibart, S. Relation of baseline microbial parameters to future periodontal attachment loss. *J. Clin. Periodontol.* **1991**, *18*, 744–750. [CrossRef]

39. Tomita, S.; Komiya-Ito, A.; Imamura, K.; Kita, D.; Ota, K.; Takayama, S.; Makino-Oi, A.; Kinumatsu, T.; Ota, M.; Saito, A. Prevalence of Aggregatibacter actinomycetemcomitans, Porphyromonas gingivalis and Tannerella forsythia in Japanese patients with generalized chronic and aggressive periodontitis. *Microb. Pathog.* **2013**, *61–62*, 11–15. [CrossRef]

40. Nomura, Y.; Otsuka, R.; Wint, W.Y.; Okada, A.; Hasegawa, R.; Hanada, N. Tooth-Level Analysis of Dental Caries in Primary Dentition in Myanmar Children. *Int. J. Environ. Res. Public Health* **2020**, *17*, 7613. [CrossRef]

41. Nakahodo, N.; Nomura, Y.; Oshiro, T.; Otsuka, R.; Kakuta, E.; Okada, A.; Inai, Y.; Takei, N.; Hanada, N. Effect of Mucosal Brushing on the Serum Levels of C-Reactive Protein for Patients Hospitalized with Acute Symptoms. *Medicina* **2020**, *56*, 549. [CrossRef]

42. Otsuka, R.; Nomura, Y.; Okada, A.; Uematsu, H.; Nakano, M.; Hikiji, K.; Hanada, N.; Momoi, Y. Properties of manual toothbrush that influence on plaque removal of interproximal surface in vitro. *J. Dent. Sci.* **2020**, *15*, 14–21. [CrossRef] [PubMed]

43. Nomura, Y.; Maung, K.; Kay Khine, E.M.; Sint, K.M.; Lin, M.P.; Win Myint, M.K.; Aung, T.; Sogabe, K.; Otsuka, R.; Okada, A.; et al. Prevalence of Dental Caries in 5- and 6-Year-Old Myanmar Children. *Int. J. Dent.* **2019**, *2019*, 5948379. [CrossRef] [PubMed]

44. Nomura, Y.; Takei, N.; Ishii, T.; Takada, K.; Amitani, Y.; Koganezawa, H.; Fukuhara, S.; Asai, K.; Uozumi, R.; Bessho, K. Factors That Affect Oral Care Outcomes for Institutionalized Elderly. *Int. J. Dent.* **2018**, *2018*, 2478408. [CrossRef]

45. Sanz-Martín, I.; Cha, J.-K.; Yoon, S.W.; Sanz-Sánchez, I.; Jung, U.W. Long-term assessment of periodontal disease progression after surgical or non-surgical treatment: A systematic review. *J. Periodontal Implant. Sci.* **2019**, *49*, 60–75. [CrossRef] [PubMed]

46. Ramseier, C.A.; Nydegger, M.; Walter, C.; Fischer, G.; Sculean, A.; Lang, N.P.; Salvi, G.E. Time between recall visits and residual probing depths predict long-term stability in patients enrolled in supportive periodontal therapy. *J. Clin. Periodontol.* **2019**, *46*, 218–230. [CrossRef]

47. Eickholz, P.; Kaltschmitt, J.; Berbig, J.; Reitmeir, P.; Pretzl, B. Tooth loss after active periodontal therapy. 1: Patient-related factors for risk, prognosis, and quality of outcome. *J. Clin. Periodontol.* **2008**, *35*, 165–174. [CrossRef]

48. König, J.; Plagmann, H.-C.; Rühling, A.; Kocher, T. Tooth loss and pocket probing depths in compliant periodontally treated patients: A retrospective analysis. *J. Clin. Periodontol.* **2002**, *29*, 1092–1100. [CrossRef]

49. Haririan, H.; Andrukhov, O.; Bertl, K.; Lettner, S.; Kierstein, S.; Moritz, A.; Rausch-Fan, X. Microbial Analysis of Subgingival Plaque Samples Compared to That of Whole Saliva in Patients with Periodontitis. *J. Periodontol.* **2014**, *85*, 819–828. [CrossRef] [PubMed]

50. Belstrøm, D.; Sembler-Møller, M.L.; Grande, M.A.; Kirkby, N.; Cotton, S.L.; Paster, B.J.; Holmstrup, P. Microbial profile comparisons of saliva, pooled and site-specific subgingival samples in periodontitis patients. *PLoS ONE* **2017**, *12*, e0182992. [CrossRef]

51. Testa, M.; Ruiz de Valladares, R.; Benito de Cárdenas, I.L. Correlation between bacterial counts in saliva and subgingival plaque. *Acta Odontol Latinoam* **1999**, *12*, 63–74.

52. Darout, I.A.; Albandar, J.M.; Skaug, N. Correlations between bacterial levels in autologous subgingival plaque and saliva of adult Sudanese. *Clin. Oral Investig.* **2002**, *6*, 210–216. [CrossRef]

53. Könönen, E.; Jousimies-Somer, H.; Asikainen, S. The most frequently isolated gram-negative anaerobes in saliva and subgingival samples taken from young women. *Oral Microbiol. Immunol.* **1994**, *9*, 126–128. [CrossRef]

54. Okada, A.; Sogabe, K.; Takeuchi, H.; Okamoto, M.; Nomura, Y.; Hanada, N. Characterization of specimens obtained by different sampling methods for evaluation of periodontal bacteria. *J. Oral Sci.* **2017**, *59*, 491–498. [CrossRef]

55. Yang, N.-Y.; Zhang, Q.; Li, J.L.; Yang, S.H.; Shi, Q. Progression of periodontal inflammation in adolescents is associated with increased number of Porphyromonas gingivalis, Prevotella intermedia, Tannerella forsythensis, and Fusobacterium nucleatum. *Int. J. Paediatr. Dent.* **2013**, *24*, 226–233. [CrossRef] [PubMed]

56. Alves, A.C.B.; Napimoga, M.H.; Klein, M.I.; Höfling, J.F.; Gonçalves, R.B. Increase in Probing Depth Is Correlated With a Higher Number of Prevotella intermedia Genotypes. *J. Periodontol.* **2006**, *77*, 61–66. [CrossRef] [PubMed]

57. Mombelli, A.; Schmid, B.; Rutar, A.; Lang, N.P. Persistence Patterns of Porphyromonas gingivalis, Prevotella intermedia/nigrescens, and Actinobacillus actinomycetemcomitans After Mechanical Therapy of Periodontal Disease. *J. Periodontol.* **2000**, *71*, 14–21. [CrossRef] [PubMed]

58. Shiloah, J.; Patters, M.R. Repopulation of Periodontal Pockets by Microbial Pathogens in the Absence of Supportive Therapy. *J. Periodontol.* **1996**, *67*, 130–139. [CrossRef]

59. Venezia, E.; Shapira, L. Use of antimicrobial agents during supportive periodontal therapy. *Oral Dis.* **2003**, *9* (Suppl. 1), 63–70. [CrossRef]
60. Yashima, A.; Morozumi, T.; Yoshie, H.; Hokari, T.; Izumi, Y.; Akizuki, T.; Mizutani, K.; Takamatsu, H.; Minabe, M.; Miyauchi, S.; et al. Biological responses following one-stage full-mouth scaling and root planing with and without azithromycin: Multicenter randomized trial. *J. Periodontal Res.* **2019**, *54*, 709–719. [CrossRef]

Journal of
Clinical Medicine

Article

Exploring the Role of Interleukin-6 Receptor Inhibitor Tocilizumab in Patients with Active Rheumatoid Arthritis and Periodontal Disease

Codrina Ancuța [1], Rodica Chirieac [2], Eugen Ancuța [3], Oana Țănculescu [4,*], Sorina Mihaela Solomon [4,*], Ana Maria Fătu [5], Adrian Doloca [6] and Cristina Iordache [5]

1   Department of Rheumatology, Faculty of Medicine, Grigore T. Popa University of Medicine and Pharmacy, 700115 Iasi, Romania; codrina_ancuta@yahoo.com
2   Sanocare Medical and Research Center, 700503 Iasi, Romania; chiriac01ro@yahoo.com
3   Research Department, Elena Doamna Clinical Hospital, 700398 Iasi, Romania; eugen01ro@yahoo.com
4   Department of Odontology-Periodontology, Fixed Prosthodontics, Faculty of Dental Medicine, Grigore T. Popa University of Medicine and Pharmacy, 700115 Iasi, Romania
5   Implantology, Removable Dentures, Dental Technology Department, Faculty of Dentistry, Grigore T. Popa University of Medicine and Pharmacy, 700115 Iasi, Romania; anafatu2007@yahoo.com (A.M.F.); ccmiiordache@yahoo.com (C.I.)
6   Department of Preventive Medicine and Interdisciplinarity, Faculty of Medicine, Grigore T. Popa University of Medicine and Pharmacy, 700115 Iasi, Romania; adrian.doloca@umfiasi.ro
*   Correspondence: oana.tanculescu@umfiasi.ro (O.Ț.); drsolomonro@gmail.com (S.M.S.)

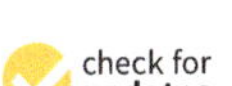

Citation: Ancuța, C.; Chirieac, R.; Ancuța, E.; Țănculescu, O.; Solomon, S.M.; Fătu, A.M.; Doloca, A.; Iordache, C. Exploring the Role of Interleukin-6 Receptor Inhibitor Tocilizumab in Patients with Active Rheumatoid Arthritis and Periodontal Disease. *J. Clin. Med.* **2021**, *10*, 878. https://doi.org/10.3390/jcm10040878

Academic Editor: Yorimasa Ogata

Received: 31 January 2021
Accepted: 18 February 2021
Published: 20 February 2021

Publisher's Note: MDPI stays neutral with regard to jurisdictional claims in published maps and institutional affiliations.

Abstract: Background: The aim of our study was to explore the influence of weekly subcutaneous administration of interleukin-6 (IL-6) receptor inhibitor tocilizumab (TCZ) on periodontal status in a local longitudinal study of patients with rheumatoid arthritis (RA) and periodontal disease (PD). Methods: We performed a 6-month prospective study in 51 patients with chronic periodontitis and moderate-to-severe RA starting TCZ in accordance with local recommendations. Extensive rheumatologic (clinical activity, inflammatory, serological biomarkers) and periodontal (visible plaque index, gingival index, bleeding on probing, probing pocket depth, clinical attachment loss) assessments were done. Changes in RA activity and periodontal status were reassessed after 3 and 6 months. Results: We demonstrated significant correlations between periodontal status, disease activity, and serologic biomarkers ($p < 0.05$). Tocilizumab significantly improved the gingival index scores and decreased the number of sites with bleeding on probing after only 3 months ($p < 0.05$), while the probing pocket depth significantly decreased after 6 months; overall, clinical attachment loss presented only slight changes without any statistical significance as well as teeth count and plaque levels ($p > 0.05$). Conclusion: IL-6 inhibition is able to improve periodontal outcomes in patients with RA and concomitant PD, which is essentially related to a dramatic decrease in serum inflammatory mediators.

Keywords: rheumatoid arthritis; periodontal disease; tocilizumab

1. Introduction

1.1. Rheumatoid Arthritis and Periodontal Disease Association

Rheumatoid arthritis (RA) is an autoimmune condition characterized by joint inflammation and destruction associated with chronic systemic inflammation accounting for significantly impaired quality of life. It is defined by the excessive activation of proinflammatory cytokines mediators, specific autoantibodies and progressive irreversible articular damage [1,2]. Considered the hallmark of RA and, perhaps, the most important step in the pathobiology of the disease, the immune response against citrullinated peptides is driven, at least in part, by antigens derived from the periodontal tissue exposed to a dysbiotic oral microbiome during periodontitis [1–4].

Chronic periodontitis is a complex condition outlined by chronic inflammation and the subsequent damage of soft collagen-rich tissues progressing to periodontal ligament and alveolar bone loss as well as a gradual increase in tooth mobility [1–9]. It is typically initiated by an infection with oral anaerobic bacteria followed by inflammatory and immune response in the gingival and periodontal microenvironment [1–3]. Autoantigens generated during citrullination induced by peptidylarginine-deaminase enzyme produced by *Porphyromonas gingivalis (P. gingivalis)*, the keystone pathogen in the oral microbial biofilm, break the immune tolerance with induction of anti-citrullinated protein antibodies (ACPA) and promote chronic inflammatory response in both periodontal and synovial/articular tissues [1,3,4,10–18].

Periodontal disease (PD) is generally associated with a broad spectrum of chronic systemic disorders including diabetes, cardiovascular, respiratory, kidney, and neurodegenerative diseases as well as immune-mediated rheumatic conditions [1–5,19]. Several epidemiological studies have already communicated that PD is more prevalent during RA and vice versa [4,20,21]. Indeed, patients with PD have an increased risk to develop RA, compared to general population, particularly those with a long history of more severe periodontitis, mostly explained by excessive protein citrullination [14,21]. Furthermore, it seems that *P. gingivalis* positive-periodontitis is more likely to occur in ACPA-positive individuals without any arthritis, suggesting that PD may precede RA [4,11,15,20,22]. On the other hand, RA patients experience a greater risk of PD, irrespective of disease duration, especially in ACPA-positive subtype [1–3,6–8,11,14,19]; moreover, they are prone to develop moderate to severe periodontitis in established compared to early disease [6–9,14,19,23]. A detailed analysis of periodontal status in first-degree relatives of RA cases discovered a higher prevalence and severity of periodontitis in ACPA-positive RA [9,14,22,23]. Altered periodontal condition during RA seems to be multifactorial, related to increased serum concentrations of proinflammatory cytokines and altered motor skills of the rheumatoid hand which can also contribute to compromised oral hygiene [1,4,5,22].

This intriguing relationship between PD and RA is roughly supported by similar pathogenic pathways in a genetically predisposed host (human leukocyte antigen HLA-haplotype DRB1, HLA-DRB1, shared epitope) triggered by common environmental risk factors (cigarette smoking) [1,2,4,10,18,24–26]. Important pathobiologic processes refer to the overexpression of proinflammatory cytokines (tumor necrosis factor alpha—TNF-α, IL-1β, IL-6 and IL-17), inflammatory mediators (prostaglandin E2, nitric oxide) and degradation enzymes (matrix metalloproteinases 1, 8, 9, and 13), osteoclast activation, and progressive articular and alveolar bone damage [1,2,4,10,18,23–26]. Considered as the cytokine signature, the aberrant activation of TNF-α and IL-6 regulates immune response and bone metabolism in RA [1,3,5,6,16]; high concentrations of both cytokines were detected in serum, synovial tissues, as well as synovial fluids [18,27], positively correlating with disease activity [28]. Different studies have also confirmed higher levels of potent IL-6 and TNF-α in inflamed gingival tissues, gingival crevicular fluid, and serum in patients with PD than in the healthy controls [17,28–33]. Moreover, increased TNF concentrations are associated with less favorable periodontal indices such as bleeding on probing (BOP), probing pocket depth (PPD), and clinical attachment loss (CAL), while serum IL-6 concentrations decreased following periodontal treatment [17,27,28,32,34]. Surprisingly, salivary levels of TNF-α, IL-6, IL-8, and IL-17A can be affected not only by periodontitis but also in RA [35,36]. Furthermore, the interplay between the subgingival biofilm, particularly PD-associated pathogens, and the host immune system may contribute to both PD and RA [1,2,18].

1.2. The Role of Different Therapies on Rheumatoid Arthritis and Periodontal Disease Outcomes

The evolving model for dynamic interrelation between RA and PD encourages the concept that standard management for RA may be effective in improving the outcomes in PD and vice versa [4,7,14,19,37–39]. Pivotal studies have already explored the role of different synthetic and biological therapies in active RA and comorbid periodontal disease, showing controversial results [5,9,12,16,17,25,28,35–39]. Overall, there is a trend to consider that TNF inhibitors, IL-6 receptor antagonist, B-cells depletive agents and, even, JAK inhibitors improve periodontal health in both RA and other arthritis (e.g., ankylosing spondylitis, psoriatic arthritis); it seems that all these drugs are ultimately effective in decreasing gingival and periodontal inflammation and, to a lesser extent, associated tissue damage [16,27,28,35–37,39–50]. Researchers even proposed a multistep approach of the sequential tissue repairing following TNF inhibitors, comprising reduced leukocytes traffic in the inflamed tissue, decreased proteolytic activity, and the normalization of osteoclast activity [1,5,9,12]. However, there are differences among anti-TNF agents as only adalimumab and etanercept significantly improved periodontal outcomes in as rapid as six months, while infliximab worsened gingival inflammation but prevented gingival bone loss [5,14]. Furthermore, according to a study by Kobayashi and colleagues, tocilizumab (TCZ) also ameliorates periodontal inflammation in RA with periodontitis as TNF inhibitors do [16,17,27,28,31,32,36,37]. Its beneficial effects were potentially explained by the decrease in TNF-α serum levels as well as immunoglobulin G and serum amyloid along with a consistent impact on serum inflammatory mediators and indirect influence on periodontal inflammation [16,17,27,28,40].

On the other hand, several papers have addressed the effect of specific periodontal therapies (e.g., non-surgical scaling and root planning) on clinical RA activity in patients with chronic periodontitis with controversial results [37,51–57]. The most recent data from the ESPERA (Experimental Study of Periodontitis and Rheumatoid Arthritis) cohort failed to demonstrate clinical improvement in established RA following aggressive and intensive periodontal treatment [57].

Considering the gap in the literature regarding the role of anti-rheumatic drugs on periodontal outcomes, the aim of our study is to assess the influence on the periodontal status of weekly subcutaneous administration of tocilizumab in a local group of patients with rheumatoid arthritis and chronic periodontitis.

2. Materials and Methods

2.1. Study Design and Population

We performed a prospective longitudinal study in fifty-one patients with moderate-to-severe RA and insufficient response to either conventional synthetic or biologic disease-modifying antirheumatic drugs (DMARDs), starting TCZ according to the local recommendation for biologic and targeted synthetic therapy aligned with European League Against Rheumatism (EULAR) consensus statement and guidelines.

We performed extensive rheumatologic and full mouth assessments at baseline (before the first administration of TCZ) as well as after 3 and 6 months of therapy.

2.2. Inclusion Criteria

The patients were aged 18 and older, able to give informed consent themselves and to participate in the study, and willing to forgo any optional examinations.

Patients fulfilled either the old 1987 American College of Rheumatology (ACR) criteria or the new 2010 classification criteria of ACR and EULAR and were followed up in one academic rheumatology department in Northeast Romania over a period of 3 years (July 2017–January 2020).

2.3. Exclusion Criteria

Several exclusion criteria were applied before enrollment in this study because of their potential interference with a correct evaluation of periodontal status, as follows: ex- or current smokers, pregnant and breastfeeding women, patients with diabetes mellitus, implants, poorly fitting fixed and/or removable prosthodontics and fewer than eight evaluable teeth, patients receiving systemic or local antimicrobials, antiplatelet drugs, any type of anti-inflammatory medication or periodontal therapy within the previous 3 months.

A total of sixty-eight patients were eligible for and received TCZ for their active RA; however, among them, seventeen had no oral issues at baseline evaluation and were excluded from the study.

2.4. Ethical Considerations

The baseline clinical documentation of periodontitis cases was collected in the context of routine check-up in the dental clinic of Sanocare Medical and Research Center.

The study was approved by the local ethics committee (Sanocare Medical and Research Center, Prot. No 15/12.12.2016) and was found to conform to the guidelines of the Declaration of Helsinki. Written informed consent regarding the use of the collected data in the context of training and research was signed by all the participants before enrollment. The data used in the study were anonymized. According to the U.S. Department of Health and Human Services (HHS) definition, this investigation is not considered human subjects research.

2.5. Rheumatologic Assessments

RA-related variables comprised clinical (tender and swollen joint count based on a 28-joint assessment, 0–10 cm visual analogue scale, VAS, pain), inflammatory tests (erythrocyte sedimentation rate, ESR, and C-reactive protein, CRP) as well as disease activity scores calculated on DAS28-CRP (Disease Activity Score on 28 joints using C-reactive protein) and SDAI (Simplified Disease Activity Index) were performed at all three visits.

DAS28-CRP was calculated with a formula that considered the tender and swollen joints, the patient's general assessment of their condition scored on a visual analogue scale (VAS), and CRP. DAS28-CRP comprises four categories: remission (DAS28-CRP < 2.3), low (2.3 $\leq$ DAS28-CRP < 2.7), moderate (2.7 $\leq$ DAS28-CRP < 4.1), and high disease activity (4.1 $\leq$ DAS28-CRP).

Designed as the numerical sum of five outcome parameters (tender and swollen joints, patient, and physician global assessment of disease activity on a 0–10 VAS and CRP level), SDAI score interpretation comprises also four categories: remission (0–3.3), low activity (3.4–11), moderate activity (11.1–26), and high RA activity (26.1–86).

Serological biomarkers (rheumatoid factor, RF, and ACPA) were evaluated only at baseline.

2.6. Periodontal Assessments

The periodontal status was recorded on a periodontal chart displaying the following clinical parameters for the entire dentition: number of present teeth, visible plaque index (VPI), gingival index (GI), bleeding on probing (BOP), probing pocket depth (PPD), and clinical attachment loss (CAL).

Clinical periodontal assessments were performed by a single trained examiner (C.I.) at the Sanocare Medical and Research Center, Iasi, who was blinded to the rheumatologic data. Access to previous assessment data was not allowed during the study. The examiner was considered calibrated when no statistically significant differences between measurements were obtained after the evaluation of 15 non-participant subjects on two occasions, one week apart. The mean values were assessed using paired *t*-test for VPI, GI, BOP, PPD, and CAL.

The periodontal evaluation was made in artificial light conditions, using a dental explorer, dental mirror, Williams probe, and air–water syringe.

VPI [58] or supragingival plaque was recorded dichotomously (present/absent) at 4 sites (mesial, distal, buccal, lingual/palatal) around each tooth. GI [59] assesses the gingival condition by gentle probing of the soft gingival wall at four sites for each tooth, as follows: 0 = absence of inflammation; 1 = mild inflammation: slight changes in color and texture, and slight edema, no bleeding on probing; 2 = moderate inflammation: redness, edema, glazing, bleeding on probing; 3 = severe inflammation: marked redness, edema, ulceration, tendency to spontaneous bleeding.

BOP evaluates gingival inflammation through bleeding observed 20 s after a probe is passed along inside the gingival sulcus or pocket. It was recorded dichotomously (present/absent) at 6 sites (mesiobuccal, mid-buccal, distobuccal, mesiolingual, mid-lingual, and distolingual) around each tooth.

PPD was measured between the gingival margin and the bottom of gingival sulcus or pocket, while CAL was measured between the cemento–enamel junction and the base of the gingival sulcus or pocket. Both parameters were determined with Williams periodontal probe and recorded on 6 sites per tooth. The recorded values were to the nearest millimeter, and every reading close to 0.5 mm was rounded to the nearest integer number.

PD was considered according to the case definition proposed by the 5th European Workshop on Periodontology in 2005 [60], as follows: level 1 (mild)—CAL $\geq$ 3 mm in 2 or more proximal sites of non-adjacent teeth and level 2 (severe)—CAL $\geq$ 5 mm in 30% or more proximal sites of teeth present; level 0 was considered—for healthy periodontal status or up to one proximal site with CAL $\geq$ 3 mm.

Patients were instructed to maintain their oral hygiene habits throughout the 6 months of follow-up; furthermore, as we intended to assess the accurate effect of TCZ on periodontal status, any periodontal treatment was avoided.

2.7. Statistical Analysis

Statistical analysis was performed with the IBM SPSS Statistics for Windows, Version 19.0. (IBM Corp., Armonk, NY, USA), with p-values less than 0.05 being considered as statistically significant; data at 3 and 6 months were summarized as means $\pm$ SD, or percentages (%) as appropriate, and correlations by Spearman rank tests; comparisons between baseline and 3 and 6 months of TCZ were assessed by Wilcoxon test.

3. Results

3.1. Baseline RA and Periodontal Assessments

Demographics, rheumatologic and periodontal characteristics, as well as RA-related drugs (concomitant glucocorticoids and immunosuppressives) taken at baseline are summarized in Table 1.

Most patients included in our study had seropositive established RA, with moderate-to-severe activity despite background medication. Eight patients (15.68%) received TCZ as their first biologic agent (bio-naïve), while the majority were bio-experienced patients, with failure (either insufficient response or adverse reactions) to previous biologics—15 (29.41%) to one biologic, 20 (39.21%) to two biologics, and 8 (15.68%) to three biologic agents.

We detected impaired oral health in all patients included in the final analysis, as follows: all had gingivitis (abnormal GI and increased prevalence of sites with BOP), and different degrees of chronic periodontitis (mainly level 1 and 2); advanced loss of attachment was reported in up to 23.52% of cases, while increased prevalence of sites with dental plaques in 21.56% of cases.

A closer look revealed a consistent positive correlation between the severity of chronic periodontitis, RA activity, and serum $r_2 = 0.71$ ACPA concentrations ($r_1 = 0.81$, $p_1 = 0.001$, $p_2 = 0.002$, respectively): the higher the RA activity and ACPA levels, the higher the PPD severity, with advanced CAL and tooth loss.

Table 1. Demographic, rheumatologic, and periodontal characteristics at baseline.

Baseline Parameters	Baseline
Demographics	
Age (years; mean $\pm$ SD)	56.3 $\pm$ 15.7
Female (*n*, %)	46 (90.1)
RA-related parameters	
Duration of RA (months; mean $\pm$ SD)	81.3 $\pm$ 68.9
DAS28-CRP (mean $\pm$ SD)	5.36 $\pm$ 1.67
SDAI (mean $\pm$ SD)	34.2 $\pm$ 16.3
Corticosteroids (*n*, %)	20 (39.21)
DMARDs (*n*, %)	46 (90.1)
ACPA levels (U/mL; mean $\pm$ SD)	239.7 $\pm$ 124.3
ACPA positivity (*n*, %)	32 (62.74)
RF levels (IU/mL; mean $\pm$ SD)	192.7 $\pm$ 85.3
RF positivity (*n*, %)	47 (92.15)
Serum CRP levels (mg/dL; mean $\pm$ SD)	15.3 $\pm$ 6.9
PD-related parameters	
Number of present teeth (mean $\pm$ SD)	23.7 $\pm$ 3.4
GI (mean $\pm$ SD)	0.98 $\pm$ 0.12
% sites with plaque (mean $\pm$ SD)	32.4 $\pm$ 16.9
% sites with BOP (mean $\pm$ SD)	10.2 $\pm$ 8.6
PPD (mm; mean $\pm$ SD)	2.8 $\pm$ 0.4
% sites with PPD $\geq$ 4 mm	12.7 $\pm$ 2.5
CAL (mm)	3.5 $\pm$ 1.2
% CAL $\geq$ 3 mm	12.5 $\pm$ 0.2

RA, rheumatoid arthritis; DAS28-CRP, Disease Activity Score on 28 joints using C-reactive protein; SDAI, Simplified Disease Activity Index; DMARDs, disease-modifying antirheumatic drugs; ACPA, anti-citrullinated protein antibodies; RF, rheumatoid factor; PD, periodontal disease; GI, gingival index; BOP, bleeding on probing; PPD, probing pocket depth; CAL, clinical attachment loss; n, number; SD, standard deviation; %, percent.

3.2. Changes in Rheumatologic and Periodontal Parameters with Tocilizumab

Changes in RA activity and periodontal status were reassessed after 3 and 6 months of TCZ; at follow-up visits, we reported significant improvement as compared to baseline ($p < 0.05$), although the results at 6 months were only slightly different from data obtained at 3 months ($p > 0.05$).

3.2.1. Changes in Rheumatologic Status

Patients displayed consistent improvements in clinical activity meaning a significant decrease in the number of tender and swollen joints, VAS pain, and morning stiffness as rapid as 3 months; as expected, clinical response was maintained 3 months later in all patients, at the final monitoring visit. Similarly, we reported a dramatic decline in inflammatory biomarkers (both ESR and CRP), as well as a considerable immunologic response, particularly for serum levels of ACPA, but also for RF (Table 2).

Table 2. Changes in rheumatoid arthritis (RA)-related parameters at 3 and 6 months after tocilizumab.

Parameter	Baseline	3 Months (V1)	6 Months (V2)	*p*-Value
DAS28-CRP (mean $\pm$ SD)	5.36 $\pm$ 1.67	3.39 $\pm$ 0.57	2.41 $\pm$ 0.19	* <0.05; ** <0.05
SDAI (mean $\pm$ SD)	34.2 $\pm$ 16.3	18.1 $\pm$ 8.2	11.1 $\pm$ 4.3	* <0.05; ** <0.05
Number of tender joints (mean $\pm$ SD)	12.31 $\pm$ 4.29	4.56 $\pm$ 1.31	3.55 $\pm$ 1.13	* <0.05; ** NS
Number of swollen joints (mean $\pm$ SD)	10.01 $\pm$ 3.37	2.85 $\pm$ 4.22	1.50 $\pm$ 2.09	* <0.05; ** NS
Pain VAS mm (mean $\pm$ SD)	82.7 $\pm$ 21.5	28.8 $\pm$ 23.2	16.3 $\pm$ 11.8	* <0.05; ** <0.05
Serum anti-CCP titer (U/mL) (mean $\pm$ SD)	239.7 $\pm$ 124.3	192.6 $\pm$ 112.4	123.6 $\pm$ 101.6	* <0.05; ** NS
Serum RF levels (IU/mL) (mean $\pm$ SD)	192.7 $\pm$ 85.3	164.8 $\pm$ 92.5	151.7 $\pm$ 89.3	* NS; ** NS
Serum CRP levels (mg/dL) (mean $\pm$ SD)	15.3 $\pm$ 6.9	4.12 $\pm$ 0.92	3.92 $\pm$ 0.34	* <0.05; ** NS

SD; standard deviation; DAS28-CRP, Disease activity score on 28 joints based on C-reactive protein; SDAI, Simplified Disease Activity Index; VAS, 0–10 cm visual analogue scale; CCP, cyclic citrullinated peptide; RF, rheumatoid factor; V, visits; * V1 compared to baseline; ** V2 compared to V1; NS, non-significant (0.05).

DAS28-CRP and SDAI strongly improved during monitoring visits reaching either low disease activity or, even, remission (EULAR responders) vs. baseline, irrespective of the severity of periodontitis.

3.2.2. Changes in Periodontal Status

Clinical data showed improvement in periodontal inflammation after only 3 months of TCZ and maintained over 6 months, as supported by an important decrease in gingival index and sites with bleeding of probing ($p < 0.05$). However, the improvement of specific periodontal parameters such as probing pocket depth becomes evident after prolonged treatment (6 months); overall, clinical attachment loss presented only slight changes without any statistical significance; teeth count and bacterial plaque scores were also not significantly influenced by medication ($p > 0.05$) (Table 3).

Table 3. Changes in PD-related parameters at 3 and 6 months after tocilizumab.

Parameter	Baseline	3 Months (V1)	6 Months (V2)	*p*-Value
GI	0.98 ± 0.12	0.85 ± 0.17	0.81 ± 0.18	* <0.05; ** NS
% sites with plaque	32.4 ± 16.9	30.5 ± 14.2	30.2 ± 15.8	* NS; ** NS
% sites with BOP	10.2 ± 8.6	7.3 ± 6.1	6.5 ± 6.8	* <0.05; ** NS
PPD (mm)	2.8 ± 0.4	2.1 ± 0.12	2.1 ± 0.09	* <0.05; ** NS
% sites with PPD ≥ 4 mm	12.7 ± 2.5	7.8 ± 3.9	6.1 ± 3.6	* <0.05; ** NS
CAL (mm)	3.5 ± 1.2	2.58 ± 0.30	2.55 ± 0.31	* <0.05; ** NS
% sites with CAL ≥ 4 mm	12.5 ± 0.2	11.2 ± 0.4	11.3 ± 0.9	* <0.05; ** NS

GI, gingival index; BOP, bleeding on probing; PPD, probing pocket depth; CAL, clinical attachment loss; * V1 compared to baseline; ** V2 compared to V1; $p > 0.05$ non-significant (NS); V1 and V2, visit 1 and 2, respectively.

No significant correlations between changes in periodontal parameters and changes in RA activity were described in our study ($p > 0.05$).

We assumed that all the modifications in the degree of local gingival and periodontal inflammation is related to IL-6 blockade as no local periodontal treatment was allowed during follow-up.

4. Discussion

We aimed to assess the influence of the IL-6 receptor inhibitor on periodontal status in active RA associated with periodontitis, assuming that TCZ might be able not only to improve clinical and biochemical RA-related parameters but also to ameliorate chronic periodontitis as a result of decreased IL-6 in the periodontal microenvironment via declining systemic inflammation.

Although our target is to demonstrate the ability of TCZ to modulate periodontal inflammation and subsequent damage, firstly, we emphasized its role in controlling RA activity. We reasonably confirmed a consistent response to TCZ in real-life settings, which was achieved in as rapid as three months and continued after six months of therapy, irrespective of background medication and clinical scenario (mono- or combined therapy, bio-naïve or bio-experienced patients); it is more than clear that even in the short-term, IL-6 blockade displays significant clinical, biological, as well as serologic disease improvement. Although we found no consistent difference in clinical response in seropositive vs. seronegative RA, we noticed a significant impact on ACPA serum concentration after six months, which is an improvement that parallels the decrease in periodontal inflammation, suggesting a role of IL-6 in both systemic and local inflammation (synovial and periodontal) and the potential implications via citrullination. Therefore, our results stand by as a proof of the effectiveness of subcutaneous TCZ in managing inflammatory and immune pathways in RA [53].

We also focused on the magnitude of compromised oral health in RA; most patients in our initial group presented a high rate of mild and severe periodontal disease, validating/reinforcing the already known risk of periodontitis in such patients, particularly in established, longstanding disease [2–4,9,37]. We have included in the final analysis only those cases with overt periodontal disease, meaning that up to 75% had at baseline altered periodontal status in a group of consecutive patients starting TCZ for their active disease. Indeed, recent reviews and meta-analyses have already discussed periodontal disease in various RA settings (independent of age, disease duration, serology profile, and disease activity) compared to general population [1,3,5,11,19–21,23].

We identified excessive gingival involvement confirmed by an increased percentage of sites with plaques and inflammation and abnormal periodontal status (e.g., increased probing depth, clinical attachment loss) supporting data from the literature [6,16,17,20,21,27,28,38,39,44,48]. Moreover, we recognized positive correlations between the severity of periodontitis, inflammatory parameters (especially CRP), serology (ACPA status and titers), and RA activity; indeed, recent studies suggest a worse periodontal status in active untreated RA, and higher CRP if RA is associated with severe periodontitis [6,11,15–17,19,27,28,37]. Finally, it seems that ACPA-positive patients had severely impaired periodontal health, while disease activity correlated with periodontitis degree as well [6,16,17,19,27,28,37].

Finally, we demonstrated that short-term tocilizumab significantly reduced gingival as well as periodontal inflammation as supported by decreased levels of gingival index, bleeding on probing, and probing pocket depth, paralleling the articular improvement. Indeed, only minor changes in clinical attachment loss were detected in our enrolled patients, and the supragingival plaque remained stable after 3 and 6 months of biological treatment ($p > 0.05$).

A closer look at recent data definitely emphasizes the dual effects of early and aggressive RA treatment with biologic and non-biologic drugs (Janus kinase inhibitors, JAK inhibitors) on articular as well as comorbid periodontal disease [5,8,12,16,17,25,27,28,37–39]. It is widely accepted that TNF and IL-6 receptor inhibitors are able to ameliorate oral health in active RA, as reflected by clinical, biological, and even serological RA biomarkers [5,16,17,27,28,38,39,44–46]. Although there are controversial effectiveness signals with TNF inhibitors in improving chronic periodontitis [3,6,11,15,19,23–26,33,38,39], all papers about anti-IL-6 therapy clearly demonstrated articular, systemic, and also periodontal benefits with TCZ without any periodontal specific treatment [16,17,27,28].

An interesting trial compared periodontal condition in patients with RA and periodontitis before and after biological therapy in two cohorts: one under tocilizumab and the other receiving medication with TNF inhibitors [16]. After 6 months, both tocilizumab and TNF inhibitors demonstrated a consistent improvement of oral health with significantly reduced periodontal inflammation (gingival index, bleeding on probing, and probing depth) compared to baseline, with similar results in both cohorts unless there was a greater decrease in gingival index and less gingival inflammation with tocilizumab; however, plaque levels remained the same irrespective of medication, while periodontal clinical attachment loss decreased only after TCZ but not after TNF inhibitors [16,17,27,28]. These observations were partially supported by the results of another study about an excessive inflammatory response against oral pathogens essentially based on high levels of IL-6 [16,17,27,28,31].

Recent meta-analyses reviewed the most important studies on TNF and non-TNF biologics in patients with RA and PD [5,20]. The critical difference between the class of TNF inhibitors and TCZ or B-cell depletive agent rituximab is that infliximab, an anti-TNF monoclonal antibody, may negatively address gingival inflammation although it may also improve alveolar bone destruction [5–7,19,37] resulting in a dissociated response for patients with severe periodontitis [5], while both tocilizumab and rituximab associate with significant a down regulation of gingival inflammation and damage in RA associated with periodontitis [16,17,27,28,50].

Additional research is necessary to clearly differentiate between the direct effects of TCZ on local periodontal inflammation and IL-6 or its receptor levels in the gingival crevicular fluids and periodontium of patients and the indirect effect via dramatically decreasing systemic inflammation, which may impact also oral health [16,17,27,28]. Indeed, numerous studies indicated a rapid and significant decline in typical inflammatory parameters (ESR and CRP), but also in serological RA biomarkers (RF, ACPA) as well as inflammatory cytokines (TNF, IL-6) and mediators (serum-amyloid A, matrixmetalloproteinases 1, 3), supporting the indirect role of TCZ in periodontitis [16,17,27,28].

In our study, we assessed specific gingival and periodontal parameters before and after short-term TCZ therapy. We demonstrated successful RA as well as periodontal outcomes with TCZ and independent of potential confounding factors (such as smoking, diabetes, hematological conditions, sex steroid hormones elevations, pharmacological agents) related to periodontal disease, as such patients were excluded from the final analysis.

We concluded that tocilizumab decreased gingival inflammation since no periodontal therapy was permitted and the dental hygiene behavior remained unchanged in our enrolled patients.

Unfortunately, we were not able to assess either the serum or gingival crevicular fluid levels of IL-6 or its receptor in all patients; we assumed that tocilizumab indirectly contributes to modulate local (gingival and periodontal) inflammation by limiting systemic inflammation. Indeed, the biofilm plaque accumulation was not consistently diminished with tocilizumab, and we were not able to depict a spectacular impact on clinical attachment loss, but we arrived to demonstrate a positive effect of IL-6 blockade on exuberant gingival inflammation.

Further studies are necessary to confirm the benefits of Il-6 inhibitors in larger populations and longer follow-ups also focusing on IL-6/IL-6 receptor levels in gingival crevicular fluid.

Author Contributions: Conceptualization, C.A., O.Ț., C.I. and S.M.S.; methodology, C.A., C.I., S.M.S. and O.Ț.; software, E.A. and A.D.; validation, C.I., A.M.F. and R.C.; formal analysis, C.A., A.D. and E.A.; investigation, C.A, R.C., C.I. and A.M.F.; resources, E.A. and A.D.; data curation, E.A.; writing—original draft preparation, C.A. and C.I.; writing—review and editing, C.A., C.I., E.A., A.D. and O.Ț.; visualization, C.A.; supervision, R.C.; project administration, E.A. All authors have read and agreed to the published version of the manuscript.

Funding: This research received no external funding

Institutional Review Board Statement: The study was conducted according to the guidelines of the Declaration of Helsinki, and approved by the Ethics Committee of Sanocare Medical and Research Center (protocol No. 15/12.12.2016).

Informed Consent Statement: Informed consent was obtained from all subjects involved in the study.

Conflicts of Interest: The authors declare no conflict of interest.

References

1. De Molon, R.S.; Rossa, C., Jr.; Thurlings, R.M.; Cirelli, J.A.; Koenders, M.I. Linkage of periodontitis and rheumatoid arthritis: Current evidence and potential biological interactions. *Int. J. Mol. Sci.* **2019**, *20*, 4541. [CrossRef]

2. Genco, R.J.; Sanz, M. Clinical and public health implications of periodontal and systemic diseases: An overview. *Periodontology 2000* **2020**, *83*, 7–13. [CrossRef] [PubMed]

3. Eezammuddeen, N.N.; Vaithilingam, R.D.; Hassan, N.H.M.; Bartold, P.M. Association between rheumatoid arthritis and periodontitis: Recent progress. *Curr. Oral Health Rep.* **2020**, *7*, 139–153. [CrossRef]

4. Sherina, N. *On the Origin of ACPA: Exploring the Role of P. Gingivalis in the Development of Rheumatoid Arthritis*; Karolinska Institutet: Stockholm, Sweden, 2019.

5. Rinaudo-Gaujous, M.; Blasco-Baque, V.; Miossec, P.; Gaudin, P.; Farge, P.; Roblin, X.; Thomas, T.; Paul, S.; Marotte, H. Infliximab induced a dissociated response of severe periodontal biomarkers in rheumatoid arthritis patients. *J. Clin. Med.* **2019**, *8*, 751. [CrossRef] [PubMed]

6. Koziel, J.; Mydel, P.; Potempa, J. The link between periodontal disease and rheumatoid arthritis: An updated review. *Curr. Rheumatol. Rep.* **2014**, *16*, 1–7. [CrossRef]

7. Rutter-Locher, Z.; Fuggle, N.; Orlandi, M.; D′Aiuto, F.; Sofat, N. Periodontal Disease and Autoimmunity: What We Have Learned from Microbiome Studies in Rheumatology. In *Periodontitis—A Useful Reference*; Arjunan, P., Ed.; IntechOpen Limited: London, UK, 2017; pp. 113–141.

8. Fuggle, N.R.; Smith, T.O.; Kaul, A.; Sofat, N. Hand to mouth: A systematic review and meta-analysis of the association between rheumatoid arthritis and periodontitis. *Front. Immunol.* **2016**, *7*, 1–10. [CrossRef] [PubMed]

9. Bartold, P.M.; Lopez-Oliva, I. Periodontitis and rheumatoid arthritis: An update 2012–2017. *Periodontology 2000* **2020**, *83*, 189–212. [CrossRef]

10. Gómez-Bañuelos, E.; Mukherjee, A.; Darrah, E.; Andrade, F. Rheumatoid arthritis-associated mechanisms of porphyromonas gingivalis and aggregatibacter actinomycetemcomitans. *J. Clin. Med.* **2019**, *8*, 1309. [CrossRef]

11. Mankia, K.; Cheng, Z.; Do, T.; Hunt, L.; Meade, J.; Kang, J.; Clerehugh, V.; Speirs, A.; Tugnait, A.; Hensor, E.M.A.; et al. Prevalence of periodontal disease and periodontopathic bacteria in anti—Cyclic citrullinated protein antibody—Positive at-risk adults without arthritis. *JAMA Netw. Open* **2019**, *2*, e195394. [CrossRef]

12. Zamri, F.; De Vries, T.J. Use of TNF inhibitors in rheumatoid arthritis and implications for the periodontal status: For the benefit of both? *Front. Immun.* **2020**, *11*, 2686. [CrossRef]

13. Eriksson, K.; Fei, G.; Lundmark, A.; Benchimol, D.; Lee, L.; Hu, Y.O.O.; Kats, A.; Saevarsdottir, S.; Catrina, A.I.; Klinge, B.; et al. Periodontal health and oral microbiota in patients with rheumatoid arthritis. *J. Clin. Med.* **2019**, *8*, 630. [CrossRef]

14. Marrotte, H. Determining the right time for the right treatment—Application to preclinical rheumatoid arthritis. *JAMA Netw. Open* **2019**, *6*, e195358. [CrossRef] [PubMed]

15. Gonzalez, S.M.; Payne, J.B.; Yu, F.; Thiele, G.M.; Erickson, A.R.; Johnson, P.G.; Schmid, M.J.; Cannon, G.W.; Kerr, G.S.; Reimold, A.M.; et al. Alveolar bone loss is associated with circulating anti-citrullinated protein antibody (ACPA) in patients with rheumatoid arthritis. *J. Periodontol.* **2015**, *86*, 222–231. [CrossRef]

16. Kobayashi, T.; Ito, S.; Kobayashi, D.; Kojima, A.; Shimada, A.; Narita, I.; Murasawa, A.; Nakazono, K.; Yoshie, H. Interleukin-6 receptor inhibitor tocilizumab ameliorates periodontal inflammation in patients with rheumatoid arthritis and periodontitis as well as tumor necrosis factor inhibitors. *Clin. Exp. Dent. Res.* **2015**, *1*, 63–73. [CrossRef] [PubMed]

17. Kobayashi, T.; Okada, M.; Ito, S.; Kobayashi, D.; Ishida, K.; Kojima, A.; Narita, I.; Murasawa, A.; Yoshie, H. Assessment of interleukin-6 Receptor inhibition therapy on periodontal condition in patients with rheumatoid arthritis and chronic periodontitis. *J. Periodontol.* **2014**, *85*, 57–67. [CrossRef]

18. Kobayashi, T.; Yoshie, H. Host responses in the link between periodontitis and rheumatoid arthritis. *Curr. Oral Health Rep.* **2015**, *2*, 1–8. [CrossRef] [PubMed]

19. Äyräväinen, L.; Leirisalo-Repo, M.; Kuuliala, A.; Ahola, K.; Koivuniemi, R.; Meurman, J.H.; Heikkinen, A.M. Periodontitis in early and chronic rheumatoid arthritis: A prospective follow-up study in Finnish population. *BMJ Open* **2017**, *7*, e011916. [CrossRef]

20. Qiao, Y.; Wang, Z.; Li, Y.; Han, Y.; Zhou, Y.; Cao, X. Rheumatoid arthritis risk in periodontitis patients: A systematic review and meta-analysis. *Jt. Bone Spine* **2020**, *87*, 556–564. [CrossRef]

21. Lee, K.H.; Choi, Y.Y. Rheumatoid arthritis and periodontitis in adults: Using the Korean National Health Insurance Service—National sample cohort. *J. Periodontol.* **2020**, *91*, 1186–1193. [CrossRef]

22. Ferreira, R.D.O.; Silva, R.D.B.; Magno, M.B.; Almeida, A.P.C.P.S.C.; Fagundes, N.C.F.; Maia, L.C.; Lima, R.R. Does periodontitis represent a risk factor for rheumatoid arthritis? A systematic review and meta-analysis. *Ther. Adv. Musculoskelet. Dis.* **2019**, *11*, 1–16.

23. Choi, I.A.; Kim, J.H.; Kim, Y.M.; Lee, J.Y.; Kim, K.H.; Lee, E.Y.; Lee, E.B.; Lee, Y.M.; Song, Y.W. Periodontitis is associated with rheumatoid arthritis. A study with longstanding rheumatoid arthritis patients in Korea. *Korean J. Intern. Med.* **2016**, *31*, 977–986. [CrossRef]

24. Ceccarelli, F.; Saccucci, M.; Di Carlo, G.; Lucchetti, R.; Pilloni, A.; Pranno, N.; Luzzi, V.; Valesini, G.; Polimeni, A. Periodontitis and rheumatoid arthritis: The same inflammatory mediators? Hindawi. *Mediat. Inflamm.* **2019**, *2019*, 6034546. [CrossRef] [PubMed]

25. Thilagar, S.; Theyagarajan, R.; Sudhakar, U.; Suresh, S.; Saketharaman, P.; Ahamed, N. Comparison of serum tumor necrosis factor-α levels in rheumatoid arthritis individuals with and without chronic periodontitis: A biochemical study. *J. Indian Soc. Periodontol.* **2018**, *22*, 116–121. [CrossRef]

26. Silvestre-Rangil, J.; Bagán, L.; Silvestre, F.-J.; Martinez-Herrera, M.; Bagán, J.-V. Periodontal, salivary and IL-6 status in rheumatoid arthritis patients. A cross-sectional study. *Med. Oral Patol. Oral Y Cir. Bucal* **2017**, *22*, 595–600. [CrossRef] [PubMed]

27. Kobayashi, T.; Yokoyama, T.; Ito, S.; Kobayashi, D.; Yamagata, A.; Okada, M.; Oofusa, K.; Narita, I.; Murasawa, A.; Nakazono, K.; et al. Periodontal and serum protein profiles inpatients with rheumatoid arthritis treated with tumor necrosis inhibitor adalimumab. *J. Periodontol.* **2014**, *85*, 1480–1488. [CrossRef] [PubMed]

28. Kobayashi, T.; Yokoyama, T.; Ishida, K.; Abe, A.; Yamamoto, K.; Yoshie, H. Serum cytokine and periodontal profiles in relation to disease activity of rheumatoid arthritis in Japanese adults. *J. Periodontol.* **2010**, *80*, 650–657. [CrossRef] [PubMed]

29. Möller, B.; Bender, P.; Eick, S.; Kuchen, S.; Maldonado, A.; Potempa, J.; Reichenbach, S.; Sculean, A.; Schwenzer, A.; Villiger, P.M.; et al. Treatment of severe periodontitis may improve clinical disease activity in otherwise treatment-refractory rheumatoid arthritis patients. *Rheumatology* **2019**, *59*, 243–245. [CrossRef]

30. Irwin, C.R.; Myrillas, T.T. The role of IL-6 in the pathogenesis of periodontal disease. *Oral Dis.* **2008**, *4*, 43–47. [CrossRef]

31. Nibali, L.; Fedele, S.; D'Aiuto, F.; Donos, N. Interleukin-6 in oral diseases: A review. *Oral Dis.* **2011**, *18*, 236–243. [CrossRef]

32. Shimada, Y.; Komatsu, Y.; Ikezawa-Suzuki, I.; Tai, H.; Sugita, N.; Yoshie, H. The effect of periodontal treatment on serum leptin, Interleukin-6, and C-Reactive protein. *J. Periodontol.* **2010**, *81*, 1118–1123. [CrossRef]

33. Takahashi, K.; Takashiba, S.; Nagai, A.; Takigawa, M.; Myoukai, F.; Kurihara, H.; Murayama, Y. Assessment of Interleukin-6 in the pathogenesis of periodontal disease. *J. Periodontol.* **1994**, *65*, 147–153. [CrossRef]

34. Vidal, F.; Figueredo, C.M.S.; Cordovil, I.; Fischer, R.G. Periodontal therapy reduces plasma levels of Interleukin-6, C-Reactive protein, and fibrinogen in patients with severe periodontitis and refractory arterial hypertension. *J. Periodontol.* **2009**, *80*, 786–791. [CrossRef]

35. Kaczynski, T.; Wronski, J.; Gluszko, P.; Kryczka, T.; Miskiewiczi, A.; Gordki, B.; Radkowski, M.; Trzemecki, D.; Grieb, P.; Gorska, R. Salivary interleukin 6, interleukin 8, interleukin 17A, and tumour necrosis factor α levels in patients with periodontitis and rheumatoid arthritis. *Cent. Eur. J. Immunol.* **2019**, *44*, 269–276. [CrossRef] [PubMed]

36. Lee, H.-J.; Kang, I.-K.; Chung, C.-P.; Choi, S.-M. The subgingival microflora and gingival crevicular fluid cytokines in refractory periodontitis. *J. Clin. Periodontol.* **1995**, *22*, 885–890. [CrossRef]
37. Ancuța, C.; Pomîrleanu, C.; Mihailov, C.; Chirieac, R.; Ancuta, E.; Iordache, C.; Bran, C.; Țanculescu, O. Efficacy of baricitinib on periodontal inflammation in patients with rheumatoid arthritis. *Jt. Bone Spine* **2020**, *87*, 235–239. [CrossRef] [PubMed]
38. Jung, G.U.; Han, J.Y.; Hwang, K.G.; Park, C.J.; Stathopoulou, P.G.; Fiorellini, J.P. Effects of conventional synthetic disease-modifying anti-rheumatic drugs on response to periodontal treatment in patients with rheumatoid arthritis. *BioMed Res. Int.* **2018**, *2018*, 1465402. [CrossRef]
39. Romero-Sanchez, C.; AM, M.; GI, L.; DM, C.; Ch P, C.; Gualtero JM, B. Is the treatment with biological or non-biological DMARDS a modifier of periodontal condition in patients with rheumatoid arthritis? *Curr. Rheumatol. Rev.* **2017**, *13*, 139–151. [CrossRef]
40. Kobayashi, T.; Ito, S.; Murasawa, A.; Ishikawa, H.; Yoshie, H. Effects of tofacitinib on the clinical features of periodontitis in patients with rheumatoid arthritis: Two case reports. *BMC Rheumatol.* **2019**, *3*, 1–4. [CrossRef]
41. Iordache, C.; Chirieac, R.; Ancuța, E.; Pomirleanu, C.; Ancuța, C. Periodontal disease in patients with ankylosing spondylitis: Myth or reality? *Studies* **2017**, *68*, 1660–1664. [CrossRef]
42. Ancuta, C.; Ancuta, E.; Chirieac, R.; Antohe, M.; Iordache, C. Anti-tumor necrosis factor alpha therapy and periodontal inflammation in rheumatoid arthritis. A clinical and biochemical approach. *Rev. Chim.* **2017**, *68*, 369–372. [CrossRef]
43. Ancuta, C.; Ancuta, E.; Chirieac, R.; Anton, C.; Surlari, Z.; Iordache, C. TNF inhibitors and periodontal inflammation in psoriatic arthritis. *Rev. Chim.* **2017**, *68*, 1914–1918. [CrossRef]
44. Mayer, Y.; Balbir-Gurman, A.; Machtei, E.E. Anti-Tumor necrosis factor-alpha therapy and periodontal parameters in patients with rheumatoid arthritis. *J. Periodontol.* **2009**, *80*, 1414–1420. [CrossRef] [PubMed]
45. Savioli, C.; Ribeiro, A.C.M.; Fabri, G.M.C.; Calich, A.L.; Carvalho, J.; Silva, C.A.; Viana, V.S.T.; Bonfá, E.; Siqueira, J.T.T. Persistent periodontal disease hampers anti-tumor necrosis factor treatment response in rheumatoid arthritis. *J. Clin. Rheumatol.* **2012**, *18*, 180–184. [CrossRef]
46. Schiefelbein, R.; Jentsch, H.F.R. Periodontal conditions during arthritis therapy with TNF-a blockers. *J. Clin. Diagn. Res.* **2018**, *12*, 27–31. [CrossRef]
47. Sikorska, D.; Orzechowska, Z.; Rutkowski, R.; Prymas, A.; Mrall-Wechta, M.; Bednarek-Hatlińska, D.; Roszak, M.; Surdacka, A.; Samborski, W.; Witowski, J. Diagnostic value of salivary CRP and IL-6 in patients undergoing anti-TNF-alpha therapy for rheumatic disease. *Inflammopharmacology* **2018**, *26*, 1183–1188. [CrossRef] [PubMed]
48. Tachibana, M.; Yonemoto, Y.; Okamura, K.; Suto, T.; Sakane, H.; Kaneko, T.; Dam, T.T.; Okura, C.; Tajika, T.; Tsushima, Y.; et al. Does periodontitis affect the treatment response of biologics in the treatment of rheumatoid arthritis? *Arthritis Res. Ther.* **2020**, *22*, 178. [CrossRef]
49. Kadkhoda, Z.; Amirzargar, A.; Esmaili, Z.; Vojdanian, M.; Akbari, S. Effect of TNF-? Blockade in gingival crevicular fluid on periodontal condition of patients with rheumatoid arthritis. *Iran. J. Immunol.* **2016**, *13*, 197–203.
50. Coat, J.; Demoersman, J.; Beuzit, S.; Cornec, D.; Devauchelle-Pensec, V.; Saraux, A.; Pers, J.-O. Anti-B lymphocyte immunotherapy is associated with improvement of periodontal status in subjects with rheumatoid arthritis. *J. Clin. Periodontol.* **2015**, *42*, 817–823. [CrossRef] [PubMed]
51. Kawalec, P.; Śladowska, K.; Malinowska-Lipien, I.; Brzostek, T.; Kózka, M. New alternative in the treatment of rheumatoid arthritis: Clinical utility of baricitinib. *Ther. Clin. Risk Manag.* **2019**, *15*, 275–284. [CrossRef] [PubMed]
52. Calderaro, D.C.; Corrêa, J.D.; Ferreira, G.A.; Barbosa, I.G.; Martins, C.C.; Silva, T.A.; Teixeira, A.L. Influence of periodontal treatment of rheumatoid arthritis: A systematic review and meta-analysis. *Rev. Bras. Reumatol.* **2017**, *57*, 238–244. [CrossRef]
53. Yang, N.-Y.; Wang, C.-Y.; Chyuan, I.-T.; Wu, K.-J.; Tu, Y.-K.; Chang, C.-W.; Hsu, P.-N.; Kuo, M.Y.-P.; Chen-Ying, W. Significant association of rheumatoid arthritis-related inflammatory markers with non-surgical periodontal therapy. *J. Formos. Med. Assoc.* **2018**, *117*, 1003–1010. [CrossRef]
54. Finckh, A.; Tellenbach, C.; Herzog, L.; Scherer, A.; Moeller, B.; Ciurea, A.; Von Muehlenen, I.; Gabay, C.; Kyburz, D.; Brulhart, L.; et al. Comparative effectiveness of antitumour necrosis factor agents, biologics with an alternative mode of action and tofacitinib in an observational cohort of patients with rheumatoid arthritis in Switzerland. *RMD Open* **2020**, *6*, e001174. [CrossRef] [PubMed]
55. Ortiz, P.; Bissada, N.F.; Palomo, L.; Han, Y.W.; Al-Zahrani, M.S.; Panneerselvam, A.; Askari, A. Periodontal therapy reduces the severity olfactive rheumatoid arthritis in patients treated with or without tumor necrosis factor inhibitors. *J. Periodontol.* **2009**, *80*, 535–540. [CrossRef]
56. Buwembo, W.; Munabi, I.G.; Kaddumukasa, M.; Kiryowa, H.; Mbabali, M.; Nankya, E.; Johnson, W.E.; Okello, E.; Sewankambo, N.K. Non-surgical oral hygiene interventions on disease activity of Rheumatoid arthritis patients with periodontitis: A randomized controlled trial. *J. Dent. Res. Dent. Clin. Dent. Prospect.* **2020**, *14*, 26–36. [CrossRef]
57. Monsarrat, P.; de Grado, G.F.; Constantin, A.; Willmann, C.; Nabet, C.; Sixou, M.; Cantagrel, A.; Barnetche, T.; Mehsen-Cetre, N.; Schaeverbeke, T. The effect of periodontal treatment on patients with rheumatoid arthritis: The ESPERA randomized controlled trial. *Jt. Bone Spine* **2019**, *86*, 600–609. [CrossRef]
58. Ainamo, J.; Bay, I. Problems and proposals for recording gingivitis and plaque. *Int. Dent. J.* **1975**, *25*, 229–235. [PubMed]
59. Löe, H.; Silness, J. Periodontal disease in pregnancy I. Prevalence and severity. *Acta Odontol. Scand.* **1963**, *21*, 533–551. [CrossRef]
60. Tonetti, M.S.; Claffey, N. Advances in the progression of periodontitis and proposal of definitions of a periodontitis case and disease progression for use in risk factor research. Group C Consensus report of the 5th European workshop in periodontology. *J. Clin. Periodontol.* **2005**, *32*, 210–213. [CrossRef]

Journal of
Clinical Medicine

Article

Chronological Gene Expression of Human Gingival Fibroblasts with Low Reactive Level Laser (LLL) Irradiation

Yuki Wada [1], Asami Suzuki [2,*], Hitomi Ishiguro [1,3], Etsuko Murakashi [1] and Yukihiro Numabe [1,3]

[1] Department of Periodontology, School of Life Dentistry at Tokyo, The Nippon Dental University, 1-9-20 Fujimi, Chiyoda-ku, Tokyo 102-8159, Japan; y-wada_d2117015@tky.ndu.ac.jp (Y.W.); hitomi-i@tky.ndu.ac.jp (H.I.); satoetsu@tky.ndu.ac.jp (E.M.); numabe-y@tky.ndu.ac.jp (Y.N.)

[2] Division of General Dentistry, The Nippon Dental University Hospital, 2-3-16 Fujimi, Chiyoda-ku, Tokyo 102-8158, Japan

[3] Dental Education Support Center, School of Life Dentistry, The Nippon Dental University, 1-9-20 Fujimi, Chiyoda-ku, Tokyo 102-8159, Japan

* Correspondence: nduh-a-suzuki@tky.ndu.ac.jp; Tel.: +81-3-3261-5511

Abstract: Though previously studies have reported that Low reactive Level Laser Therapy (LLLT) promotes wound healing, molecular level evidence was uncleared. The purpose of this study is to examine the temporal molecular processes of human immortalized gingival fibroblasts (HGF) by LLLT by the comprehensive analysis of gene expression. HGF was seeded, cultured for 24 h, and then irradiated with a Nd: YAG laser at 0.5 W for 30 s. After that, gene differential expression analysis and functional analysis were performed with DNA microarray at 1, 3, 6 and 12 h after the irradiation. The number of genes with up- and downregulated differentially expression genes (DEGs) compared to the nonirradiated group was large at 6 and 12 h after the irradiation. From the functional analysis results of DEGs, Biological Process (BP) based Gene Ontology (GO), BP 'the defense response' is considered to be an important process with DAVID. Additionally, the results of PPI analysis of DEGs involved in the defense response with STRING, we found that the upregulated DEGs such as CXCL8 and NFKB1, and the downregulated DEGs such as NFKBIA and STAT1 were correlated with multiple genes. We estimate that these genes are key genes on the defense response after LLLT.

Keywords: Low reactive Level Laser Therapy (LLLT); human gingival fibroblasts (HGF); microarray; differentially gene expression (DEGs); gene ontology; biological processes (BP); protein–protein interaction (PPI)

Citation: Wada, Y.; Suzuki, A.; Ishiguro, H.; Murakashi, E.; Numabe, Y. Chronological Gene Expression of Human Gingival Fibroblasts with Low Reactive Level Laser (LLL) Irradiation. *J. Clin. Med.* **2021**, *10*, 1952. https://doi.org/10.3390/jcm10091952

Academic Editors: Gianrico Spagnuolo and Yorimasa Ogata

Received: 19 April 2021
Accepted: 29 April 2021
Published: 1 May 2021

Publisher's Note: MDPI stays neutral with regard to jurisdictional claims in published maps and institutional affiliations.

1. Introduction

Periodontal disease is a chronic multifactorial inflammatory disease caused by genetic, immune, environmental, microbial factors and lifestyles, with anaerobic bacteria in the oral cavity as the main causative organism. Periodontal disease is generally treated with nonsurgical therapy, that is performed with hand or power-driven instrumentation. In recent years, combination therapies with scaler and laser, or laser alone have attracted attention [1–3].

The laser therapy methods currently used for treatment of periodontal diseases can be broadly divided into two types: High reactive Level Laser Therapy (HLLT) and Low reactive Level Laser Therapy (LLLT).

HLLT is an application of laser intensity that produces an irreversible reaction (photobiological destruction reaction) beyond the cell survival region and is used for tissue incision and transpiration [4–6].

On the other hand, LLLT is a treatment that applies laser intensity to generate a reversible reaction (photobiologically active reaction) within the cell survival threshold. LLLT are expected to have anti-inflammatory effects [7–12]; pain relief [13], improvement/promotion of blood flow [14], activation of cells in tissues, wound healing by proliferation [15] and tissue regeneration without causing tissue degeneration with low-power

155

laser irradiation conditions [16–18]. In recent years, the promotion of wound healing with LLLT has been one of the highlights.

Wound healing is thought to progress in the process of hemorrhagic coagulation phase, inflammatory phase, proliferative phase, reconstruction phase after injury by external stimulus. During the hemorrhagic coagulation phase, blood is coagulated by platelets, which is one of the coagulation factors, and growth factors such as platelet-derived growth factor (PDGF) and cytokines are released from the platelets. During the inflammatory phase, factors such as Nuclear Factor-κB (NF-κB) cause infiltration of inflammatory cells such as neutrophils and macrophages. Then, the release of growth factors and cytokines such as transforming growth factor-β (TGF-β) and fibroblast growth factor (FGF) is observed [19]. During the proliferative phase, it promotes the migration and proliferation of fibroblasts and keratinocytes [20]. Extracellular matrix is synthesized from fibroblasts and serves as a scaffold for cell migration and adhesion. During maturity, scar tissue formation occurs.

In the study of LLLT, TGF-β1 is closely involved in cell differentiation, migration, and adhesion by LLLT. In addition, it is thought to be involved in a wide range of areas such as ontogeny, tissue reconstruction, wound healing, inflammation/immunity, and cancer infiltration/metastasis. It has also been reported that the expression of NF-κB is increased [21].

Previous studies reported that the effects of low-reaction level laser irradiation on periodontium-derived cultured cells have been mainly on cell proliferation and cell transport ability related to wound healing. However, the elucidation of the mechanism at the molecular level leading to the promotion of wound healing by laser irradiation has been insufficient. It is considered that there may be a series of processes related to various wound healing by LLLT by the approach from biological processes (BP). There are few studies on mechanism analysis at the gene level using microarrays by LLLT for HGF, and only a limited number of studies have analyzed gene expression over time [22,23]. In order to clarify the effect of laser irradiation by LLLT, it is important to analyze and consider changes in gene expression and changes in BP of differentially expression genes (DEGs) over time in order to understand the mechanism at the molecular level.

In this study, HGF was irradiated with LLLT, and gene expression fluctuations at 1, 3, 6, and 12 h after irradiation were analyzed using a DNA microarray. In addition, we focused on the defense response, which showed remarkable changes in gene expression over time in relation to wound healing obtained from the results of vast amounts of analytical data, and to investigate for the mechanism from the expression change genes and BP due to the photobiological effects of laser. The aim of this study was to elucidate the changes of gene expression on the wound healing, especially defense response, over time after irradiation.

2. Materials and Methods

2.1. Cell Culture

Human immortalized gingival fibroblasts (HGF; Applied Biological Material, Richmond, BC, Canada) were used, and 10% Fetal Bovine Serum (Moregate, Bulimba, Australia), 50 U/mL Penicilin G, 50 μg/mL Amphotericin. The study was carried out by culturing in D-MEM/F-12 medium (Life Technologies Corporation, Grand Island, NY, USA) under 37 °C, and 5% CO_2 conditions. The HGF at the time of irradiation was in the logarithmic growth phase.

2.2. Dental Laser Device and Laser Irradiation Stent

A dental Nd: YAG laser: impulse dental laser (Incisive Japan Co., Ltd., Tokyo, Japan) was used as a dental laser device, and an ultrafine fiber with a diameter of 320 nm was used for the laser light guide tip. Stents (Gikousha, Kanagawa, Japan) were prepared to uniformly irradiate cells with laser, attached to a handpiece, and used for research. Irradiation conditions were found to be significantly different in cell proliferation curvature in previous studies, irradiation output conditions 0.5 W (100 mJ, 5 pps), irradiation time

30 s, irradiation distance from the tip of the fiber guide to each well plate. Laser irradiation was performed with a distance of 20 mm to the bottom [24].

2.3. Microarray Analysis

Total RNA was extracted from HGF with RNeasy® Plus Micro Kit (QIAGEN, Valencia, CA, USA) before laser irradiation 1, 3, 6 and 12 h after irradiation on a 96-well plate. cDNA was synthesized from total RNA using the SuperScript™ VILO™ cDNA Synthesis Kit (Invitrogen, Carlsbad, CA, USA). After synthesis, the cDNA was fragmented and biotin labeled. Biotin-labeled cDNA was added to the GeneChipTM Human Gene 2.0 ST Array (Thermo Fisher Scientific Inc., Waltham, MA, USA) and hybridized with a probe (GeneChipTM Hybridization, Wash, and Stain Kit; Thermo Fisher Scientific Inc., Waltham, MA, USA). Phycoerythrin staining was performed, and the fluorescence signal was measured with a GeneChip scanner (Scanner 3000 7 G; Thermo Fisher Scientific Inc., Waltham, MA, USA). After normalization, Expression Gene was analyzed by SST-RMA algorithm.

2.4. Data Analysis of Differentially Expressed Genes (DEGs)

DEGs were extracted with Affymetrix® Expression ConsoleTM (Thermo Fisher Scientific Inc. Waltham, MA, USA). The cutoff values were fold change (FC) $\geq$ |1.5| and p-value < 0.05. The nonirradiated group (control), the irradiated group (test) were defined as upregulated DEGs with significantly increased expression, and downregulated DEGs with significantly decreased expression with respect to the control group.

2.5. Functional Analysis of Differentially Expressed Genes

A functional analysis of DEGs was performed based on Gene Ontology (GO) with the database for annotation, visualization and integrated discovery (DAVID). Enrichment analysis on the BP was performed on the DEGs at 1, 3, 6, and 12 h after irradiation. The cutoff value was a modified Fisher exact p-value < 0.1, total count $\leq$ 2. The DEGs contained in the upregulated and the downregulated regions at each irradiation time were analyzed.

2.6. Protein–Protein Interaction (PPI) of Up- or Downregulated DEGs

We focused on the defense response which is an important response on the wound healing process. PPI analysis of up- or downregulated DEGs, which were involved in defense response and continuously observed as up- or downregulated at 6 and 12 h, and related expression fluctuations were observed at all times after irradiation and Search Tool for the Retrieval of Interacting Genes (STRING) was performed. Furthermore, from the analysis of PPI, variation in the expression of genes correlated with other DEGs over time was analyzed.

3. Results

3.1. Extraction of DEGs

In order to compare the gene expression after laser irradiation on time course, the DEGs were extracted. DEGs were extracted under the condition that the cutoff value was FC $\geq$ |1.5| and p-value < 0.05. Control and test were defined as upregulated DEGs with significantly increased expression and downregulated DEGs with significantly decreased expression with respect to the control group. At 1 h after the irradiation, 83 upregulated and 50 downregulated genes were extracted (Supplementary Tables S1 and S2). At 3 h after, 46 upregulated genes and 32 downregulated genes (Supplementary Tables S3 and S4), at 6 h after, 362 upregulated and 549 downregulated genes (Supplementary Tables S5 and S6) and at 12 h after, 253 upregulated genes and 413 downregulated were extracted (Supplementary Tables S7 and S8).

The number of DEGs was large 6 and 12 h after the irradiation. At 6 h, the number of DEGs was the highest in both the upregulated gene and downregulated gene groups.

3.2. Functional Analysis on GO

From the results of functional analysis with DAVID, the number of BP related with each DEG after the irradiation was 13 BP on upregulated and 35 BP on downregulated DEGs at 1 h after, 6 BP on upregulated and 68 BP on downregulated DEGs at 3 h after, 212 BP on upregulated and 288 BP on downregulated DEGs at 6 h after, and 84 BP on upregulated and 425 BP on downregulated DEGs at 12 h after. The number of BP was small at 1 and 3 h, and the number of BP was large at 6 and 12 h. Tables 1–8 show the top BPs for each irradiation time.

BPs on upregulated DEGs at 1 h after irradiation are, for example, GO: 0050867 ~ positive regulation of cell activation, GO: 0050865 ~ regulation of cell activation (Table 1), BPs on downregulated DEGs are, for example, GO: 0032774 ~ RNA biosynthetic process, GO: 0007267 ~ cell–cell signaling (Table 2).

At 3 h after the irradiation, BPs on upregulated DEGs are, for example, GO: 0055085 ~ transmembrane transport, GO: 0006820 ~ anion transport (Table 3), BPs on down-]regulated DEGs are, for example, GO: 0032774 ~ RNA biosynthetic process, GO: 0016070 ~ RNA metabolic process (Table 4).

At 6 h, BPs on upregulated DEGs are, for example, GO: 0008283 ~ cell proliferation, GO: 0007155 ~ cell adhesion GO: 0022610 ~ biological adhesion, GO: 0042127 ~ regulation of cell proliferation, GO: 0006952 ~ defense response, GO: 0060429 ~ epithelium development (Table 5), BP on downregulated DEGs are, for example, GO: 0034645 ~ cellular macromolecule biosynthetic process, GO: 0019438 ~ aromatic compound biosynthetic process, GO: 0006325 ~ chromatin organization, GO: 0007049 ~ cell cycle (Table 6).

At 12 h, BPs on upregulated DEGs are, for example, GO: 0006955 ~ immune response, GO: 0006952 ~ defense response, GO: 0009605 ~ response to external stimulus, GO: 0048584 ~ positive regulation of response to stimulus (Table 7), BPs on downregulated DEGs are, for example, GO: 0010468 BP such as ~ regulation of gene expression, GO: 0007155 ~ cell adhesion, GO: 0022610 ~ biological adhesion (Table 8).

Table 1. The functional analysis of the upregulated genes at 1 h after Low Reactive Level Laser (LLL) irradiation.

Gene Ontology (GO) ID and Terms on Biological Process (BP)	Count	%	*p*-Value
GO:0061024~membrane organization	6	6.67	4.45×10^{-2}
GO:0002696~positive regulation of leukocyte activation	4	4.44	2.15×10^{-2}
GO:0050867~positive regulation of cell activation	4	4.44	2.31×10^{-2}
GO:0002694~regulation of leukocyte activation	4	4.44	6.00×10^{-2}
GO:0072657~protein localization to membrane	4	4.44	6.66×10^{-2}
GO:0050865~regulation of cell activation	4	4.44	7.06×10^{-2}
GO:0008037~cell recognition	3	3.33	3.60×10^{-2}
GO:0072659~protein localization to plasma membrane	3	3.33	5.98×10^{-2}
GO:1990778~protein localization to cell periphery	3	3.33	6.97×10^{-2}
GO:0007009~plasma membrane organization	3	3.33	9.88×10^{-2}
GO:0006910~phagocytosis, recognition	2	2.22	6.52×10^{-2}
GO:2000243~positive regulation of reproductive process	2	2.22	8.66×10^{-2}
GO:0006911~phagocytosis, engulfment	2	2.22	9.18×10^{-2}

Count: genes involved in the term; percentage (%): involved genes/total genes; *p*-value: modified Fisher exact *p*-value.

Table 2. The functional analysis of the downregulated genes at 1 h after LLL irradiation.

Gene Ontology (GO) ID and Terms on Biological Process (BP)	Count	%	*p*-Value
GO:0032774~RNA biosynthetic process	10	18.52	1.27×10^{-2}
GO:0034654~nucleobase-containing compound biosynthetic process	10	18.52	2.59×10^{-2}
GO:0018130~heterocycle biosynthetic process	10	18.52	2.79×10^{-2}
GO:0019438~aromatic compound biosynthetic process	10	18.52	2.85×10^{-2}
GO:0016070~RNA metabolic process	10	18.52	3.87×10^{-2}
GO:0034645~cellular macromolecule biosynthetic process	10	18.52	6.07×10^{-2}
GO:0010467~gene expression	10	18.52	8.40×10^{-2}
GO:0007267~cell–cell signaling	5	9.26	7.48×10^{-2}
GO:0006614~SRP-dependent cotranslational protein targeting to membrane	3	5.56	4.74×10^{-3}
GO:0006613~cotranslational protein targeting to membrane	3	5.56	5.44×10^{-3}
GO:0045047~protein targeting to ER	3	5.56	5.54×10^{-3}
GO:0072599~establishment of protein localization to endoplasmic reticulum	3	5.56	5.96×10^{-3}
GO:0000184~nuclear-transcribed mRNA catabolic process, nonsense-mediated decay	3	5.56	7.67×10^{-3}
GO:0070972~protein localization to endoplasmic reticulum	3	5.56	8.28×10^{-3}
GO:0019083~viral transcription	3	5.56	1.60×10^{-2}
GO:0006413~translational initiation	3	5.56	1.74×10^{-2}
GO:0006612~protein targeting to membrane	3	5.56	1.76×10^{-2}
GO:0019080~viral gene expression	3	5.56	1.79×10^{-2}
GO:0000956~nuclear-transcribed mRNA catabolic process	3	5.56	2.05×10^{-2}
GO:0044033~multiorganism metabolic process	3	5.56	2.20×10^{-2}
GO:0006402~mRNA catabolic process	3	5.56	2.33×10^{-2}
GO:0006401~RNA catabolic process	3	5.56	2.91×10^{-2}
GO:0006364~rRNA processing	3	5.56	3.36×10^{-2}
GO:0016072~rRNA metabolic process	3	5.56	3.52×10^{-2}
GO:0042254~ribosome biogenesis	3	5.56	5.00×10^{-2}
GO:0090150~establishment of protein localization to membrane	3	5.56	5.91×10^{-2}
GO:0034655~nucleobase-containing compound catabolic process	3	5.56	6.25×10^{-2}
GO:0034470~ncRNA processing	3	5.56	7.17×10^{-2}
GO:0046700~heterocycle catabolic process	3	5.56	7.17×10^{-2}
GO:0044270~cellular nitrogen compound catabolic process	3	5.56	7.38×10^{-2}
GO:0019439~aromatic compound catabolic process	3	5.56	7.54×10^{-2}
GO:1901361~organic cyclic compound catabolic process	3	5.56	8.23×10^{-2}
GO:0019058~viral life cycle	3	5.56	8.55×10^{-2}
GO:0022613~ribonucleoprotein complex biogenesis	3	5.56	9.37×10^{-2}
GO:0072657~protein localization to membrane	3	5.56	9.67×10^{-2}

Count: genes involved in the term; percentage (%): involved genes/total genes; *p*-value: modified Fisher exact *p*-value.

Table 3. The functional analysis of the upregulated genes at 3 h after LLL irradiation.

Gene Ontology (GO) ID and Terms on Biological Process (BP)	Count	%	*p*-Value
GO:0055085~transmembrane transport	5	0.30	5.87×10^{-2}
GO:1901615~organic hydroxy compound metabolic process	3	0.18	7.85×10^{-2}
GO:0006820~anion transport	3	0.18	9.85×10^{-2}
GO:0051180~vitamin transport	2	0.12	3.81×10^{-2}
GO:0006767~water-soluble vitamin metabolic process	2	0.12	9.58×10^{-2}

Count: genes involved in the term; percentage (%): involved genes/total genes; *p*-value: modified Fisher exact *p*-value.

Table 4. The functional analysis of the downregulated genes at 3 h after LLL irradiation.

Gene Ontology (GO) ID and Terms on Biological Process (BP)	Count	%	*p*-Value
GO:0032774~RNA biosynthetic process	6	15.79	3.8×10^{-2}
GO:0034654~nucleobase-containing compound biosynthetic process	6	15.79	5.85×10^{-2}
GO:0018130~heterocycle biosynthetic process	6	15.79	6.14×10^{-2}
GO:0019438~aromatic compound biosynthetic process	6	15.79	6.21×10^{-2}
GO:0016070~RNA metabolic process	6	15.79	7.52×10^{-2}
GO:0044085~cellular component biogenesis	5	13.16	5.62×10^{-2}
GO:0006614~SRP-dependent cotranslational protein targeting to membrane	4	10.53	1.52×10^{-5}
GO:0006613~cotranslational protein targeting to membrane	4	10.53	1.88×10^{-5}
GO:0045047~protein targeting to ER	4	10.53	1.94×10^{-5}
GO:0072599~establishment of protein localization to endoplasmic reticulum	4	10.53	2.17×10^{-5}
GO:0000184~nuclear-transcribed mRNA catabolic process, nonsense-mediated decay	4	10.53	3.20×10^{-5}
GO:0070972~protein localization to endoplasmic reticulum	4	10.53	3.60×10^{-5}
GO:0019083~viral transcription	4	10.53	1.01×10^{-4}
GO:0006413~translational initiation	4	10.53	1.15×10^{-4}
GO:0006612~protein targeting to membrane	4	10.53	1.17×10^{-4}
GO:0019080~viral gene expression	4	10.53	1.20×10^{-4}
GO:0000956~nuclear-transcribed mRNA catabolic process	4	10.53	1.48×10^{-4}
GO:0044033~multiorganism metabolic process	4	10.53	1.66×10^{-4}
GO:0006402~mRNA catabolic process	4	10.53	1.83×10^{-4}
GO:0006401~RNA catabolic process	4	10.53	2.60×10^{-4}
GO:0006364~rRNA processing	4	10.53	3.26×10^{-4}
GO:0016072~rRNA metabolic process	4	10.53	3.51×10^{-4}
GO:0042254~ribosome biogenesis	4	10.53	6.20×10^{-4}
GO:0090150~establishment of protein localization to membrane	4	10.53	8.15×10^{-4}
GO:0034655~nucleobase-containing compound catabolic process	4	10.53	8.94×10^{-4}
GO:0034470~ncRNA processing	4	10.53	1.12×10^{-3}
GO:0046700~heterocycle catabolic process	4	10.53	1.12×10^{-3}
GO:0044270~cellular nitrogen compound catabolic process	4	10.53	1.18×10^{-3}
GO:0019439~aromatic compound catabolic process	4	10.53	1.22×10^{-3}
GO:1901361~organic cyclic compound catabolic process	4	10.53	1.41×10^{-3}
GO:0019058~viral life cycle	4	10.53	1.51×10^{-3}
GO:0022613~ribonucleoprotein complex biogenesis	4	10.53	1.76×10^{-3}
GO:0072657~protein localization to membrane	4	10.53	1.85×10^{-3}
GO:0034660~ncRNA metabolic process	4	10.53	2.87×10^{-3}
GO:0006412~translation	4	10.53	4.01×10^{-3}
GO:0072594~establishment of protein localization to organelle	4	10.53	4.47×10^{-3}
GO:0043043~peptide biosynthetic process	4	10.53	4.49×10^{-3}
GO:0016071~mRNA metabolic process	4	10.53	4.58×10^{-3}
GO:1902582~single-organism intracellular transport	4	10.53	5.04×10^{-3}
GO:0006605~protein targeting	4	10.53	5.16×10^{-3}
GO:0043604~amide biosynthetic process	4	10.53	5.92×10^{-3}
GO:0006518~peptide metabolic process	4	10.53	7.94×10^{-3}
GO:0044802~single-organism membrane organization	4	10.53	9.30×10^{-3}
GO:0033365~protein localization to organelle	4	10.53	1.04×10^{-2}
GO:0006396~RNA processing	4	10.53	1.07×10^{-2}
GO:0016032~viral process	4	10.53	1.26×10^{-2}
GO:0044764~multiorganism cellular process	4	10.53	1.28×10^{-2}
GO:0044265~cellular macromolecule catabolic process	4	10.53	1.29×10^{-2}
GO:0043603~cellular amide metabolic process	4	10.53	1.36×10^{-2}
GO:0044403~symbiosis, encompassing mutualism through parasitism	4	10.53	1.37×10^{-2}

Count: genes involved in the term; percentage (%): involved genes/total genes; *p*-value: modified Fisher exact *p*-value.

Table 5. The functional analysis of the upregulated genes at 6 h after LLL irradiation.

Gene Ontology (GO) ID and Terms on Biological Process (BP)	Count	%	*p*-Value
GO:0034645~cellular macromolecule biosynthetic process	72	19.20	4.50×10^{-3}
GO:0010467~gene expression	67	17.87	9.63×10^{-2}
GO:0016070~RNA metabolic process	65	17.33	1.49×10^{-2}
GO:0010468~regulation of gene expression	64	17.07	4.90×10^{-3}
GO:0051171~regulation of nitrogen compound metabolic process	64	17.07	6.75×10^{-3}
GO:0019219~regulation of nucleobase-containing compound metabolic process	61	16.27	5.24×10^{-3}
GO:0010556~regulation of macromolecule biosynthetic process	60	16.00	8.42×10^{-3}
GO:2000112~regulation of cellular macromolecule biosynthetic process	59	15.73	7.09×10^{-3}
GO:0034654~nucleobase-containing compound biosynthetic process	59	15.73	4.02×10^{-2}
GO:0018130~heterocycle biosynthetic process	59	15.73	4.99×10^{-2}
GO:0019438~aromatic compound biosynthetic process	59	15.73	5.24×10^{-2}
GO:0051252~regulation of RNA metabolic process	58	15.47	3.04×10^{-3}
GO:0097659~nucleic acid-templated transcription	57	15.20	5.26×10^{-3}
GO:0032774~RNA biosynthetic process	57	15.20	1.10×10^{-2}
GO:0006355~regulation of transcription, DNA-templated	54	14.40	8.34×10^{-3}
GO:1903506~regulation of nucleic acid-templated transcription	54	14.40	9.42×10^{-3}
GO:2001141~regulation of RNA biosynthetic process	54	14.40	1.04×10^{-2}
GO:0006351~transcription, DNA-templated	53	14.13	1.32×10^{-2}
GO:0010646~regulation of cell communication	47	12.53	7.96×10^{-3}
GO:0023051~regulation of signaling	47	12.53	1.07×10^{-2}
GO:0009893~positive regulation of metabolic process	46	12.27	1.53×10^{-2}
GO:0010604~positive regulation of macromolecule metabolic process	43	11.47	2.02×10^{-2}
GO:0009966~regulation of signal transduction	42	11.20	1.56×10^{-2}
GO:0009892~negative regulation of metabolic process	40	10.67	1.19×10^{-2}
GO:0007166~cell surface receptor signaling pathway	40	10.67	2.78×10^{-2}
GO:0031325~positive regulation of cellular metabolic process	40	10.67	6.31×10^{-2}
GO:0010605~negative regulation of macromolecule metabolic process	39	10.40	5.48×10^{-3}
GO:0031324~negative regulation of cellular metabolic process	37	9.87	1.76×10^{-2}
GO:0006366~transcription from RNA polymerase II promoter	35	9.33	1.02×10^{-3}
GO:0006357~regulation of transcription from RNA polymerase II promoter	35	9.33	1.17×10^{-3}
GO:0010628~positive regulation of gene expression	33	8.80	1.19×10^{-3}
GO:0071310~cellular response to organic substance	32	8.53	6.76×10^{-2}
GO:0051173~positive regulation of nitrogen compound metabolic process	31	8.27	8.31×10^{-3}
GO:0048584~positive regulation of response to stimulus	31	8.27	4.62×10^{-2}
GO:0009891~positive regulation of biosynthetic process	30	8.00	1.39×10^{-2}
GO:0031328~positive regulation of cellular biosynthetic process	29	7.73	1.93×10^{-2}
GO:0006468~protein phosphorylation	29	7.73	4.20×10^{-2}
GO:0008283~cell proliferation	29	7.73	4.43×10^{-2}
GO:0008219~cell death	29	7.73	7.43×10^{-2}
GO:0010557~positive regulation of macromolecule biosynthetic process	28	7.47	1.29×10^{-2}
GO:0045935~positive regulation of nucleobase-containing compound metabolic process	28	7.47	2.02×10^{-2}
GO:0012501~programmed cell death	28	7.47	6.56×10^{-2}
GO:0007155~cell adhesion	27	7.20	4.37×10^{-2}
GO:0022610~biological adhesion	27	7.20	4.53×10^{-2}
GO:2000026~regulation of multicellular organismal development	27	7.20	4.97×10^{-2}
GO:0006915~apoptotic process	27	7.20	5.76×10^{-2}
GO:0051254~positive regulation of RNA metabolic process	26	6.93	9.98×10^{-3}
GO:1902531~regulation of intracellular signal transduction	26	6.93	7.47×10^{-2}
GO:0048585~negative regulation of response to stimulus	25	6.67	1.26×10^{-2}
GO:0009890~negative regulation of biosynthetic process	25	6.67	3.44×10^{-2}
GO:0042127~regulation of cell proliferation	25	6.67	4.68×10^{-2}
GO:0048646~anatomical structure formation involved in morphogenesis	24	6.40	3.17×10^{-3}
GO:0010558~negative regulation of macromolecule biosynthetic process	24	6.40	3.25×10^{-2}

Table 5. *Cont.*

Gene Ontology (GO) ID and Terms on Biological Process (BP)	Count	%	*p*-Value
GO:0031327~negative regulation of cellular biosynthetic process	24	6.40	4.89×10^{-2}
GO:0010629~negative regulation of gene expression	24	6.40	5.04×10^{-2}
GO:0045893~positive regulation of transcription, DNA-templated	23	6.13	3.43×10^{-2}
GO:1903508~positive regulation of nucleic acid-templated transcription	23	6.13	3.43×10^{-2}
GO:1902680~positive regulation of RNA biosynthetic process	23	6.13	3.94×10^{-2}
GO:0002682~regulation of immune system process	23	6.13	4.19×10^{-2}
GO:0006952~defense response	23	6.13	9.61×10^{-2}
GO:0010648~negative regulation of cell communication	22	5.87	1.69×10^{-2}
GO:0023057~negative regulation of signaling	22	5.87	1.74×10^{-2}
GO:0080134~regulation of response to stress	22	5.87	3.76×10^{-2}
GO:2000113~negative regulation of cellular macromolecule biosynthetic process	22	5.87	4.90×10^{-2}
GO:0051240~positive regulation of multicellular organismal process	22	5.87	9.32×10^{-2}
GO:0051094~positive regulation of developmental process	21	5.60	1.41×10^{-2}

Count: genes involved in the term; percentage (%): involved genes/total genes; *p*-value: modified Fisher exact *p*-value.

Table 6. The functional analysis of the downregulated genes at 6 h after LLL irradiation.

Gene Ontology (GO) ID and Terms on Biological Process (BP)	Count	%	*p*-Value
GO:0034645~cellular macromolecule biosynthetic process	158	0.26	1.20×10^{-6}
GO:0010467~gene expression	147	0.24	2.91×10^{-3}
GO:0019438~aromatic compound biosynthetic process	141	0.23	4.17×10^{-6}
GO:0018130~heterocycle biosynthetic process	138	0.23	1.59×10^{-5}
GO:0034654~nucleobase-containing compound biosynthetic process	135	0.22	3.51×10^{-5}
GO:0016070~RNA metabolic process	132	0.22	2.23×10^{-3}
GO:0051171~regulation of nitrogen compound metabolic process	128	0.21	1.10×10^{-3}
GO:0010468~regulation of gene expression	126	0.21	1.36×10^{-3}
GO:0032774~RNA biosynthetic process	121	0.20	1.21×10^{-4}
GO:2000112~regulation of cellular macromolecule biosynthetic process	121	0.20	3.10×10^{-4}
GO:0010556~regulation of macromolecule biosynthetic process	121	0.20	9.90×10^{-4}
GO:0019219~regulation of nucleobase-containing compound metabolic process	118	0.20	2.92×10^{-3}
GO:0097659~nucleic acid-templated transcription	113	0.19	8.72×10^{-4}
GO:0006351~transcription, DNA-templated	109	0.18	8.23×10^{-4}
GO:0006355~regulation of transcription, DNA-templated	107	0.18	1.77×10^{-3}
GO:1903506~regulation of nucleic acid-templated transcription	107	0.18	2.18×10^{-3}
GO:2001141~regulation of RNA biosynthetic process	107	0.18	2.60×10^{-3}
GO:0051252~regulation of RNA metabolic process	107	0.18	7.16×10^{-3}
GO:0009892~negative regulation of metabolic process	90	0.15	9.30×10^{-6}
GO:0010605~negative regulation of macromolecule metabolic process	86	0.14	4.06×10^{-6}
GO:0031324~negative regulation of cellular metabolic process	85	0.14	1.02×10^{-5}
GO:0044085~cellular component biogenesis	85	0.14	1.03×10^{-2}
GO:0043933~macromolecular complex subunit organization	83	0.14	1.72×10^{-4}
GO:0009893~positive regulation of metabolic process	83	0.14	5.25×10^{-2}
GO:0010604~positive regulation of macromolecule metabolic process	81	0.13	2.37×10^{-2}
GO:0022607~cellular component assembly	79	0.13	6.13×10^{-3}
GO:0051276~chromosome organization	67	0.11	7.48.E-12
GO:0033554~cellular response to stress	66	0.11	6.62×10^{-5}
GO:0032268~regulation of cellular protein metabolic process	65	0.11	6.56×10^{-2}
GO:0031327~negative regulation of cellular biosynthetic process	63	0.10	1.62×10^{-6}
GO:0009890~negative regulation of biosynthetic process	63	0.10	2.75×10^{-6}
GO:0051172~negative regulation of nitrogen compound metabolic process	63	0.10	2.91×10^{-6}
GO:0010558~negative regulation of macromolecule biosynthetic process	61	0.10	1.79×10^{-6}
GO:0010629~negative regulation of gene expression	61	0.10	7.71×10^{-6}

Table 6. *Cont.*

Gene Ontology (GO) ID and Terms on Biological Process (BP)	Count	%	*p*-Value
GO:2000113~negative regulation of cellular macromolecule biosynthetic process	60	0.10	4.10×10^{-7}
GO:0071822~protein complex subunit organization	60	0.10	4.18×10^{-4}
GO:0045934~negative regulation of nucleobase-containing compound metabolic process	59	0.10	2.74×10^{-6}
GO:0065003~macromolecular complex assembly	59	0.10	1.21×10^{-3}
GO:0006461~protein complex assembly	56	0.09	9.86×10^{-5}
GO:0070271~protein complex biogenesis	56	0.09	1.00×10^{-4}
GO:0007049~cell cycle	56	0.09	3.16×10^{-3}
GO:0006325~chromatin organization	54	0.09	4.01×10^{-13}
GO:1903507~negative regulation of nucleic acid-templated transcription	53	0.09	2.84×10^{-6}
GO:1902679~negative regulation of RNA biosynthetic process	53	0.09	4.28×10^{-6}
GO:0051253~negative regulation of RNA metabolic process	53	0.09	1.24×10^{-5}
GO:0045892~negative regulation of transcription, DNA-templated	52	0.09	1.93×10^{-6}
GO:0034622~cellular macromolecular complex assembly	47	0.08	4.38×10^{-6}
GO:0031399~regulation of protein modification process	47	0.08	9.08×10^{-2}
GO:0044248~cellular catabolic process	46	0.08	8.76×10^{-2}
GO:0022402~cell cycle process	44	0.07	1.65×10^{-2}

Count: genes involved in the term; percentage (%): involved genes/total genes; *p*-value: modified Fisher exact *p*-value.

Table 7. The functional analysis of the upregulated genes at 12 h after LLL irradiation.

Gene Ontology (GO) ID and Terms on Biological Process (BP)	Count	%	*p*-Value
GO:0007166~cell surface receptor signaling pathway	28	10.73	2.37×10^{-2}
GO:0006955~immune response	24	9.20	3.39×10^{-4}
GO:0006952~defense response	23	8.81	6.37×10^{-4}
GO:0009605~response to external stimulus	23	8.81	2.57×10^{-2}
GO:0048584~positive regulation of response to stimulus	22	8.43	3.73×10^{-2}
GO:0003008~system process	21	8.05	5.96×10^{-2}
GO:0002682~regulation of immune system process	19	7.28	6.30×10^{-3}
GO:0050776~regulation of immune response	18	6.90	1.70×10^{-4}
GO:0016192~vesicle-mediated transport	17	6.51	5.14×10^{-2}
GO:0045087~innate immune response	16	6.13	6.31×10^{-4}
GO:0007186~G-protein coupled receptor signaling pathway	16	6.13	2.69×10^{-2}
GO:0051707~response to other organism	15	5.75	1.72×10^{-3}
GO:0043207~response to external biotic stimulus	15	5.75	1.72×10^{-3}
GO:0009607~response to biotic stimulus	15	5.75	2.76×10^{-3}
GO:0050877~neurological system process	15	5.75	6.02×10^{-2}
GO:0006897~endocytosis	14	5.36	6.77×10^{-4}
GO:0002252~immune effector process	14	5.36	1.83×10^{-3}
GO:0001775~cell activation	14	5.36	1.00×10^{-2}
GO:0002684~positive regulation of immune system process	14	5.36	1.51×10^{-2}
GO:0009617~response to bacterium	13	4.98	3.80×10^{-4}
GO:0050778~positive regulation of immune response	13	4.98	2.71×10^{-3}
GO:0045321~leukocyte activation	13	4.98	5.47×10^{-3}
GO:0051094~positive regulation of developmental process	13	4.98	7.29×10^{-2}
GO:0002768~immune response-regulating cell surface receptor signaling pathway	12	4.60	9.96×10^{-5}
GO:0002764~immune response-regulating signaling pathway	12	4.60	1.12×10^{-3}
GO:0046649~lymphocyte activation	12	4.60	4.91×10^{-3}
GO:0002449~lymphocyte mediated immunity	11	4.21	2.83×10^{-5}
GO:0002443~leukocyte mediated immunity	11	4.21	1.79×10^{-4}
GO:0002250~adaptive immune response	11	4.21	6.62×10^{-4}
GO:0098542~defense response to other organism	11	4.21	2.23×10^{-3}
GO:0042742~defense response to bacterium	10	3.83	9.53×10^{-5}
GO:0002460~adaptive immune response based on somatic recombination of immune receptors built from immunoglobulin superfamily domains	10	3.83	1.65×10^{-4}

Table 7. *Cont.*

Gene Ontology (GO) ID and Terms on Biological Process (BP)	Count	%	*p*-Value
GO:0002429~immune response-activating cell surface receptor signaling pathway	10	3.83	1.01×10^{-3}
GO:0002757~immune response-activating signal transduction	10	3.83	7.63×10^{-3}
GO:0002253~activation of immune response	10	3.83	1.40×10^{-2}
GO:0016064~immunoglobulin mediated immune response	9	3.45	2.44×10^{-5}
GO:0019724~B cell mediated immunity	9	3.45	2.66×10^{-5}
GO:0042113~B cell activation	9	3.45	3.08×10^{-4}
GO:0006959~humoral immune response	9	3.45	3.25×10^{-4}
GO:0051251~positive regulation of lymphocyte activation	9	3.45	8.51×10^{-4}
GO:0002696~positive regulation of leukocyte activation	9	3.45	1.42×10^{-3}
GO:0050867~positive regulation of cell activation	9	3.45	1.70×10^{-3}
GO:0051249~regulation of lymphocyte activation	9	3.45	7.68×10^{-3}
GO:0002694~regulation of leukocyte activation	9	3.45	1.59×10^{-2}
GO:0050865~regulation of cell activation	9	3.45	2.28×10^{-2}
GO:0006958~complement activation, classical pathway	8	3.07	5.50×10^{-6}
GO:0002455~humoral immune response mediated by circulating immunoglobulin	8	3.07	1.24×10^{-5}
GO:0006956~complement activation	8	3.07	1.65×10^{-5}
GO:0072376~protein activation cascade	8	3.07	6.86×10^{-5}
GO:0006909~phagocytosis	8	3.07	3.11×10^{-3}

Count: genes involved in the term; percentage (%): involved genes/total genes; *p*-value: modified Fisher exact *p*-value.

Table 8. The functional analysis of the downregulated genes at 12 h after LLL irradiation.

Gene Ontology (GO) ID and Terms on Biological Process (BP)	Count	%	*p*-Value
GO:0034645~cellular macromolecule biosynthetic process	102	23.83	6.82×10^{-2}
GO:0051171~regulation of nitrogen compound metabolic process	94	21.96	2.85×10^{-2}
GO:0010556~regulation of macromolecule biosynthetic process	90	21.03	1.75×10^{-2}
GO:0010468~regulation of gene expression	90	21.03	6.33×10^{-2}
GO:0018130~heterocycle biosynthetic process	90	21.03	8.55×10^{-2}
GO:0034654~nucleobase-containing compound biosynthetic process	89	20.79	8.55×10^{-2}
GO:0019219~regulation of nucleobase-containing compound metabolic process	88	20.56	3.19×10^{-2}
GO:2000112~regulation of cellular macromolecule biosynthetic process	87	20.33	2.26×10^{-2}
GO:0097659~nucleic acid-templated transcription	82	19.16	2.93×10^{-2}
GO:0032774~RNA biosynthetic process	82	19.16	6.02×10^{-2}
GO:0051252~regulation of RNA metabolic process	81	18.93	3.76×10^{-2}
GO:0006351~transcription, DNA-templated	80	18.69	2.03×10^{-2}
GO:1903506~regulation of nucleic acid-templated transcription	80	18.69	2.30×10^{-2}
GO:2001141~regulation of RNA biosynthetic process	80	18.69	2.58×10^{-2}
GO:0006355~regulation of transcription, DNA-templated	79	18.46	2.76×10^{-2}
GO:0023051~regulation of signaling	77	17.99	1.04×10^{-3}
GO:0010646~regulation of cell communication	76	17.76	1.04×10^{-3}
GO:0009966~regulation of signal transduction	71	16.59	6.57×10^{-4}
GO:0065009~regulation of molecular function	71	16.59	2.42×10^{-3}
GO:0035556~intracellular signal transduction	69	16.12	8.62×10^{-4}
GO:0009893~positive regulation of metabolic process	68	15.89	3.21×10^{-2}
GO:0010604~positive regulation of macromolecule metabolic process	67	15.65	1.20×10^{-2}
GO:0006796~phosphate-containing compound metabolic process	67	15.65	5.25×10^{-2}
GO:0006793~phosphorus metabolic process	67	15.65	5.42×10^{-2}
GO:0031325~positive regulation of cellular metabolic process	66	15.42	1.53×10^{-2}
GO:0044085~cellular component biogenesis	65	15.19	3.58×10^{-2}
GO:0050790~regulation of catalytic activity	63	14.72	7.74×10^{-4}
GO:0022607~cellular component assembly	63	14.72	9.06×10^{-3}
GO:0007166~cell surface receptor signaling pathway	59	13.79	5.38×10^{-2}
GO:0051128~regulation of cellular component organization	57	13.32	9.42×10^{-3}

Table 8. *Cont.*

Gene Ontology (GO) ID and Terms on Biological Process (BP)	Count	%	*p*-Value
GO:0016310~phosphorylation	56	13.08	6.37×10^{-3}
GO:0051246~regulation of protein metabolic process	55	12.85	7.26×10^{-2}
GO:0033554~cellular response to stress	53	12.38	1.80×10^{-4}
GO:0007399~nervous system development	52	12.15	2.06×10^{-2}
GO:0031324~negative regulation of cellular metabolic process	52	12.15	6.82×10^{-2}
GO:0032268~regulation of cellular protein metabolic process	51	11.92	9.24×10^{-2}
GO:1902531~regulation of intracellular signal transduction	50	11.68	5.12×10^{-4}
GO:0044093~positive regulation of molecular function	50	11.68	2.35×10^{-3}
GO:0006928~movement of cell or subcellular component	48	11.21	4.02×10^{-3}
GO:0009605~response to external stimulus	47	10.98	6.26×10^{-2}
GO:0043085~positive regulation of catalytic activity	46	10.75	7.29×10^{-4}
GO:0007049~cell cycle	46	10.75	2.94×10^{-3}
GO:0006357~regulation of transcription from RNA polymerase II promoter	46	10.75	1.51×10^{-2}
GO:1902589~single-organism organelle organization	45	10.51	2.52×10^{-3}
GO:0006366~transcription from RNA polymerase II promoter	45	10.51	2.03×10^{-2}
GO:0008219~cell death	44	10.28	8.02×10^{-2}
GO:0031399~regulation of protein modification process	43	10.05	1.20×10^{-2}
GO:0048585~negative regulation of response to stimulus	41	9.58	1.17×10^{-3}
GO:0007155~cell adhesion	41	9.58	3.80×10^{-2}
GO:0022610~biological adhesion	41	9.58	3.98×10^{-2}

Count: genes involved in the term; percentage (%): involved genes/total genes; *p*-value: modified Fisher exact *p*-value.

3.3. PPI of Up- or Downregulated DEGs

As for the upregulated DEGs, there were DEGs involved in the defense response at all irradiation times. From the functional analysis of DAVID, we focused on the defense response related to wound healing, which is the BP of the DEGs of the upregulated DEGs at 6 and 12 h after the irradiation. Among the genes involved in the defense response, HCP5, DAPK3, IGLC2 and IGHV1 OR21-1 at 1 h after the irradiation for the upregulated DEGs, HP, IGHE and TPSAB1 at 3 h, CAKM2 B, IGHM, SEMA7 A, CXCL8 and TNFAIP6 at 6 h. IL34, ITGA2, SERPINE1, INHBA, PRDM1, LYZL4, PVR, BMP6, NFKB1, FOXP1, PER1, IGHG1, IGHA1, IGHA2, LDLR, SLC25 A6, PF4 and TGM2. At 12 h, it was IGHM, ITIH4, SEMA7 A, EDN1, HIST1 H2 BJ, KCNJ8, TNFAIP6, CCL21, CARD9, IFI6, SERPINB9, LYZL2, DEFB108 B, ECSIT, IGHG1, IGKC, IGHD, CHRFAM7 A and SLAMF6.

As the results of PPI analysis of DEGs involved in the defense response with STRING, CXCL8 was a gene associated with multiple genes in the upregulated DEGs of the defense response. The relationships with SERPINE1, PF4, NFKB1, TNFAIP6, EDN1, CCL21, ITIH4 and HP were confirmed. In addition, HP, ITIH4, TNFAIP6 and NFKB1 were also associated with multiple genes (Figure 1). In the downregulated DEGs, STAT1 was a gene associated with multiple genes. In particular, it was associated with SMAD3, IL15, CCL22, NAIP, NRIH3, LY96, NFKBIA, PSMB8, PSMB9, PSMB10, OSA2, OAS3, IRF2, BCL6 and DDX58. In addition, NFKBIA, TLR3, IL15 and IRF2 were genes associated with multiple genes (Figure 2).

Figure 3 shows the changes in the expression of major genes related to other genes over time.

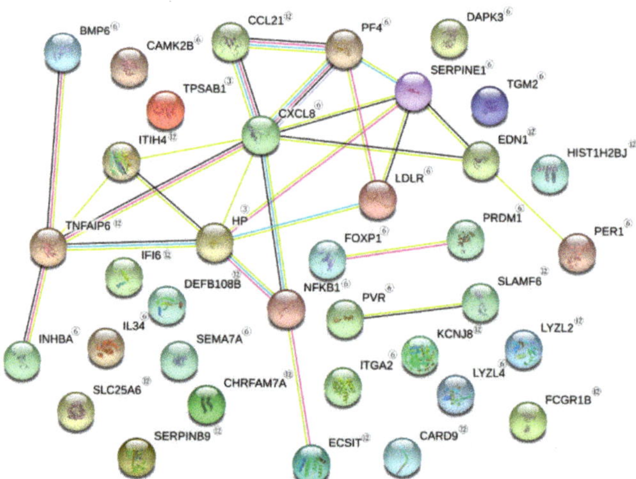

Figure 1. PPI of upregulated DEGs. Search tool STRING analysis of interacting genes and proteins reveals a protein–protein interaction PPI network in defense response by LLLT. PPI of upregulated DEGs related to 'defense response' Related genes after the irradiation at 3 h; ③, 6 h; ⑥ 12 h; ⑫ Red line: indicates the presence of fusion evidence; green line: neighborhood evidence; blue line: cooccurrence evidence; purple line: experimental evidence; yellow line: text mining evidence; light blue line: database evidence; black line: coexpression evidence.

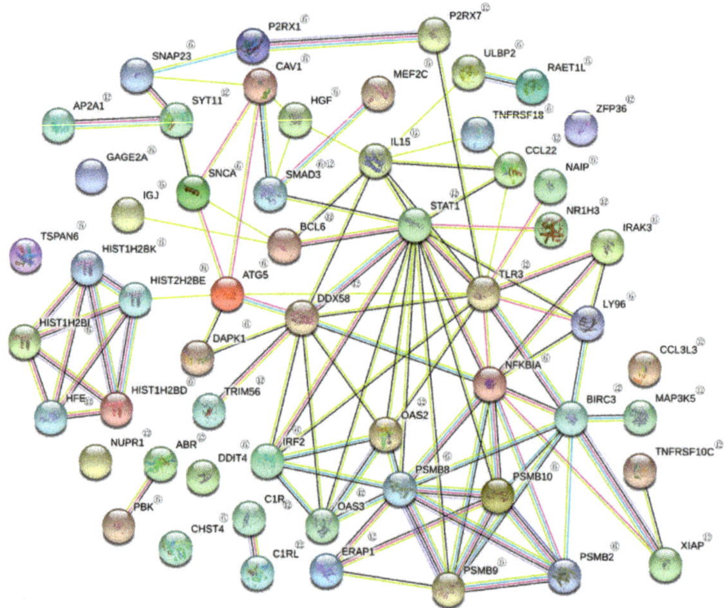

Figure 2. PPI of downregulated DEGs. Search tool STRING analysis of interacting genes and proteins reveals a protein–protein interaction network between proteins in defense response by LLLT. PPI of downregulated DEGs related to 'defense response.' Related genes after the irradiation at 3 h; ③, 6 h; ⑥ 12 h; ⑫.

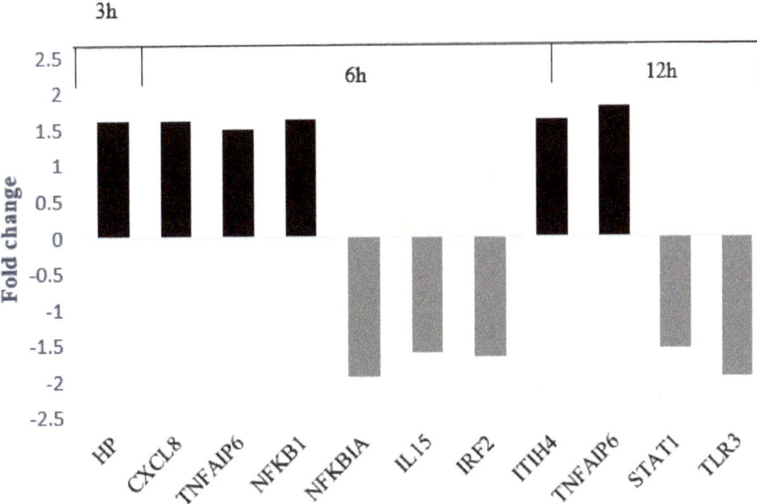

Figure 3. The FC of the key genes.

4. Discussion

Laser treatment has been studied for clinical application in many fields including medical and dentistry. In the field of dentistry, lasers are used for various purposes such as promoting wound healing, gingival incision, caries removal and sterilization/disinfection effect [25]. It has been reported that LLLT promotes wound healing and bone formation in the oral cavity by utilizing the bioactivating effect of chronic periodontitis.

In addition, at the cellular and tissue levels, there are research reports related to promotion of wound healing, such as cell proliferation of fibroblasts [15,20], osteoblasts [22,26] and vascular endothelial cells [18,27] by LLLT. However, clinical application in medical and dental treatments has not been carried out much. This is because the molecular biological findings are not clear.

Intracellular biological effects of LLLT include physiological activity by photoreceptors, changes in intracellular signal cascades, and changes in genes. The intracellular photoreceptor of LLLT is cytochrome c oxidase, which is a transmembrane protein complex that is an electron transport chain enzyme found in mitochondria. Promotion of cell proliferation by increasing cytochrome c oxidase activity and ATP is one of the representative mechanisms in LLLT research [28–31].

As a method for elucidating the mechanism of wound healing by LLLT, DNA microarrays are considered to be useful because they can examine thousands to tens of thousands of gene expressions at a time. From the obtained gene expression data, DEGs are extracted using bioinformatics analysis tools [32,33]. Furthermore, based on GO, by analyzing the biological process (BP) of DEGs, it is possible to estimate what is happening at each time by knowing the known functions of the gene contained in DEGs. It is also possible to infer what is about to happen at that time by performing chronological analysis. In addition, by searching for protein–protein interaction (PPI), it is possible to search for relationships at the molecular level [34,35]. The method plays a major role in elucidating the effects of proteins controlled by LLLT-stimulated genes at the molecular level.

In this study, we focused on the defense response from a huge amount of microarray data and analyzed the chronological changes in gene expression and the function of the genes after LLLT to HGF and the function of the genes.

The gene expression reactions related to wound healing with LLLT were remarkable 6–12 h after the irradiation. Analysis of the gene expression changes within these times were considered important for investigating the molecular mechanism of effects on HGF with LLLT. In particular, among the DEGs, those that are common over time and those with a significantly large expression fluctuation amount were considered largely affected by LLLT.

Analysis of BP over time revealed that upregulated were activated from 1 to 3 h after the initial irradiation, and downregulated BP was involved in RNA metabolism and activity. In both cases, the number of applicable DEGs for BP was 10 or less. It was suggested that there was no cohesive expression fluctuation as a function.

At 6 h after the irradiation, upregulated DEGs were observed related to cell proliferation, adhesion and defense reaction. Additionally, downregulated DEGs were observed in many BPs involved in RNA metabolism, activity, cell polymer production, and metabolism.

At 12 h after the irradiation, upregulated DEGs were observed with many BPs involved in defense reaction, immune reaction and response to external stimuli, and downregulated DEGs were observed with many BPs involved in RNA metabolism, activity, cell polymer production and metabolism. In the upregulated group, many BPs associated with wound healing were observed at 6–12 h after irradiation. Additionally, in downregulated, similar BP such as RNA metabolism were observed from 1 to 12 h after irradiation.

The BP 'the defense response' focused on in this study belongs to the BP 'the response to stress of response to stimulus' in GO. The response to stimulus is the process by which the state or activity of a cell or organism changes as a result of stimulation. The response to stress also causes motility, secretion, enzyme production, gene expression, etc., as a result of impaired homeostasis of the organism or cell due to extrinsic factors (temperature, humidity, ionizing radiation). The defense response, which belongs to response to stress, is a reaction caused in response to the presence of foreign substances or the occurrence of injuries, and is an important BP involved in the restriction, prevention/recovery of damage to living organisms.

In the BP 'defense response', the protein encoded by CXCL8, which is a downregulated DEG, is called interleukin-8 (IL-8). IL-8 is secreted by mononuclear macrophages, neutrophils, eosinophils, T lymphocytes, epithelial cells, and fibroblasts. IL-8 is also known as a neutrophil chemotactic factor with two major functions. It induces chemotaxis of target cells to the infected site. IL-8 is also known as a strong promoter of angiogenesis. IL-8 expression is regulated by the transcription factor NF-κB [36–42].

In particular, the upregulated DEGs NFKB1 and the downregulated DEGs NFKBIA are genes involved in NF-κB. These are one of the genes involved in the existing mechanism. NFKB1 is a transcriptional regulator that is activated by various intracellular and extracellular stimuli such as cytokines, oxidant free radicals, UV irradiation and bacterial or viral products. Activated NFKB stimulates the expression of genes involved in biological functions associated with many biological processes such as inflammation, immunity, differentiation and cells [43]. NFKBIA is a member of a family of cellular proteins that function to inhibit NF-κB transcription factors and IκBα masks the nuclear localization signals (NLS) of NF-κB proteins and inactivates them in the cytoplasm. It inhibits NF-κB by isolating it into a state [44]. From this result, it can be seen that NFKB1 increases and NFKBIA decreases when irradiation is performed, so that the activity of the pathway containing NF-κB occurs. Additionally, the defense response of BP is considered to be strongly related to the NF-κB pathway.

In a study by Chen et al., NF-κB activity was observed 1–10 h after irradiation [21].

In this study as well, changes in the expression of NFKB1 and NFKBIA were observed 6 h after irradiation, and the results of analysis from the viewpoint of BP also suggest that the movement of NF-κB due to irradiation is an important process in the mechanism of LLLT. In this study, we focused on the defense response. Further, we also will need to focus on wound healing-related processes such as BP 'the immune response'.

By adding analysis more time points, it is possible to analyze detailed time-series processes after LLL irradiation. As future developments, we plan to study other BPs, collect LLLT microarray data at different time points, and analyze the effects of laser irradiation on fibroblasts and molecular-level processes during the healing process. We will lead to the elucidation of molecular evidence in 'the response to stress of response to stimulus' by LLLT.

5. Conclusions

The time points of 1, 3, 6 and 12 h after LLL irradiation were compared over time. The most DEGs after the LLL irradiation on HGF were showed at 6 h upregulated gene. The number of DEGs peaked 6 h after irradiation and slightly decreased at 12 h after irradiation. From the time-dependent functional analysis, the upregulated DEGs were involved in BPs of cell proliferation, adhesion, and defense response related to wound healing from 6 h. In addition, defense response is one of the important mechanisms in BP after the irradiation. We found that the upregulated DEGs such as CXCL8 and NFKB1, and the downregulated DEGs such as NFKBIA and STAT1 were correlated with multiple genes from these PPI. From these results, irradiation of LLLT showed fluctuations in the expression of genes related to BP defense response.

Supplementary Materials: The following are available online at https://www.mdpi.com/article/10.3390/jcm10091952/s1, Table S1: DEGs of the up-regulated genes at 1 h after LLL irradiation, Table S2: DEGs of the down-regulated genes at 1 h after LLL irradiation, Table S3: DEGs of the up-regulated genes at 3 h after LLL irradiation, Table S4: DEGs of the down-regulated genes at 3 h after LLL irradiation, Table S5: DEGs of the up-regulated genes at 6 h after LLL irradiation, Table S6: DEGs of the down-regulated genes at 6 h after LLL irradiation, Table S7: DEGs of the up-regulated genes at 12 h after LLL irradiation, Table S8: DEGs of the down-regulated genes at 12 h after LLL irradiation.

Author Contributions: Conceptualization, Y.W., A.S., H.I., E.M. and Y.N.; methodology, Y.W., A.S., H.I., E.M. and E.M.; software, Y.W. and A.S.; validation, Y.W., A.S., H.I., and E.M.; formal analysis, Y.W., A.S. and H.I.; investigation, Y.W., A.S., H.I., and E.M.; resources, Y.W., A.S., H.I., and E.M.; data curation, Y.W., A.S., H.I., and E.M.; writing—original draft preparation, Y.W., A.S., H.I., E.M. and Y.N.; writing—review and editing, Y.W., A.S., H.I., E.M. and Y.N.; visualization, Y.W. and A.S.; supervision, Y.N.; project administration, A.S., H.I., and E.M.; funding acquisition, H.I. and E.M. All authors have read and agreed to the published version of the manuscript.

Funding: This research was funded by grants from the JSPS KAKENHI (grant numbers: JP18 K09585).

Institutional Review Board Statement: Not applicable.

Informed Consent Statement: Not applicable.

Conflicts of Interest: The authors declare no conflict of interest.

References

1. Lopes, B.M.V.; Marcantonio, R.A.C.; Thompson, G.M.A.; Neves, L.H.M.; Theodoro, L.H. Short-Term Clinical and Immunologic Effects of Scaling and Root Planing With Er:YAG Laser in Chronic Periodontitis. *J. Periodontol.* **2008**, *79*, 1158–1167. [CrossRef] [PubMed]
2. Schwarz, F.; Sculean, A.; Georg, T.; Reich, E. Periodontal Treatment With an Er:YAG Laser Compared to Scaling and Root Planing. A Controlled Clinical Study. *J. Periodontol.* **2001**, *72*, 361–367. [CrossRef]
3. Crespi, R.; Capparè, P.; Toscanelli, I.; Gherlone, E.; Romanos, G.E. Effects of Er:YAG Laser Compared to Ultrasonic Scaler in Periodontal Treatment: A 2-Year Follow-Up Split-Mouth Clinical Study. *J. Periodontol.* **2007**, *78*, 1195–1200. [CrossRef]
4. White, J.M.; Goodis, H.E.; Rose, C.L. Use of the Pulsed Nd:YAG Laser for Intraoral Soft Tissue Surgery. *Lasers Surg. Med.* **1991**, *11*, 455–461. [CrossRef]
5. Lauritano, D.; Lucchese, A.; Gabrione, F.; Di Stasio, D.; Silvestre Rangil, J.; Carinci, F. The Effectiveness of Laser-Assisted Surgical Excision of Leukoplakias and Hyperkeratosis of Oral Mucosa: A Case Series in A Group of Patients. *Int. J. Environ. Res. Public. Health* **2019**, *16*, 210. [CrossRef]
6. Pick, R.M.; Colvard, M.D. Current Status of Lasers in Soft Tissue Dental Surgery. *J. Periodontol.* **1993**, *64*, 589–602. [CrossRef]
7. Aimbire, F.; Santos, F.V.; Albertini, R.; Castro-Faria-Neto, H.C.; Mittmann, J.; Pacheco-Soares, C. Low-Level Laser Therapy Decreases Levels of Lung Neutrophils Anti-Apoptotic Factors by a NF-KappaB Dependent Mechanism. *Int. Immunopharmacol.* **2008**, *8*, 603–605. [CrossRef]
8. Aimbire, F.; Ligeiro de Oliveira, A.P.; Albertini, R.; Corrêa, J.C.; Ladeira de Campos, C.B.; Lyon, J.P.; Silva, J.A.; Costa, M.S. Low Level Laser Therapy (LLLT) Decreases Pulmonary Microvascular Leakage, Neutrophil Influx and IL-1beta Levels in Airway and Lung from Rat Subjected to LPS-Induced Inflammation. *Inflammation* **2008**, *31*, 189–197. [CrossRef]

9. Albertini, R.; Aimbire, F.S.C.; Correa, F.I.; Ribeiro, W.; Cogo, J.C.; Antunes, E.; Teixeira, S.A.; De Nucci, G.; Castro-Faria-Neto, H.C.; Zângaro, R.A.; et al. Effects of Different Protocol Doses of Low Power Gallium-Aluminum-Arsenate (Ga-Al-As) Laser Radiation (650 Nm) on Carrageenan Induced Rat Paw Ooedema. *J. Photochem. Photobiol. B* **2004**, *74*, 101–107. [CrossRef] [PubMed]

10. Albertini, R.; Villaverde, A.B.; Aimbire, F.; Bjordal, J.; Brugnera, A.; Mittmann, J.; Silva, J.A.; Costa, M. Cytokine MRNA Expression Is Decreased in the Subplantar Muscle of Rat Paw Subjected to Carrageenan-Induced Inflammation after Low-Level Laser Therapy. *Photomed. Laser Surg.* **2008**, *26*, 19–24. [CrossRef] [PubMed]

11. Albertini, R.; Aimbire, F.; Villaverde, A.B.; Silva, J.A.; Costa, M.S. COX-2 MRNA Expression Decreases in the Subplantar Muscle of Rat Paw Subjected to Carrageenan-Induced Inflammation after Low Level Laser Therapy. *Inflamm. Res. Off. J. Eur. Histamine Res. Soc. Al* **2007**, *56*, 228–229. [CrossRef]

12. Albertini, R.; Villaverde, A.B.; Aimbire, F.; Salgado, M.A.C.; Bjordal, J.M.; Alves, L.P.; Munin, E.; Costa, M.S. Anti-Inflammatory Effects of Low-Level Laser Therapy (LLLT) with Two Different Red Wavelengths (660 Nm and 684 Nm) in Carrageenan-Induced Rat Paw Edema. *J. Photochem. Photobiol. B* **2007**, *89*, 50–55. [CrossRef] [PubMed]

13. Aoki, A.; Mizutani, K.; Schwarz, F.; Sculean, A.; Yukna, R.A.; Takasaki, A.A.; Romanos, G.E.; Taniguchi, Y.; Sasaki, K.M.; Zeredo, J.L.; et al. Periodontal and Peri-Implant Wound Healing Following Laser Therapy. *Periodontol. 2000* **2015**, *68*, 217–269. [CrossRef] [PubMed]

14. Kami, T.; Yoshimura, Y.; Nakajima, T.; Ohshiro, T.; Fujino, T. Effects of Low-Power Diode Lasers on Flap Survival. *Ann. Plast. Surg.* **1985**, *14*, 278–283. [CrossRef]

15. Ogita, M.; Tsuchida, S.; Aoki, A.; Satoh, M.; Kado, S.; Sawabe, M.; Nanbara, H.; Kobayashi, H.; Takeuchi, Y.; Mizutani, K.; et al. Increased Cell Proliferation and Differential Protein Expression Induced by Low-Level Er:YAG Laser Irradiation in Human Gingival Fibroblasts: Proteomic Analysis. *Lasers Med. Sci.* **2015**, *30*, 1855–1866. [CrossRef] [PubMed]

16. Kipshidze, N.; Nikolaychik, V.; Keelan, M.H.; Shankar, L.R.; Khanna, A.; Kornowski, R.; Leon, M.; Moses, J. Low-Power Helium: Neon Laser Irradiation Enhances Production of Vascular Endothelial Growth Factor and Promotes Growth of Endothelial Cells in Vitro. *Lasers Surg. Med.* **2001**, *28*, 355–364. [CrossRef] [PubMed]

17. Yu, H.S.; Chang, K.L.; Yu, C.L.; Chen, J.W.; Chen, G.S. Low-Energy Helium-Neon Laser Irradiation Stimulates Interleukin-1 Alpha and Interleukin-8 Release from Cultured Human Keratinocytes. *J. Investig. Dermatol.* **1996**, *107*, 593–596. [CrossRef]

18. Khanna, A.; Shankar, L.R.; Keelan, M.H.; Kornowski, R.; Leon, M.; Moses, J.; Kipshidze, N. Augmentation of the Expression of Proangiogenic Genes in Cardiomyocytes with Low Dose Laser Irradiation in Vitro. *Cardiovasc. Radiat. Med.* **1999**, *1*, 265–269. [CrossRef]

19. Gkogkos, A.S.; Karoussis, I.K.; Prevezanos, I.D.; Marcopoulou, K.E.; Kyriakidou, K.; Vrotsos, I.A. Effect of Nd:YAG Low Level Laser Therapy on Human Gingival Fibroblasts. *Int. J. Dent.* **2015**, *2015*. [CrossRef]

20. Misa, O.; Etsuko, M.; Hitomi, I.; Hiroko, T.-I.; Yukihiro, N. Effect of Low-Level Nd: YAG Laser Irradiation for Wound Healing on Human Gingival Fibroblasts. Ph.D. Thesis, The Nippon Dental University, Tokyo, Japan, 2016.

21. Chen, A.C.-H.; Arany, P.R.; Huang, Y.-Y.; Tomkinson, E.M.; Sharma, S.K.; Kharkwal, G.B.; Saleem, T.; Mooney, D.; Yull, F.E.; Blackwell, T.S.; et al. Low-Level Laser Therapy Activates NF-KB via Generation of Reactive Oxygen Species in Mouse Embryonic Fibroblasts. *PLoS ONE* **2011**, *6*, e22453. [CrossRef]

22. Ohsugi, Y.; Aoki, A.; Mizutani, K.; Katagiri, S.; Komaki, M.; Noda, M.; Takagi, T.; Kakizaki, S.; Meinzer, W.; Izumi, Y. Evaluation of Bone Healing Following Er:YAG Laser Ablation in Rat Calvaria Compared with Bur Drilling. *J. Biophotonics* **2019**, *12*, e201800245. [CrossRef] [PubMed]

23. Kong, S.; Aoki, A.; Iwasaki, K.; Mizutani, K.; Katagiri, S.; Suda, T.; Ichinose, S.; Ogita, M.; Pavlic, V.; Izumi, Y. Biological Effects of Er:YAG Laser Irradiation on the Proliferation of Primary Human Gingival Fibroblasts. *J. Biophotonics* **2018**, *11*. [CrossRef] [PubMed]

24. Naoya, Y.; Etsuko, M.; Hiroko, I.; Yukihiro, N. The Effect of Nd:YAG Laser Irradiation on Human Gingival Fibroblasts—A Study of the Irradiation Output and Distance. *J. Jpn. Soc. Laser Dent.* **2013**, *24*, 72–82. (In Japanese) [CrossRef]

25. Akiyama, F.; Aoki, A.; Miura-Uchiyama, M.; Sasaki, K.M.; Ichinose, S.; Umeda, M.; Ishikawa, I.; Izumi, Y. In Vitro Studies of the Ablation Mechanism of Periodontopathic Bacteria and Decontamination Effect on Periodontally Diseased Root Surfaces by Erbium:Yttrium-Aluminum-Garnet Laser. *Lasers Med. Sci.* **2011**, *26*, 193–204. [CrossRef]

26. Pyo, S.-J.; Song, W.-W.; Kim, I.-R.; Park, B.-S.; Kim, C.-H.; Shin, S.-H.; Chung, I.-K.; Kim, Y.-D. Low-Level Laser Therapy Induces the Expressions of BMP-2, Osteocalcin, and TGF-B1 in Hypoxic-Cultured Human Osteoblasts. *Lasers Med. Sci.* **2013**, *28*, 543–550. [CrossRef] [PubMed]

27. Li, Y.; Xu, Q.; Shi, M.; Gan, P.; Huang, Q.; Wang, A.; Tan, G.; Fang, Y.; Liao, H. Low-Level Laser Therapy Induces Human Umbilical Vascular Endothelial Cell Proliferation, Migration and Tube Formation through Activating the PI3K/Akt Signaling Pathway. *Microvasc. Res.* **2020**, *129*, 103959. [CrossRef]

28. Karu, T.I.; Pyatibrat, L.V.; Afanasyeva, N.I. Cellular Effects of Low Power Laser Therapy Can Be Mediated by Nitric Oxide. *Lasers Surg. Med.* **2005**, *36*, 307–314. [CrossRef] [PubMed]

29. Smith, K.C. The Photobiological Basis of Low Level Laser Radiation Therapy. *Laser Ther.* **1991**, *3*, 19–24. [CrossRef]

30. Silveira, P.C.L.; Streck, E.L.; Pinho, R.A. Evaluation of Mitochondrial Respiratory Chain Activity in Wound Healing by Low-Level Laser Therapy. *J. Photochem. Photobiol. B Biol.* **2007**, *86*, 279–282. [CrossRef]

31. Lim, J.; Sanders, R.A.; Snyder, A.C.; Eells, J.T.; Henshel, D.S.; Watkins, J.B. Effects of Low-Level Light Therapy on Streptozotocin-Induced Diabetic Kidney. *J. Photochem. Photobiol. B Biol.* **2010**, *99*, 105–110. [CrossRef]

32. Huang, D.W.; Sherman, B.T.; Lempicki, R.A. Bioinformatics Enrichment Tools: Paths toward the Comprehensive Functional Analysis of Large Gene Lists. *Nucleic Acids Res.* **2009**, *37*, 1–13. [CrossRef] [PubMed]
33. Huang, D.W.; Sherman, B.T.; Lempicki, R.A. Systematic and Integrative Analysis of Large Gene Lists Using DAVID Bioinformatics Resources. *Nat. Protoc.* **2009**, *4*, 44–57. [CrossRef] [PubMed]
34. Zeidán-Chuliá, F.; Gursoy, M.; de Oliveira, B.-H.N.; Gelain, D.P.; Könönen, E.; Gursoy, U.K.; Moreira, J.C.F.; Uitto, V.-J. Focussed Microarray Analysis of Apoptosis in Periodontitis and Its Potential Pharmacological Targeting by Carvacrol. *Arch. Oral Biol.* **2014**, *59*, 461–469. [CrossRef]
35. Zeidán-Chuliá, F.; Rybarczyk-Filho, J.L.; Gursoy, M.; Könönen, E.; Uitto, V.-J.; Gursoy, O.V.; Cakmakci, L.; Moreira, J.C.F.; Gursoy, U.K. Bioinformatical and in Vitro Approaches to Essential Oil-Induced Matrix Metalloproteinase Inhibition. *Pharm. Biol.* **2012**, *50*, 675–686. [CrossRef]
36. Van Damme, J.; Rampart, M.; Conings, R.; Decock, B.; Van Osselaer, N.; Willems, J.; Billiau, A. The Neutrophil-Activating Proteins Interleukin 8 and Beta-Thromboglobulin: In Vitro and in Vivo Comparison of NH2-Terminally Processed Forms. *Eur. J. Immunol.* **1990**, *20*, 2113–2118. [CrossRef]
37. Schutyser, E.; Struyf, S.; Proost, P.; Opdenakker, G.; Laureys, G.; Verhasselt, B.; Peperstraete, L.; Van de Putte, I.; Saccani, A.; Allavena, P.; et al. Identification of Biologically Active Chemokine Isoforms from Ascitic Fluid and Elevated Levels of CCL18/Pulmonary and Activation-Regulated Chemokine in Ovarian Carcinoma. *J. Biol. Chem.* **2002**, *277*, 24584–24593. [CrossRef] [PubMed]
38. Hébert, C.A.; Luscinskas, F.W.; Kiely, J.M.; Luis, E.A.; Darbonne, W.C.; Bennett, G.L.; Liu, C.C.; Obin, M.S.; Gimbrone, M.A.; Baker, J.B. Endothelial and Leukocyte Forms of IL-8. Conversion by Thrombin and Interactions with Neutrophils. *J. Immunol.* **1990**, *145*, 3033–3040.
39. Ungureanu, D.; Vanhatupa, S.; Kotaja, N.; Yang, J.; Aittomaki, S.; Jänne, O.A.; Palvimo, J.J.; Silvennoinen, O. PIAS Proteins Promote SUMO-1 Conjugation to STAT1. *Blood* **2003**, *102*, 3311–3313. [CrossRef]
40. Zhang, Y.; Mao, D.; Roswit, W.T.; Jin, X.; Patel, A.C.; Patel, D.A.; Agapov, E.; Wang, Z.; Tidwell, R.M.; Atkinson, J.J.; et al. PARP9-DTX3L Ubiquitin Ligase Targets Host Histone H2BJ and Viral 3C Protease to Enhance Interferon Signaling and Control Viral Infection. *Nat. Immunol.* **2015**, *16*, 1215–1227. [CrossRef]
41. Chen, K.; Liu, J.; Liu, S.; Xia, M.; Zhang, X.; Han, D.; Jiang, Y.; Wang, C.; Cao, X. Methyltransferase SETD2-Mediated Methylation of STAT1 Is Critical for Interferon Antiviral Activity. *Cell* **2017**, *170*, 492–506.e14. [CrossRef]
42. Liu, B.; Liao, J.; Rao, X.; Kushner, S.A.; Chung, C.D.; Chang, D.D.; Shuai, K. Inhibition of Stat1-Mediated Gene Activation by PIAS1. *Proc. Natl. Acad. Sci. USA* **1998**, *95*, 10626–10631. [CrossRef] [PubMed]
43. Beinke, S.; Robinson, M.J.; Hugunin, M.; Ley, S.C. Lipopolysaccharide Activation of the TPL-2/MEK/Extracellular Signal-Regulated Kinase Mitogen-Activated Protein Kinase Cascade Is Regulated by IkappaB Kinase-Induced Proteolysis of NF-KappaB1 P105. *Mol. Cell. Biol.* **2004**, *24*, 9658–9667. [CrossRef] [PubMed]
44. Scherer, D.C.; Brockman, J.A.; Chen, Z.; Maniatis, T.; Ballard, D.W. Signal-Induced Degradation of I Kappa B Alpha Requires Site-Specific Ubiquitination. *Proc. Natl. Acad. Sci. USA* **1995**, *92*, 11259–11263. [CrossRef] [PubMed]

MDPI
St. Alban-Anlage 66
4052 Basel
Switzerland
Tel. +41 61 683 77 34
Fax +41 61 302 89 18
www.mdpi.com

Journal of Clinical Medicine Editorial Office
E-mail: jcm@mdpi.com
www.mdpi.com/journal/jcm